IONOSPHERIC EFFECTS OF SOLAR FLARES

ASTROPHYSICS AND
SPACE SCIENCE LIBRARY

A SERIES OF BOOKS ON THE RECENT DEVELOPMENTS

OF SPACE SCIENCE AND OF GENERAL GEOPHYSICS AND ASTROPHYSICS

PUBLISHED IN CONNECTION WITH THE JOURNAL

SPACE SCIENCE REVIEWS

VOLUME 46

A. P. MITRA

Radio Science Division, National Physical Laboratory, New Delhi-12, India

IONOSPHERIC EFFECTS OF SOLAR FLARES

D. REIDEL PUBLISHING COMPANY

DORDRECHT-HOLLAND / BOSTON-U.S.A.

Library of Congress Catalog Card Number 74–76480

ISBN-13: 978-94-010-2233-0   e-ISBN-13: 978-94-010-2231-6
DOI: 10.1007/ 978-94-010-2231-6

Published by D. Reidel Publishing Company,
P.O. Box 17, Dordrecht, Holland

Sold and distributed in the U.S.A., Canada and Mexico
by D. Reidel Publishing Company, Inc.
306 Dartmouth Street, Boston,
Mass. 02116, U.S.A.

TABLE OF CONTENTS

PREFACE

Sudden Ionospheric Disturbances resulting from an interaction of the Solar Flare radiation with the constituents of the upper atmosphere constitute one of the three major aspects of ground level monitoring of solar flares – the other two being optical observations of flares, and the observations of solar bursts in radio wavelengths. SIDs, therefore, form a major part of flare monitoring programme in many observatories.

Unlike the other two, however, the ionospheric effects of flares provide one major additional source of interest – the reaction of the ionospheric plasma to an impulsive ionization. The high atmosphere provides a low pressure laboratory without walls in which a host of reactions occur between electrons, ions and neutral particles. The resulting products and their distributions may bear no resemblance to those of the primary neutral constituents or their direct ionization products. The variations with the time of the day, with season and with solar activity that form the bulk of the ionospheric measurements are too slow to allow any insight into the nature of these ionospheric reactions whose lifetimes are often very short. The relaxation time of the ionospheric ionization is only a few minutes or fraction of a minute in the lower ionosphere and in the E-region and is about 30 min to an hour at 300 km. The flares provide a sudden short impulse comparable to these time scales.

Conventional techniques of recording ionospheric effects of solar flares involve the use of signal strengths of terrestrial transmitters, of radio noise coming from our Galaxy, or of atmospherics, or measuring the sudden changes in the Earth's magnetic field. In addition, there is now an increasing use of more quantitative techniques, involving the use of partial reflection, wave interaction, multifrequency absorption, low and very low frequency propagation data, incoherent-scatter, satellite radio beacon transmission and rocket-borne measurements of flare time electron density. These have provided information on: (i) the statistical properties of SIDs, (ii) the distribution of the ionization enhancements, as well as the nature of growth and decay of these enhancements, and in some cases on ion chemistry, and (iii) the relationship between these enhancements with the incident XUV fluxes and their spectral characteristics.

It appears that the two most important applications of quantitative SID studies are currently: (a) the ability of the D-region SIDs to serve as a probe of the basic chemical processes operating in the D-region, and (b) the ability of the F-region SIDs to successfully monitor the time variations in EUV flux and any relative changes in fluxes in the competing bands.

The SIDs constitute the immediate effects of Solar Flares. There is a whole sequence of effects following the SIDs, beginning from energetic protons that spiral down the Earth's magnetic field and enter into the polar ionosphere causing Polar Cap Absorption Events (PCAs) to the slower ions and electrons that arrive 20–40 h later causing magnetic and ionospheric storms and the auroras. The latter are not discussed in this work, except the PCAs in a short summary chapter at the end. The question of prediction of solar flares and of its various hazards are also not discussed, since this is not directly relevant. The purpose of this volume is to give a detailed account of the immediate ionospheric effects.

Acknowledgements

The material of this volume was originally prepared as an invited article for a journal. Revisions introduced for the book version have essentially been minimal and were intended to provide completeness in presentation. Dr B. M. Reddy of this laboratory went through the first as well as the present version of this manuscript and helped greatly in its preparation. Dr B. C. N. Rao read the complete manuscript and gave several useful suggestions. I have drawn heavily from the works of Professors A. J. Ferraro and H. S. Lee and of Dr John Rowe, some of which are unpublished. I have also depended very heavily on the works of my ex-students Dr C. V. Subrahmanyam, Dr (Mrs) Mirjana Karabin and Dr S. D. Deshpande, whose contributions to SID studies have been considerable. I wish also to record my debt of gratitude to Mrs Santosh Aggarwal and Mr P. K. Pasricha for assistance in the preparation of the manuscript, and to Mr P. C. Rana, Mr K. L. Malik, Mrs S. A. Joseph, Mr C. L. Kesavan, Mr D. B. Sharma and Mr M. M. Sabharwal, in typing this manuscript and in the preparation of the diagrams. I wish also to thank Dr Michael Mendillo (U.S.A.) and Professor John V. Evans (U.S.A.) for providing August 72 results in advance; and Dr G. N. Taylor for using in Figure 112 his unpublished profiles for the flares of 11 December 1968 and 19 August 1970. To Dr Richard F. Donnelly (U.S.A.) I owe a special debt of gratitude: much of the material presented in this work on SFDs comes from his works; he has also kindly taken the trouble of proofreading and reviewing this book.

PUBLISHERS	PUBLICATIONS
Air Force Cambridge Research Laboratories, U.S.A.	*AFCRL Reports*
American Geophysical Union	*Journal of Geophysical Research* *Radio Science*
Communications Research Centre, Canada	*CRC Reports*
ESSA Research Laboratories	*Technical Reports*
Institute of Physics and the Physical Society, London	*Proc. Phys. Soc., London*
MIT Press	*Annals of IQSY*

Macmillan Journals Limited, England	*Nature (Physical Sciences)*
North Holland Publishing Co., Amsterdam	*Space Research* *Solar Flares and Space Research*
Pergamon Press, England	*Journal of Atmospheric and Terrestrial Physics* *Planetary and Space Science*
Pennsylvania State University, U.S.A.	*Ionospheric Research Laboratory Scientific Reports*
Radio Research Laboratories, Japan	*Journal of the Radio Research Laboratories*
D. Reidel Publishing Co., Dordrecht	*Solar Physics* *Space Science Reviews* *Intercorrelated Satellite Observations Related to Solar Events* *Solar-Terrestrial Physics/1970*
Science Council of Japan	*Solar Activity Chart*
Svenska Geofysiska Foreningen, Sweden	*Tellus*
University of Chicago Press, U.S.A.	*Astrophysical Journal*
Universitetsforlaget, Tromsö	*Magnetosphere-Ionosphere Interactions*
World Data Centre A	*Upper Atmosphere Geophysics Reports (UAG Series)*

A. P. MITRA

CHAPTER 1

INTRODUCTION

The Sun acts as a formidable source of radiation during a solar flare, throwing out increased electromagnetic radiation in the X-rays and in the ultraviolet and a stream of particles of a wide spectrum of energies, from the solar cosmic rays ($E \sim 1$ GeV) to sub-relativistic protons of energies between 1–1000 MeV that spiral along the Earth's magnetic field lines into the polar regions within some 80 min to 4 h, down to the slower energy cloud of ions and electrons that envelope the Earth some 20–40 h later. In their interaction with the Earth's atmosphere, several effects are produced. The increase in the XUV radiation produces immediate increase in the ionospheric ionization of varying degrees at different heights, together called the Sudden Ionospheric Disturbances or SIDs. These increases are also seen as a sudden brightening of the Hα radiation, for long the only way of monitoring such solar eruptions, and over the entire radio spectrum as radio bursts, much of which can penetrate to the

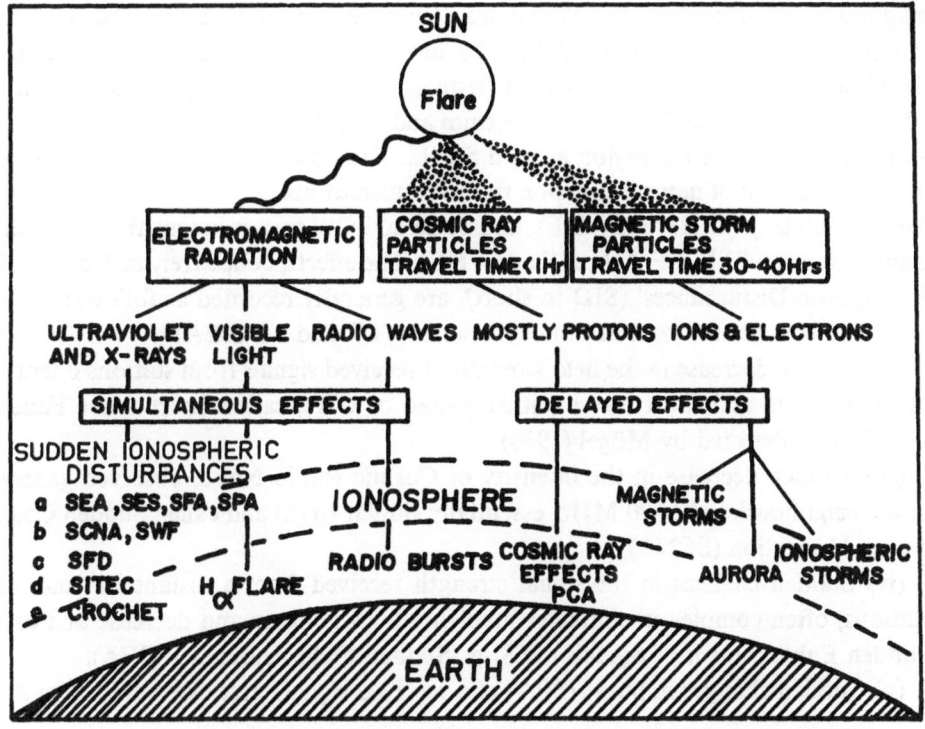

Fig. 1. An illustrative sequence of ionospheric effects associated with solar flares.

ground and are monitored by radio astronomical observatories (Figure 1). The solar cosmic ray particles are recorded at the ground by neutron monitors as Ground Level Events (GLE); such events are relatively few. The sub-relativistic particles (mainly protons) entering the polar regions spiralling down the Earth's field lines cause increased absorption in the polar cap region, called Polar Cap Absorption Events (PCAs). Slower particles causing magnetic storms also cause complex changes in the ionospheric ionization, principally in the F-region and above. These are the ionospheric storms; there are usually considerable structural changes in the ionization profile. In the higher latitudes there is, in addition, the generation of brilliant aurorae.

This work is concerned primarily with the effects caused by the increased electromagnetic radiation and, therefore, deal principally with the Sudden Ionospheric Disturbances. Nevertheless, the immediate particle effects causing PCAs are discussed, in the last chapter, mainly for completeness, but also because of their similarity with SIDs in probing the ion chemistry of the ionosphere.

Changes in the electromagnetic spectrum have been best monitored in the X-rays, with a number of satellites; but more recently increases in some of the ultraviolet regions have also been recorded. The enhancement is the most severe in the region $\lambda < 10$ Å; these are also the wavelengths that penetrate deep into the lower ionosphere (heights below 100 km). These are thus the heights most affected.

There is also evidence for some enhancement of the ionization above this level both in the E-region (Bibl, 1951) and the F-region (Minnis and Buzzard, 1958), but the magnitudes are considerably less and are, therefore, more difficult to observe. These ionization changes necessarily affect the ionospherically propagated and received signal intensities over a large range of frequencies, the effect being an increase in the VLF and the lower part of the LF spectrum and a decrease at higher frequencies with a further decrease in the region around 5 kHz. The observational techniques usually involve reception of natural emission from thunderstorms (at ELF, VLF and LF) and from the Galaxy (at HF and VHF), and pulse and CW transmissions from terrestrial transmitters (at VLF, LF, MF, HF and VHF). The effects, collectively called, 'Sudden Ionospheric Disturbances' (SID in short), are generally recorded as follows:

(i) Sudden enhancement of atmospherics at VLF and LF (SEA)*.

(ii) Sudden decrease in the field strength of received signals from stations operating in the medium and short waves (either pulsed or CW), called Short Wave Fadeout (SWF), first detected by Mögel (1930).

(iii) Sudden decrease in the intensity of Cosmic Radio Noise, observed generally in the neighbourhood of 20 MHz, essentially similar to (ii) and called Sudden Cosmic Noise Absorption (SCNA).

(iv) Sudden changes in the signal strength received from a distant LF and VLF stations, often complex in character involving both increase and decrease and called Sudden Enhancement of Signals (SES) or Sudden Field Anomalies (SFA).

(v) Sudden changes in phase (SPA) of a received signal, first extensively measured

---

* This is a misnomer. The effect observed at certain frequencies is a decrease in intensity (such as at frequencies below 10 kHz).

by Bracewell and Straker (1949) on 16 kHz Rugby transmission, and later observed at several frequencies.

(vi) Sudden frequency shifts observed with highly stable frequency transmissions, first reported by Watts and Davies (1960), and called Sudden Frequency Deviations (SFD); these originate at heights above 100 km, the exact level being determined by the frequency used, and the range of path.

In addition, flare effects are also observed on f_{min}, the minimum frequency at which reflections are observable at any time on ionograms, and in the critical frequencies

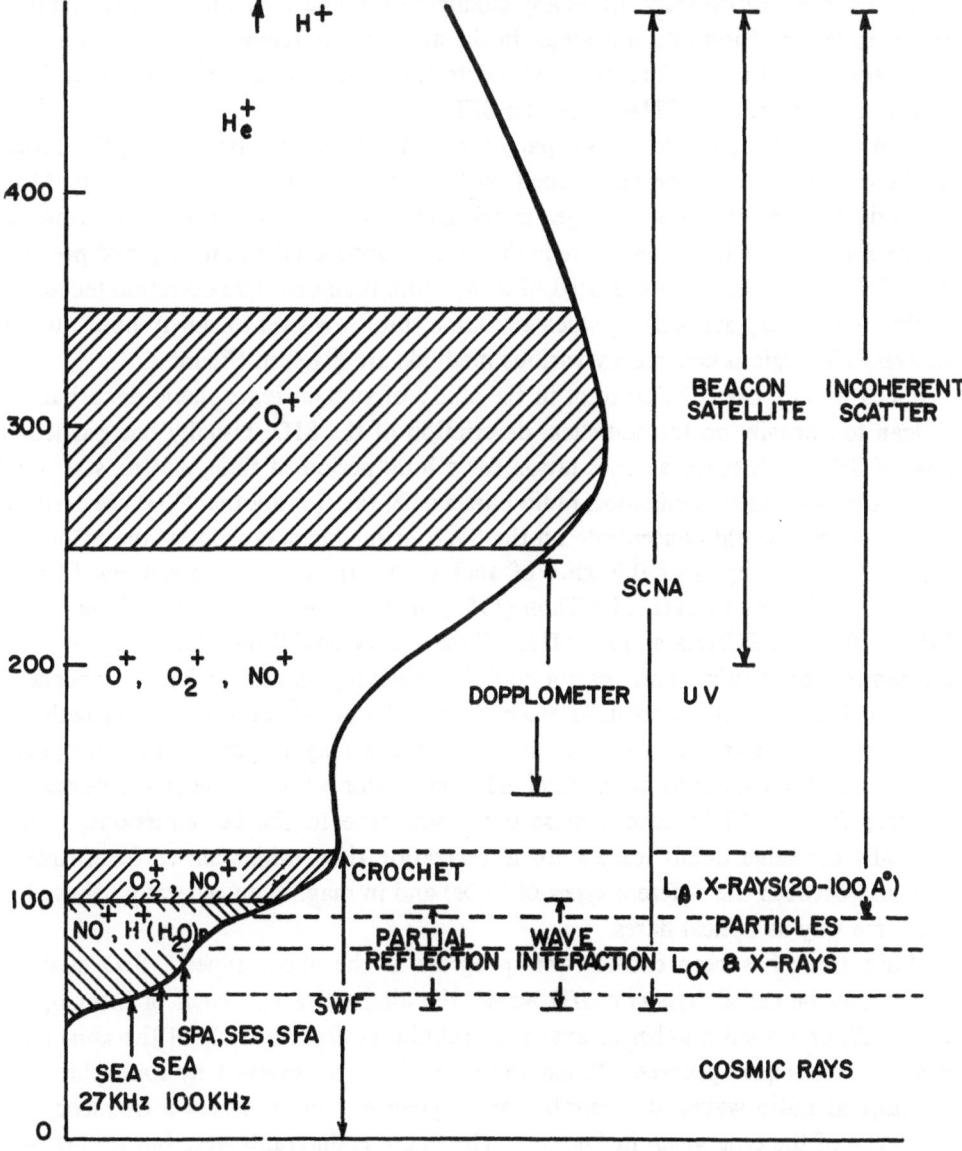

Fig. 2. A diagrammatic picture of the altitude ranges explored by different SID techniques.

of the E- and F-regions, especially in f_oF_2 (Dieminger and Geisweid, 1951; Shapley and Knecht, 1957; and Minnis and Buzzard, 1958).

Recording of SIDs generally have two objectives: (i) flare patrol and (ii) determination of the magnitudes of ionization changes in the various regions of the ionosphere and the subsequent mechanisms controlling the decay of this excess ionization.

These two objectives are not always compatible. A technique which is satisfactory for flare patrol is not always satisfactory for the study of ionospheric physics. Thus, while a shortwave fade-out is generally a very efficient detector of solar flares, it cannot be used with advantage for any quantitative estimates of the ionospheric profile changes. Conversely, while any change in f_oE and f_oF_2 can be directly interpreted in terms of ionization changes in the appropriate regions, these are not easy to observe and are, therefore, poor flare detectors. Routine methods that serve both objectives well are the SCNA, SPA and SFD.

Even when the objective is principally the study of ionospheric physics, the choice of the techniques must be determined by the region sought to be examined. Most methods involve ionization changes in the D-region – those providing information on the F-region are few: these include the SFD, ionosonde data on f_oF_2, and possibly the SCNA. There is thus some need of using more than one flare detection technique at the same place, particularly when the same station wishes to satisfy both the objectives. The regions covered by various methods are given in Figure 2.

It has usually been the custom, in the study of sudden ionospheric disturbances, to lean too heavily on the statistical description of the SID. Usually one particular type of SID is observed at one station for a long period of time, and its statistical properties *vis-à-vis* Hα emissions, radio bursts or X-rays (or another SID) are derived. Thus a station would concentrate on the SWF or magnetic crochet or the SCNA or SEA. Classic examples of this kind of analysis are those of Bracewell and Straker (1949) on SPAs at 16 kHz, of Ellison (1953) on SEAs at 27 kHz and of Shain and Mitra (1954) on SCNAs at 18.3 MHz. These works established the values of these techniques for routine flare patrol and for studying flare-associated ionospheric changes. Details of the percentage association with optical flares, time of growth and decay of the effects and correlations between the magnitudes of the ionospheric effects and Hα emissions were obtained. The major value of these works was to indicate that the SIDs have a close correspondence to the Hα emissions, follow generally the same trends (except for a 'relaxation time' which varied from case to case and between the different types of SIDs) and in magnitude generally behaved in the same way as optical flares.

While there has been considerable progress in the above type of SID analysis, which may be called '*synoptic SID work*', the alternative approach of bringing together all, or a good number of available techniques, for the study of the *same flare*, has not seen equal progress. When the same flare is observed by ionospherically propagated radio waves at a number of frequencies, one observes a frequency dependence of the type given in Figure 3. The figure is diagrammatic, but it illustrates how sensitive the SID is to the frequency of the exploring radio wave. There are two

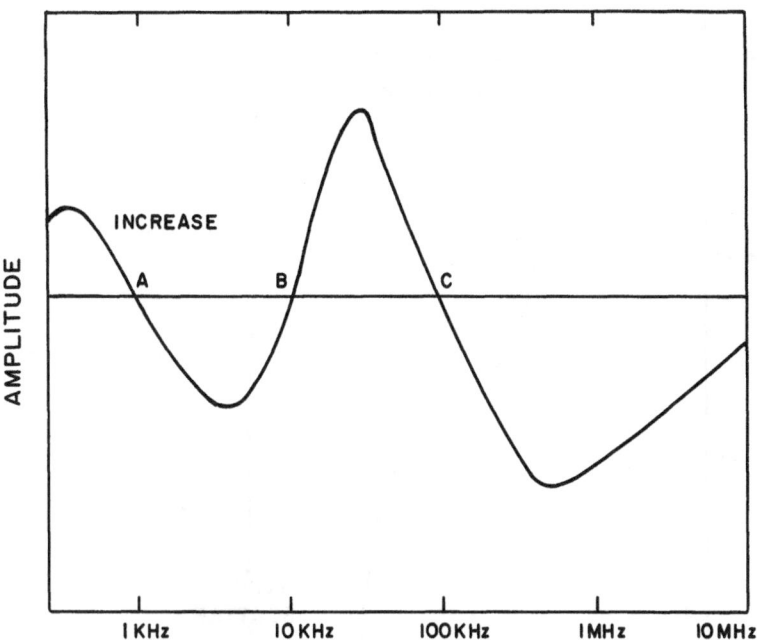

Fig. 3. Effect of solar flares on signal strengths at various frequencies.

frequency regions over which the effect of a solar flare is to cause a decrease in signal intensity – one is in the VLF below about 10 kHz and the other in the medium and shortwaves. In between these two ranges (and also in the ELF range) there are zones of signal enhancement. The three transition frequencies A, B and C are not definitely known. The first is believed to be around 1 kHz, the second around 10 kHz (Sao *et al.*, 1970), and the third, which, unlike the first two, must depend on the angle of incidence of the radio wave, is almost certainly below 75 kHz for vertical incidence and is probably around 50 kHz.

The effects we have so far discussed, and the ones most commonly observed, are indirect ones; these arise because of an enhancement in ionization which progressively decreases with height from a factor of 5–10 around 70–80 km to about 50–100% in the *E*-region and about 1–20% in the *F*-region. Figure 4 gives a rough indication of the different degrees of ionization enhancement at different levels. Attempts to determine these ionization changes quantitatively are relatively more limited; amongst the more important efforts are those by the Pennsylvania State University which uses a high-power wave interaction technique and obtains quickrun profiles of the *D*-region ionization during the entire course of the flare (Rowe *et al.*, 1970); those by Belrose, and his colleagues who use a partial-reflection technique (Belrose, 1969; Montbriand and Belrose, 1972); a remarkable series of profile determinations from 100 to about 300 km by the incoherent-scatter equipment at Arecibo for the two large flares occurring on May 21 and 23, 1967 (Thome and Wagner, 1971) and more recent determination at Malvern for *E*-region heights (Taylor and Watkins, 1970), at Millstone

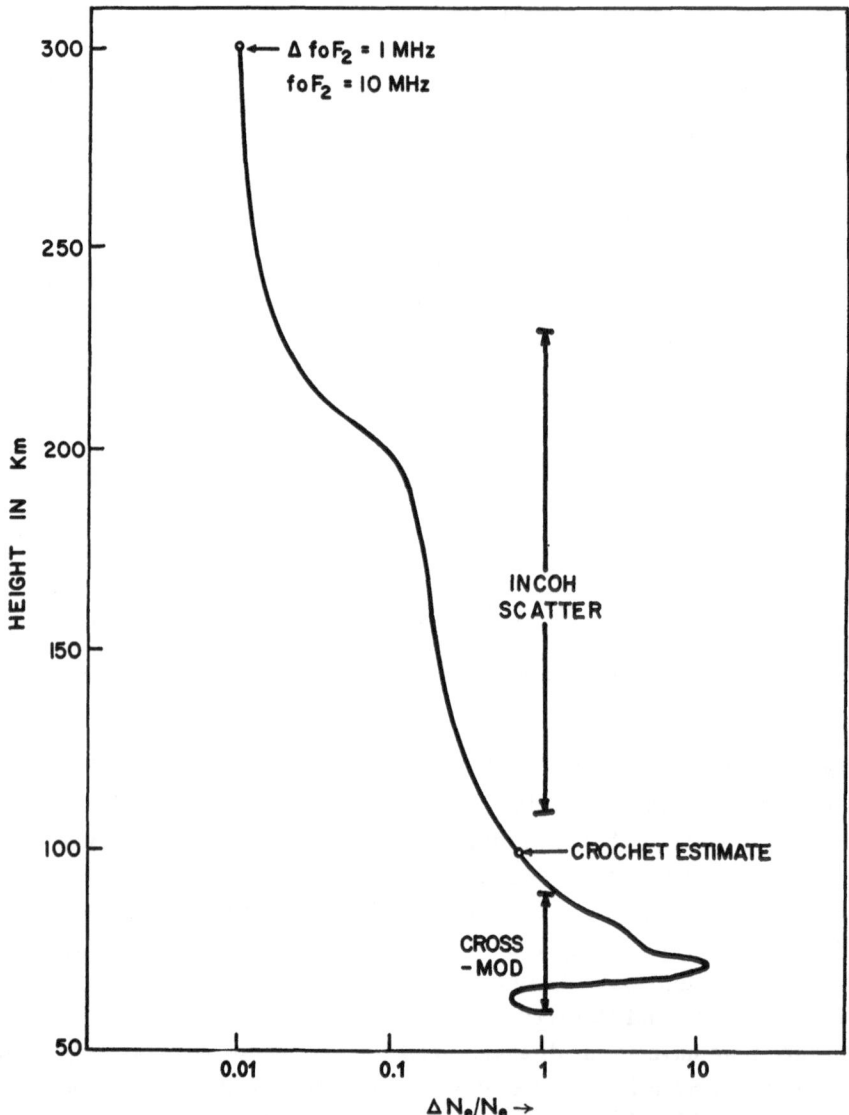

Fig. 4.   Representative diagram showing ionization enhancements at various levels in the ionosphere
during a moderate solar flare. Sources of information for different heights are indicated.

Hill, U.S.A. and Chatanika, Alaska, for F-region heights, and two series of rocket
firings by Somayajulu and Aikin (1969, 1970) into the flares occurring on January 15,
1968 and August 21, 1968. Since such works are necessarily limited, there have been
several attempts to use the more conventional SID technique for profile-studies. This
was done by May (1966) for the flare of October 7, 1948 with VLF observations, by
Deshpande and Mitra (1972c) with multifrequency SCNA observations and with
combination of SCNA and LF/VLF observations; and for the F-region by Donnelly
(1968, 1969) with SFDs at different frequencies.

In any physical study of flare-associated ionospheric effects, it is desirable to have additionally the following information:

(1) The complete time history and the spectral distribution of the ionizing flux, along with changes in the spectral distribution.

(2) The nature and concentrations of the atmospheric constituents ionized, and

(3) The nature of the effective loss rate.

It is, however, very rare to have all this information for any one event. Firstly, while entire time histories are now recorded for one or more bands (e.g. 0–3 Å, 0–8 Å, 8–20 Å, 44–60 Å and 2–12 Å), it is not easy to build a reliable spectral distribution out of these measurements, partly because the detector response itself is a function of the spectral temperature, and partly because this temperature is not constant over the band. Secondly, difficulties arise specially in the D-region where much of the pre-flare ionization is controlled by the minor constituent NO for which conflicting information existed until recently, and the relative contributions of the different ionization sources are uncertain. Thirdly, evidence exists to show that there are sharp decreases in the D-region loss rates during a flare and that these changes may well vary from flare to flare. However, at heights above 100 km no information exists for any change in the loss rate.

CHAPTER 2

DESCRIPTION OF SID TECHNIQUES AND OBSERVATIONS

Over the years the equipments that have been evolved fall in two categories: those that are relatively simple and inexpensive and easy to operate in solar and ionospheric observatories, and those that are extremely complex and sophisticated facilities, available only to a few stations in the world. In the first category we have the SEA, SES/ SFA, SPA, SCNA, SFD and the satellite radio beacon method; in the second category, use is made of cross-modulation, partial reflection and incoherent-scatter. The flare programmes in the latter category are usually only a marginal part of the total activity of the facility, and frequently needs pre-planning and change in operational schedules.

There is some confusion in published literature about the nomenclature of the SIDs. The different nomenclatures and the ones used in this work (and recommended for general uses) are given below:

ELF/VLF/LF		Currently used	Recommended
Atmospherics	(a) Sudden Decrease in Atmospherics	SDA	
	(b) Sudden Increase in Atmospherics	SEA	SDA/SEA
Transmitters	(a) Sudden Field Anomalies	SFA	
	(b) Sudden Increase in Long Wave Signals	SIL	SFA
	(c) Sudden Enhancement in Signals	SES	
	Phase Sudden Phase Anomalies	SPA	SPA

MF/HF/VHF		Currently used	Recommended
Transmitters	Short Wave Fadeout	SWF	SWF
Cosmic Noise	Sudden Cosmic Noise Absorption	SCNA	SCNA
Satellite-Radio Beacons	Sudden Increase in Total Electron Content	SITEC	SITEC

The magnitude of an event is given on a subjective scale from 1^- (the least important) by 3^+ (the most important) in a 9-step scale. The definiteness of an event is described by the following scale:

5 = definite	2 = fair
4 = reasonably definite	1 = possible
3 = reasonable	0 = questionable

2.1. The Sudden Enhancement of Atmospherics (SEAs)

The Sudden Enhancement of Atmospherics (SEAs) were discovered by Bureau (1937) who found that these enhancements were characteristic of the spectral range 7 km to 16 km (Bureau, 1947), being most pronounced at a wavelength of about 11 km (27 kHz) – a result later verified by Ellison (1950), Sachdev (1958) and others. Enhancement of atmospherics does not, however, extend over a wide frequency range. At frequencies below 10 kHz a *decrease* has been observed (Kamada, 1961), with another region of enhancement below 1 kHz. SEAs are recorded in the VLF by many observatories as a means of routine flare patrol.

Recordings of atmospherics can be made fairly easily and cheaply. A block diagram of the atmospherics recorder is shown in Figure 5. The output meter scale is usually

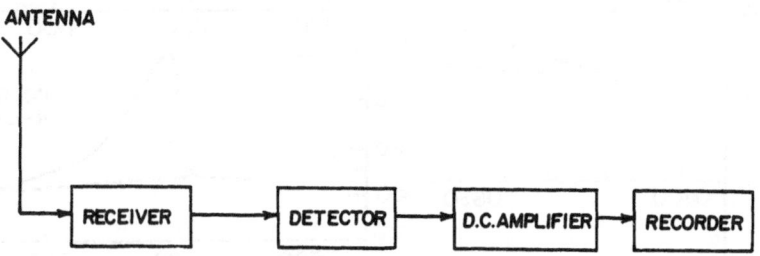

Fig. 5.   Block diagram of an atmospherics recorder.

made logarithmic to prevent the equipment for saturation during local thunderstorms. The receiver in this simple system consists of two tuned amplifier stages followed by a cathode follower before feeding into the detector stage. It is necessary to provide for a wave trap circuit at the input of the receiver to eliminate interference from local, strong, medium-wave broadcast stations. The antennas used are generally of the inverted *L*-types. In the Delhi equipments the antenna for 100 kHz observations is 38 m long and 11 m high and for 27 kHz, 38 m long and 17 m high. The feed-in is by means of a very short length of transmission lines.

Observations of atmospherics have been extended to ELF range through observations of the *magnetic* field intensities of atmospherics. Sao *et al.* (1966) have reported such measurements at 600, 260, 100, 30 and 9 Hz, i.e. the frequency range of slow tail atmospherics and Schumann resonances. They used a *buried* coil placed in the N–S direction, buried to avoid effects due to wind-induced vibration or the electric charges of rain drops. All these frequencies showed enhancements (as evident in

Figure 8). The largest enhancements were on 600 and 260 Hz, the enhancements decreasing in magnitude and occasionally vanishing at 100 Hz.

Typical smoothed records of atmospherics enhancements on 27, 30 and 100 kHz recorded at New Delhi are given in Figure 6a. Some raw SEA records observed at New Delhi are shown in Figure 6b. For 10, 21 and 27 kHz atmospherics Sakurai (1968) has classified them in three types A, B and C which are schematically shown in Figure 7. Most cases observed in Japan are of type C, which are further classified

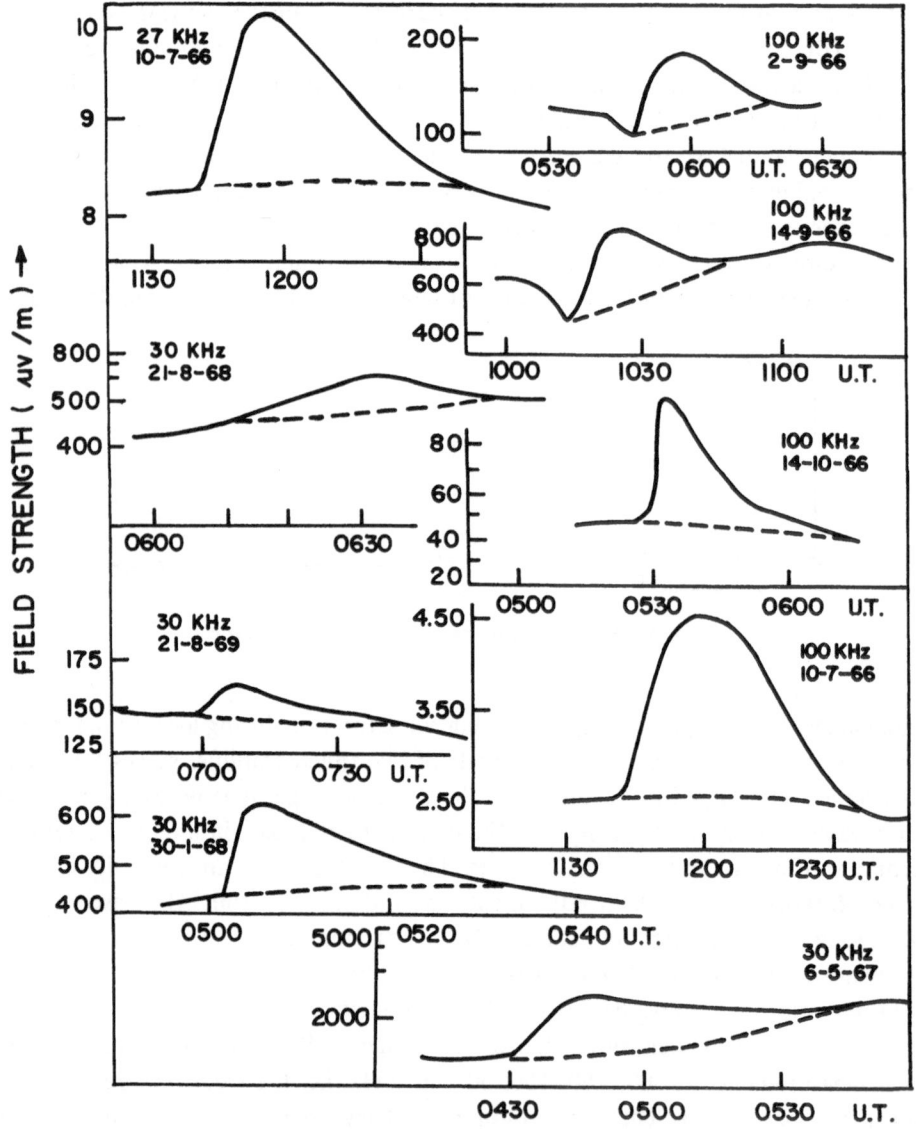

Fig. 6a.   Typical examples of sudden enhancements of atmospherics (SEAs) on 27, 30 and 100 kHz (from observations recorded at Delhi, India).

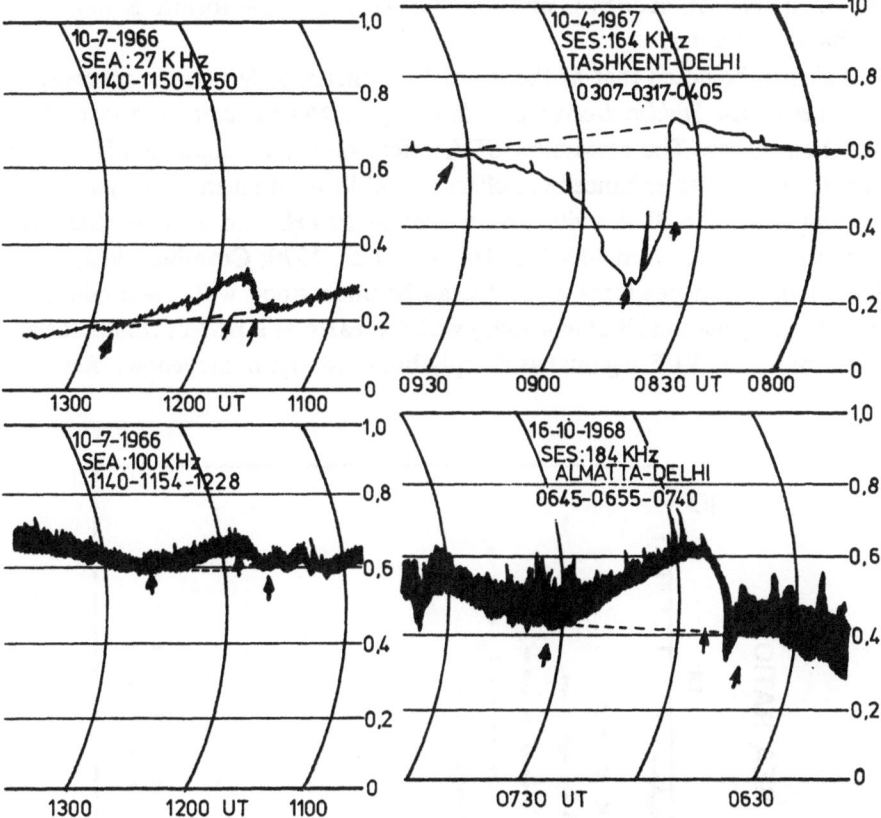

Fig. 6b. Examples of actual recordings of SEA at 27 and 100 kHz (on the same day 10 July 1966) and SES observations on 164 kHz (Tashkent-Delhi) and 184 kHz (Alma Ata–Delhi) (from observations recorded at NPL, Delhi).

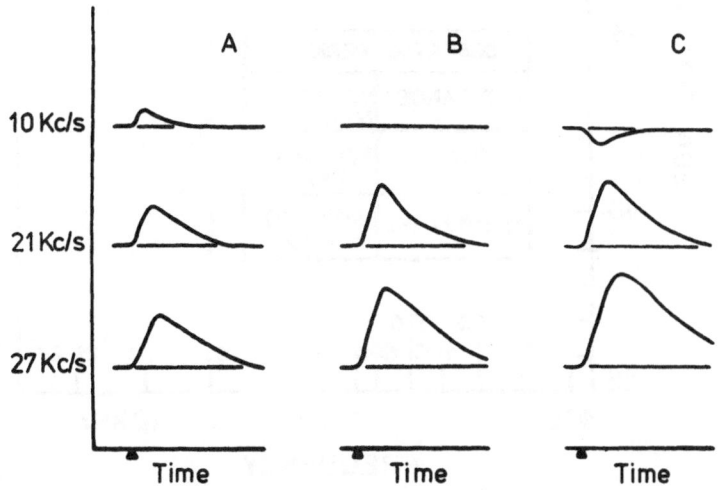

Fig. 7. Developing patterns of representative SEAs at 10, 21 and 27 kHz (after Sakurai, 1968).

into two sub-classes, sudden-C and slow-C; the rise of the former is much sharper than that of the latter.

It is also to be noted that at the lower frequencies, a *decrease* in intensity occurs (sometimes called *Sudden Decrease in Intensity*, SDA) as against an increase in the higher frequencies. The maximum SDA effect occurs at about 5 kHz, thereafter switching again to an enhancement effect at still lower frequencies (Figure 8).

The upper SEA to SDA switchover is around 10 kHz and the lower SDA to SEA switchover occurs between 800–5000 Hz (Sao *et al.*, 1970; Crombie, 1965).

Such frequency dependence of SEAs can be understood with the mode theory of VLF radio propagation (Budden, 1961) with the earth as a perfect conductor and the ionosphere (in the VLF regions) as sharply bounded and homogenous. According to

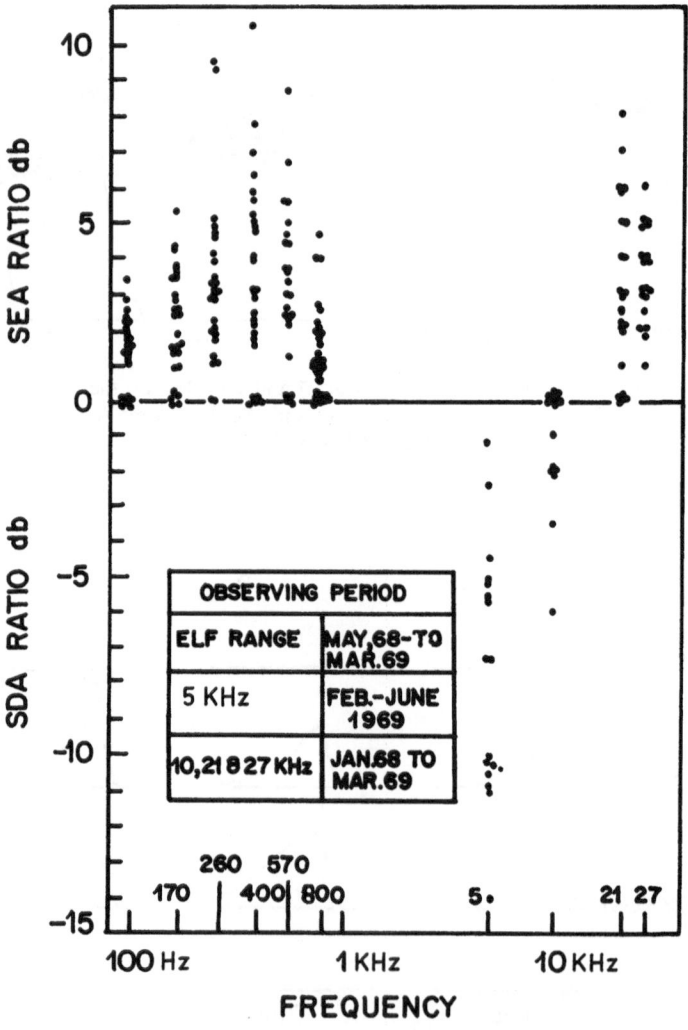

Fig. 8.   Flare effects of atmospherics in the ELF (after Sao *et al.*, 1970).

this theory, the cut-off frequency f_c in the VLF radio propagation within the Earth-ionosphere cavity is a function of the cavity height h. The cut-off wavelength λ_c corresponding to f_c is given by h/n, in which an integer n gives the nth harmonic in the propagation mode. Since during a flare there is lowering of height of the lowest portion of the ionosphere, this causes a decrease in the cavity height h, and consequently the cut-off frequency moves to higher frequencies, causing a decrease of the atmospherics field intensity at lower frequencies.

In the ELF range the mode theory interpretation assumes *two* sharply-bounded homogenous layers, the lower layer with refractive-index n_1 located at height h_1 and the upper layer with refractive-index n_2 located at height $h_1 + \varDelta$. Sao *et al.* (1970) find that observational data at 260 and 570 Hz can be reproduced for a distance of 1000 km with the following parameters:

$$
\text{Normal} \begin{cases}
\text{Noon:} & (\omega_{r1})_n = 5 \times 10^4 \, \text{s}^{-1}, \quad (\omega_{r2})_n = 1 \times 10^5 \, \text{s}^{-1} \\
& \varDelta_n = 25 \text{ km and } h_{1n} = 65 \text{ km} \\
\text{a.m. and p.m.} & (\omega_{r1})_n = 2 \times 10^4 \, \text{s}^{-1}, \quad (\omega_{r2})_n = 2 \times 10^5 \, \text{s}^{-1} \\
& \varDelta_n = 25 \text{ km and } h_{1n} = 70 \text{ km}
\end{cases}
$$

$$
\text{Disturbed} \begin{cases}
(\omega_{r1})_f = 1 \times 10^4 \, \text{s}^{-1}, \quad (\omega_{r2})_f = 2 \times 10^5 \, \text{s}^{-1} \\
\varDelta_f = 25 \text{ km}, \; h_{1f} = 60 \text{ km}.
\end{cases}
$$

Here ω_r is the conductivity parameter, discussed in further detail in later sections.

2.2. Sudden Enhancement of Signal Strengths (SES), Sudden Field Anomalies (SFA)

This is a modification of SEA in that, instead of using atmospherics, one uses transmissions from stable transmitters in the VLF, and in some cases, in the LF. The disadvantage in the atmospherics is that the sources are uncertain and variable, and the recorded noise is an integrated effect of many such sources. Consequently, it is difficult to use such records for the study of ionospheric physics, although it is quite useful for the detection of flares. A distinct advantage arises if one can select and monitor transmissions at LF and VLF from transmitters at known distances. In the VLF and for very long distances in the LF, enhancements are observed in the signal strengths; in the LF the resulting change is often quite complex.

This technique was first applied by the Cambridge Group (Bracewell and Straker, 1949; Bain *et al.*, 1952) which recorded the signal strengths of GBR (16 kHz) at Cambridge (90 km), at Aberdeen (a distance of 535 km) and at Edinburgh (a distance of 418 km), thus involving both steep and oblique incidence propagation. At steep incidence it is usually the conversion coefficient $_{\|}R_{\perp}$ which is increased; at oblique incidence it is the reflection coefficient $_{\|}R_{\|}$. There is a well-established relationship (Bracewell *et al.*, 1951) between the maximum decrease in $_{\|}R_{\perp}$ at 16 kHz and the maximum decrease in h; a value of 0.6 for $_{\|}R_{\perp}/_{\|}R_{\perp}$ corresponds to $\varDelta h = 6$ km. The flare effects, called by Bain (1953) as *Sudden Field Anomalies (SFAs)*, were characterised mostly by a rapid increase of signal strength, followed by a slow decrease, but on several occasions the onset consisted of a decrease in signal strength, or of a decrease followed by an increase, such as in Figure 9. These unusual SFAs all occurred in the winter months from November to March. An interesting feature was that the nature of changes was quite different at Aberdeen and Edinburgh. As will be noted in Figure 9, all effects recorded at Edinburgh began with a decrease in the resultant signal strength, although the largest anomalies contained an increase to a maximum after this decrease. There appeared to be slight increase in $_{\|}R_{\|}$ during an SID on 16 kHz observed at a distance of 1350 km (Belrose, 1957).

Although most of the SFAs were observed by simple amplitude recording, on several occasions the phase of the received signal at Aberdeen was also measured

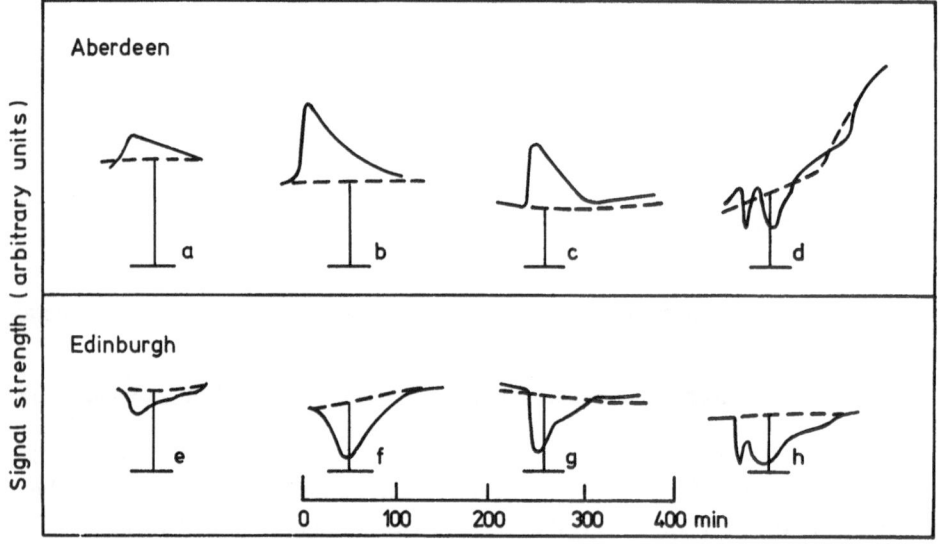

Fig. 9.   Differences in sudden field anomalies between GBR – Aberdeen (535 km) and GBR–Edinburgh (419 km) (after Bain, 1953).

simultaneously. A locus in the complex plane obtained at such a time showed that the change in the signal strength during the anomaly was due mainly to a change in the phase of the sky-wave; the amplitude of the sky-wave was scarcely affected. It was also evident that the anomaly was nearly entirely due to changes in the once-reflected component of the sky-wave.

The shapes of summer SFAs at Aberdeen and Edinburgh could be explained qualitatively on the basis of a sky-wave for which the phase lag on the ground wave decreased during the build-up of the anomaly. At Aberdeen the sky-wave normally lagged on the ground wave by a phase angle of about 150° during the summer day. A decrease in this phase lag due to an SID caused the resultant signal strength to increase. At Edinburgh, nearer the transmitter, the phase lag of the sky wave on the ground wave was estimated to be greater and around 238°. A further small decrease in phase lag, therefore, caused a decrease in signal strength. However, if the phase changes were sufficiently great to drive the sky wave beyond the position of antiphase with the ground wave, an increase of signal strength followed the decrease.

Subsequently, such recordings were introduced at other stations. In Boulder (U.S.A.) both phase (Section 2.3) and amplitude variations of phase-stabilized VLF transmissions were recorded for many years over six long-range propagation paths from NBA (18 kHz), Balboa, Panama; NPG (18.6 kHz), Seattle, Washington; GBR (16 kHz), Rugby, England. The network covered approximately one quarter of the Earth's ionosphere and also a comparison of the east–west and north–south paths. More recently, phase and amplitude perturbations of 100 kHz LORAN-C signals are being received at Boulder, Colorado, from Nantucket, Massachusetts; Cape Fear, North Carolina; and Jupiter Inlet, Florida; as also VLF transmissions on 24 kHz from Panama, Canal Zone; on 21.4 kHz from Annapolis, Maryland; and 18.6 kHz from Seattle, Washington. Observations recorded for the large flare event of May 23, 1967 are shown in Figure 10. The lowest curve in Figure 10 gives the 27 kHz atmospheric radio noise received from an omni-directional antenna. The amplitude enhancements were in the ratios 1, 0.2 and 0.1 as against phase ratios of 1, 3 and 1.

In Japan extensive observations with the LORAN-C chain were started at Hiraiso

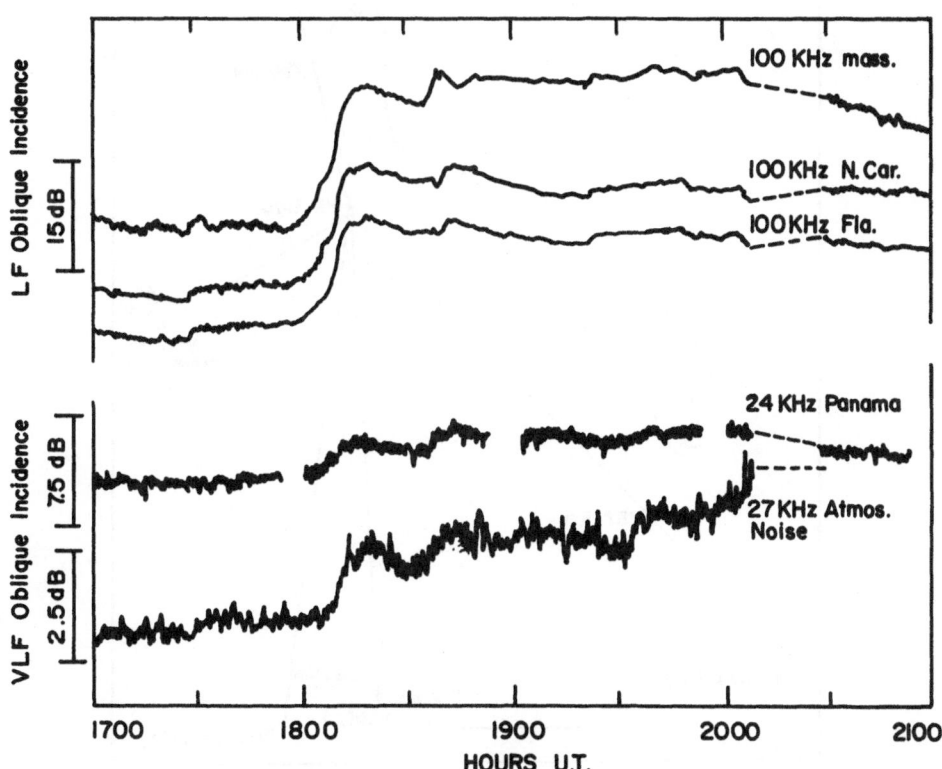

Fig. 10.   Boulder recordings of VLF –SEAs for the event of May 23, 1967 for different paths (from Jean and Large, 1967).

(36°22′N, 140°38′E) in March 1970, as a part of Japan IASY programme. The LORAN-C chain covering the North-West Pacific area (SS-3) consists of a master and four slave stations. The geometrical arrangement of the LORAN-C transmitters and the receiver at Hiraiso is shown in Figure 11. The master station at Iwojima transmits a group of nine pulses of width 300 μs each. Each is separated by 1 ms from the preceding pulse except the ninth pulse; for the latter the separation is 2 ms to allow identification. Each slave station transmits 8 pulses identical with the master station minus the ninth pulse. These slave pulse groups are delayed by an appropriate amount (coding delay) after the reception of the master signals. It turns out that the total transmission delay at a slave transmitter with respect to the master pulses is equal to the sum of the propagation delay and the coding delay. A 27 m vertical

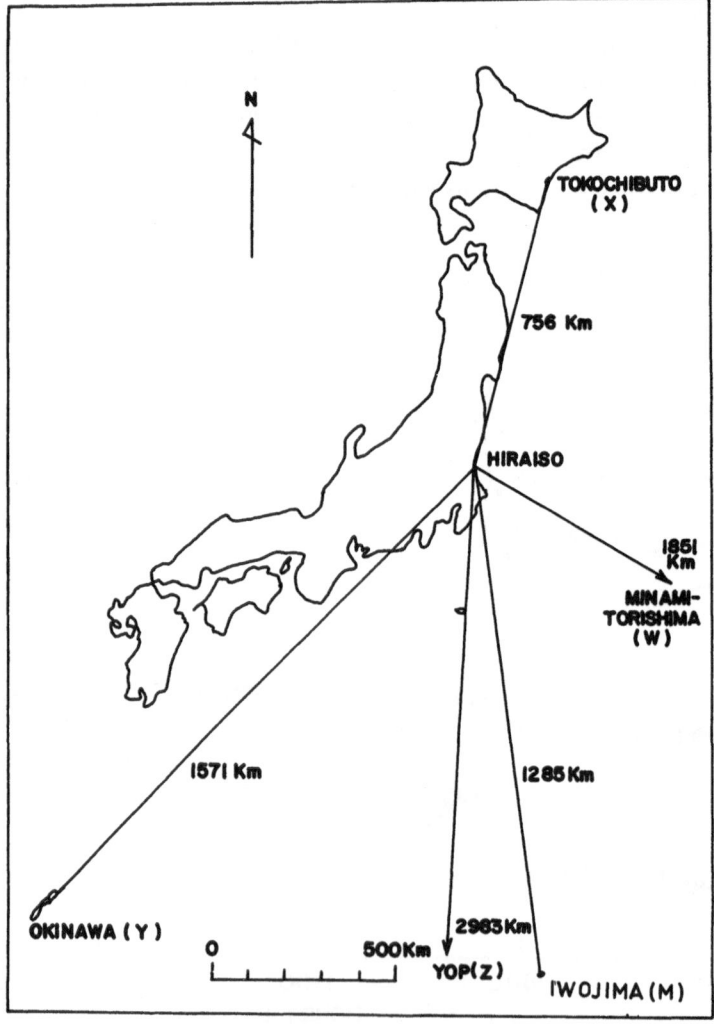

Fig. 11.   Geometric arrangement of Loran-C transmitters and receivers in Japan
(after Wakai *et al.*, 1973).

antenna was used. The path distances in the Japanese arrangement range from 756 km to 2983 km.

A simple VLF equipment measuring the relative phase and field strength of frequency-stabilized VLF transmissions is described below. In this arrangement, the received signal is compared with a voltage derived from a highly stable 100 kHz oscillator in a way indicated in the block diagram given in Figure 12. The signal is

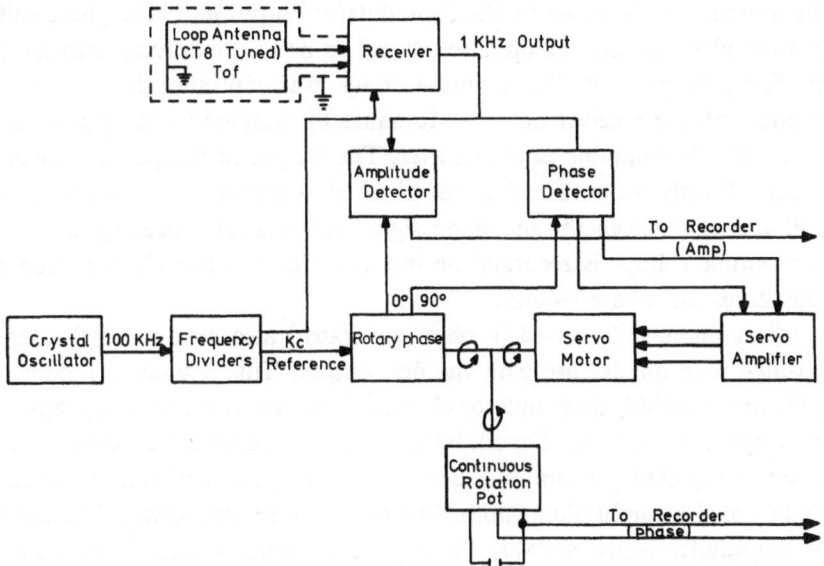

Fig. 12. Block diagram of a VLF phase and amplitude measuring equipment.

received in a balanced, electrostatistically shielded loop antenna. The antenna and transmission line are resonated at the receiver input terminals to the received frequency f. The receiver may be a transistorized superheterodyne converting the incoming frequency f to a signal of 1 kHz using a local oscillator frequency of $(f-1)$ kHz, derived from the 100 kHz reference oscillator. The 1 kHz signal from the VLF receiver is then compared in two coherent detectors with a 1 kHz reference voltage, also derived from the reference oscillator. If the receiver output signal is represented as:

$$E_2 = E_{20} \sin (2\pi f_1 t + \phi)$$

and the reference signal

$$E_1 = E_{10} \sin (2\pi f_1 t)$$

then the demodulator output, which is essentially the product of E_1 and E_2 can be represented as

$$E_0 = \tfrac{1}{2}E_1 E_2 \cos \phi + H - F \text{ terms}.$$ (1)

This is passed through a low pass filter and provides an output that is directly proportional to the phase of the incoming signal.

It is desired in such measurements to record both phase and amplitude simultaneously. The output voltage of the phase demodulator is amplified and used to derive a servo-motor which in turn drives a rotary phase-shifter. The output of the phase shifter changes the phase of its output signal by 360° for a 360° rotation of its shaft, the sign of the phase shift depending on the direction of shaft rotation. If ϕ is not 90°, the servo-motor is driven by the demodulator output until the phase-shifter has changed the phase of the commutation signal to be in quadrature with the receiver output. Thus, the phase of the commutator signal is continuously adjusted to be in quadrature with the receiver output. Mechanically coupled to the phase-shifter is a 360° continuously rotatable potentiometer. The output of the potentiometer sliding arm is thus directly proportional to the phase-shifter shaft position and thus to the phase difference between the incoming signal and the reference signal. The potentiometer output voltage is recorded on one channel of a two-channel recorder and this constitutes the 'phase record'.

The rotary phase-shifter used (a phase generator) also produces a further output signal which is in quadrature with the first output. This is used as a commutating signal for the amplitude demodulator channel. Since the servo system keeps the commutation signal of the phase demodulator in quadrature with the receiver output, the commutation signal of the amplitude demodulator is in phase with the receiver output. The output of the amplitude demodulator is, therefore, proportional to the receiver output amplitude. After suitable filtering, this output signal is recorded on the

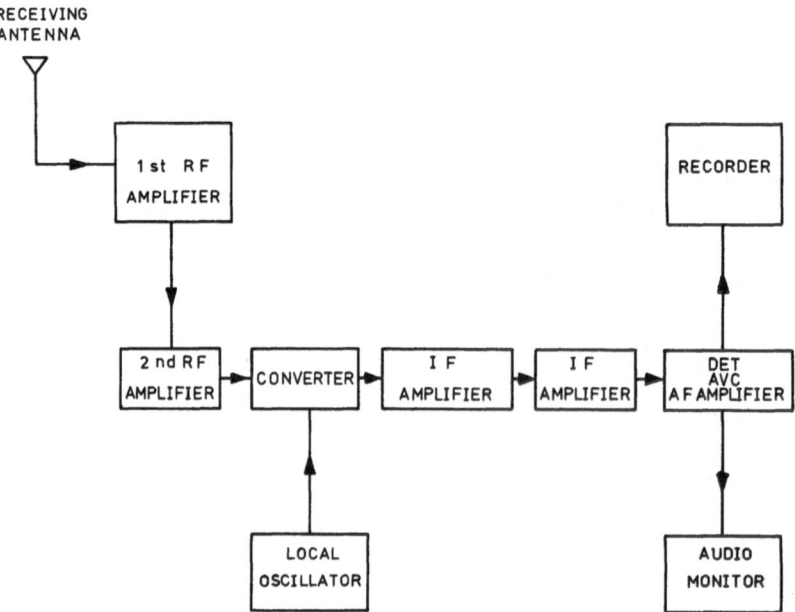

Fig. 13.   Block diagram of a simple 164 kHz receiver receiving transmissions from Tashkent.

other trace of the two channel recorder, which constitutes the 'amplitude' record.

A much simpler equipment can be used when merely the signal strength is recorded. The block diagram of such an equipment is given in Figure 13. In India much work has been done with the 164 kHz transmission from Radio Tashkent. The reception was initially started by S. N. Mitra (1964) at the All India Radio, New Delhi, at a distance of 1630 km. The coordinates and other details of the Tashkent station, as well as of a few other stations now extensively used, are given in Table I. The effect observed is an increase in the signal strength, but often more complex structures are seen. Although the effect was termed SIL by S. N. Mitra, we will call this effect also SES or SFA, since the effect is of the same kind as the effects observed elsewhere with different transmissions. Similar recordings were later initiated at the National Physical Laboratory, New Delhi (later with additions of transmissions on 182 kHz from Alma Ata and 236 kHz from Leningrad), and also at Ahmedabad. 164 kHz signals are also recorded at Sofia in Bulgaria from station Allouis, and Kühlungsborn in German Democratic Republic (Entzian, 1964) regularly records signal strengths on 185 kHz from station 'Stimme der DDR' and on 245 kHz from Kalundborg (Table I).

TABLE I

Some LF and VLF Transmitters used for routine SFA/SES recording

Frequency (1)	Transmitting station (2)	Transmitter details (3)	Path and distance (4)	Receiving organizations (5)
164 kHz	Tashkent	150 kW	Tashkent–Delhi (1600 km)	(i) National Physical Lab., New Delhi. (ii) All India Radio, New Delhi.
164 kHz	Tashkent	150 kw	Tashkent–Ahmedabad	Physical Research Lab., Ahmedabad.
164 kHz	Allouis	150 kW	Allouis–Sofia	Geophysical Institute, Sofia.
182 kHz	Alma Ata	~100 kW	Alma Ata–Delhi (1565 km)	National Physical Laboratory, New Delhi
245 kHz	Kalundborg		Kalundborg–Kühlungsborn (Refl. point at 54.9°N, 11.4°E) 180 km	Heinrich-Hertz Institute, Berlin.
185 kHz	'Stimme der DDR'		182 km (Refl. point at 53.5°N, 12.6°E)	Heinrich-Hertz Institute, Berlin.

Three different types of flare effects are observed on the 164 kHz field intensity records. These are illustrated in Figure 14. The types are:

Type 1: Positive Effect: A rapid increase of signal strength and a gradual recovery (Figure 14a).

Type 2: Negative Effect: A sudden decrease and then a gradual recovery of the field strength (Figure 14b).

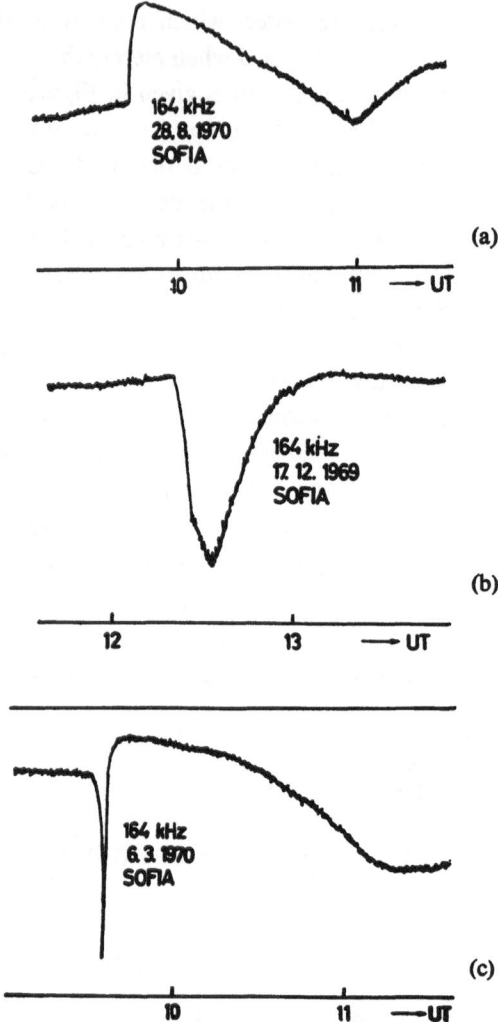

Fig. 14. Three major types of SES observed on 164 kHz in Allouis-Sofia path
(after Letfus *et al.*, 1973).

Type 3: An intermediate type in which there is an initial dip in signal strength followed by an increase. The initial fall takes place within 3 or 4 min, but, in some cases,
may take a longer time (Figure 14c).

These different types have been recognized and discussed by various authors. They
were discussed as early as 1959 by Haubert (1959); by S. N. Mitra (1964), and most
recently by Letfus *et al.* (1973). The exact nomenclature is, however, somewhat different. While type 1 is type A with S. N. Mitra, type 2 is type C and type 3 is type B.
With Letfus *et al.*, the negative effect is called type 1 and the positive effect is type 2.
It is, however, more appropriate to call the positive effect type 1; firstly because
traditionally a positive effect is expected; secondly at lower frequencies the positive
effect is more frequent and a similar classification reduces confusion; and thirdly

even at 164 kHz, negative and the mixed effects are seen most often in winter, when chemical and dynamical conditions are known to be anomalous in middle and high latitudes (Letfus *et al.*, 1973).

Letfus *et al.* introduced also another type observed in winter called type 4, in which two consecutive flares occur, the second one occurring before the ionospheric effect of the first one has subsided. The first effect is identical with types 2 or 3, and the second one is of type 1, normally associated with summer.

Classification of SFA at 164 kHz

	Present classification	S. N. Mitra's classification	Classification by Letfus *et al.*
Type 1	Positive Effect	Type A	Type 2
Type 2	Negative Effect	Type C	Type 1
Type 3	Intermediate Effect	Type B	Type 3

Subrahmanyam (1967) and Deshpande (1971) reporting on the 164 kHz observations at NPL, Delhi, on the Tashkent – Delhi path, have identified further complexities in the structures outlined in the above types. These are shown in a series of records in Figures 15–17. They point out that in the intermediate type 3, the maximum intensity sometimes exceeds the original intensity (Figure 15a), sometimes equals the original (Figure 15b), and sometimes is less than the original intensity (Figure 15c). In type 2 events, the attenuation is sometimes so large as to reduce the received field strength to base level. The average time necessary to reach this maximum attenuation starting from the beginning of the effect is about 1.5 h. The recovery from minimum to normal takes place, on the average, in about 40 min.

At relatively long distances for which such measurements are made, the field strength, under typical conditions, is a vector sum of the ground wave and first-hop sky-wave, and is given by:

$$E \approx E_G \sqrt{\frac{4\varrho}{\eta}} \sqrt{\frac{1}{2}\left(\frac{\eta}{2\varrho} + \frac{2\varrho}{\eta}\right) + \cos\phi} \, . \tag{2}$$

where E_G is the magnitude of the ground wave signal strength at the receiving station, ϱ is the ionospheric reflection coefficient, η is the ground attenuation factor and $\phi = (2\pi/\lambda)\Delta\gamma$ is the phase change of the sky wave with respect to the ground wave, $\Delta\gamma$ being the difference in path length of the sky and ground wave.

In this arrangement the solar flare effect is a combination of both the amplitude and phase changes of the sky wave. The phase changes are transient in nature before the amplitude changes take control. The change in the reflection coefficient during a flare effect is of secondary importance at these large distances and its effect on the field strength is small (Ramanamurthy, 1970). For propagation over distances of 1000–1500 km and levels of reflection in the region 70–80 km and for radio waves in the range of 100–200 kHz, a complete phase sweep (2π) occurs for height variations

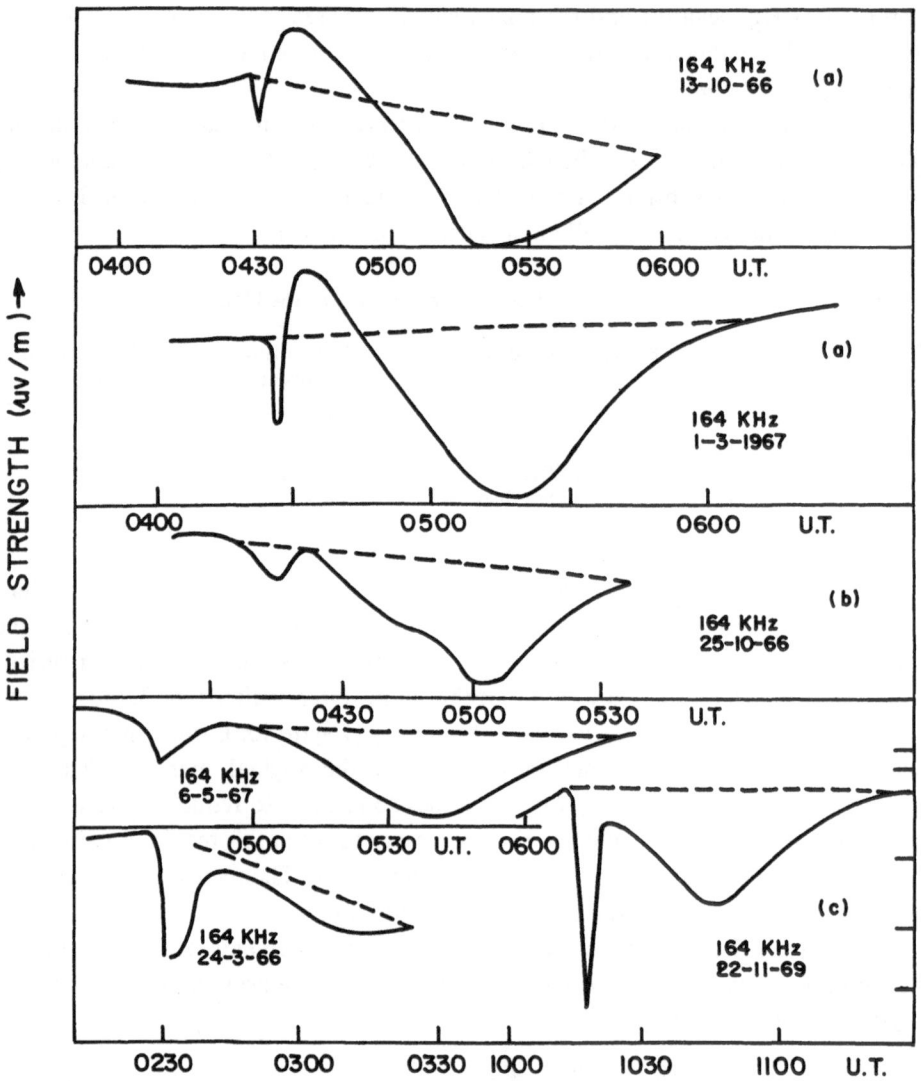

Fig. 15.   Intermediate Type SES observed at Delhi with Tashkent radio transmission at 164 kHz. Different subtypes are indicated.

$\Delta h = 4$–6 km. These waves thus react much more sensitively to phase height variations than VLF waves and allow a more accurate determination of the variations of the level of reflection that occurs during a flare.

2.3. Sudden Phase Anomalies (SPA)

To measure phase variations in VLF signals propagated by the ionosphere, a phase reference signal must be used at the receiving point. Ideally, this would be the signal as observed at the transmitting antenna. For short distances (upto a few hundred

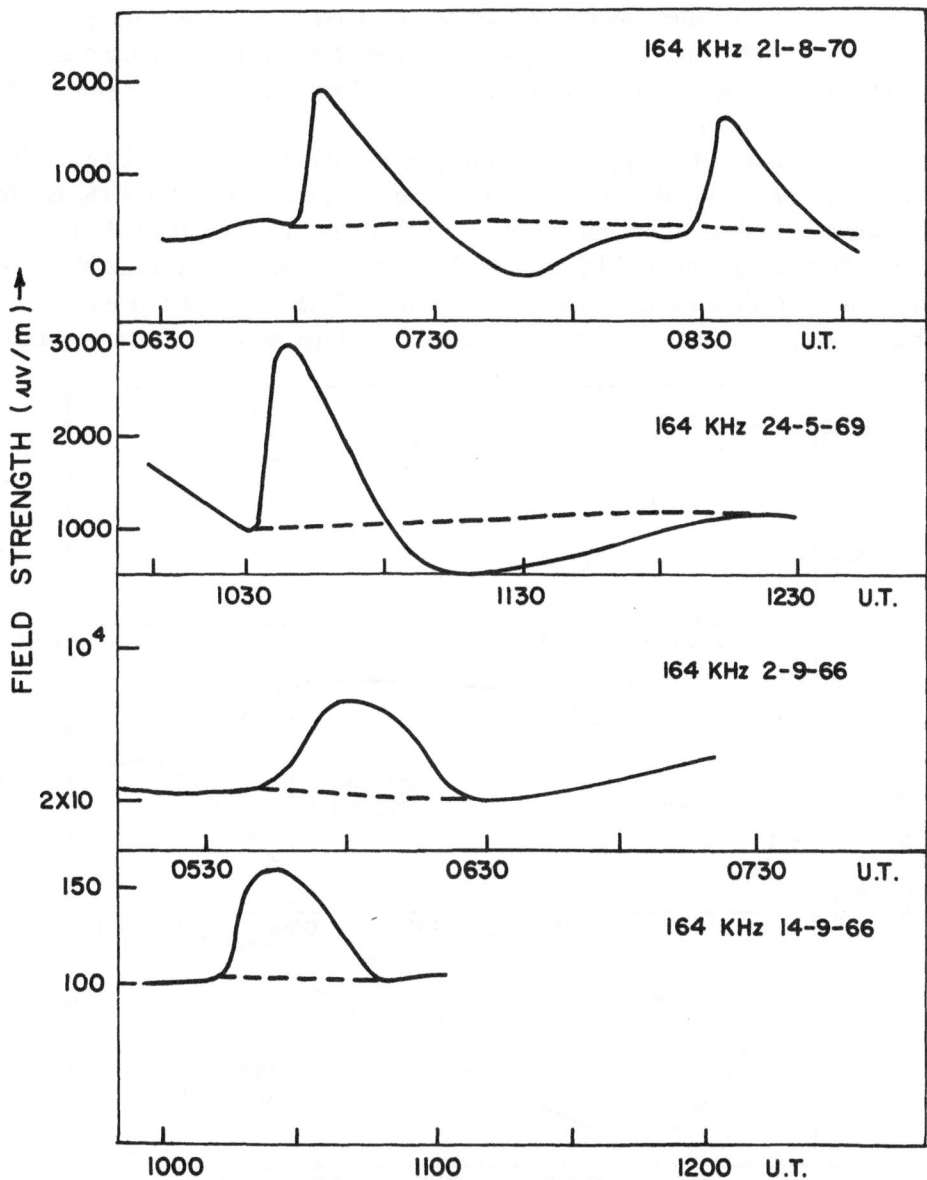

Fig. 16.   Positive SES observed at Delhi with Tashkent radio transmission at 164 kHz. Different subtypes are indicated.

kilometers) the ground wave signal for which the phase velocity is essentially constant can be separated from the sky-wave signal and is frequently used as the reference signal. For larger distances, the reference signal can be transmitted to the receiver by commercial telephone lines. This method has the advantage that the actual frequency transmitted need not be very stable and any transmitter may be used.

This procedure cannot, however, be used for distances beyond 1000 km. For such distances, the transmitted frequency must be extremely constant, and the phase reference signal at the receiver is then provided by another extremely stable oscillator. Ideally, the two oscillators should be of identical frequency or, at the worst, both should drift at the same constant rate.

An SPA is observed as an advance in the phase of the downcoming sky wave. The disturbances have a rapid onset and a slower recovery, and as in other SIDs, last for time which vary from a few minutes upto a few hours. An estimate of the intensity of the effect can be obtained in terms of the number of degrees of phase shift. The length of the path must, of course, be considered. Observations are given either in terms of the phase change $\Delta\phi$ or of an equivalent change Δh in the phase height of a

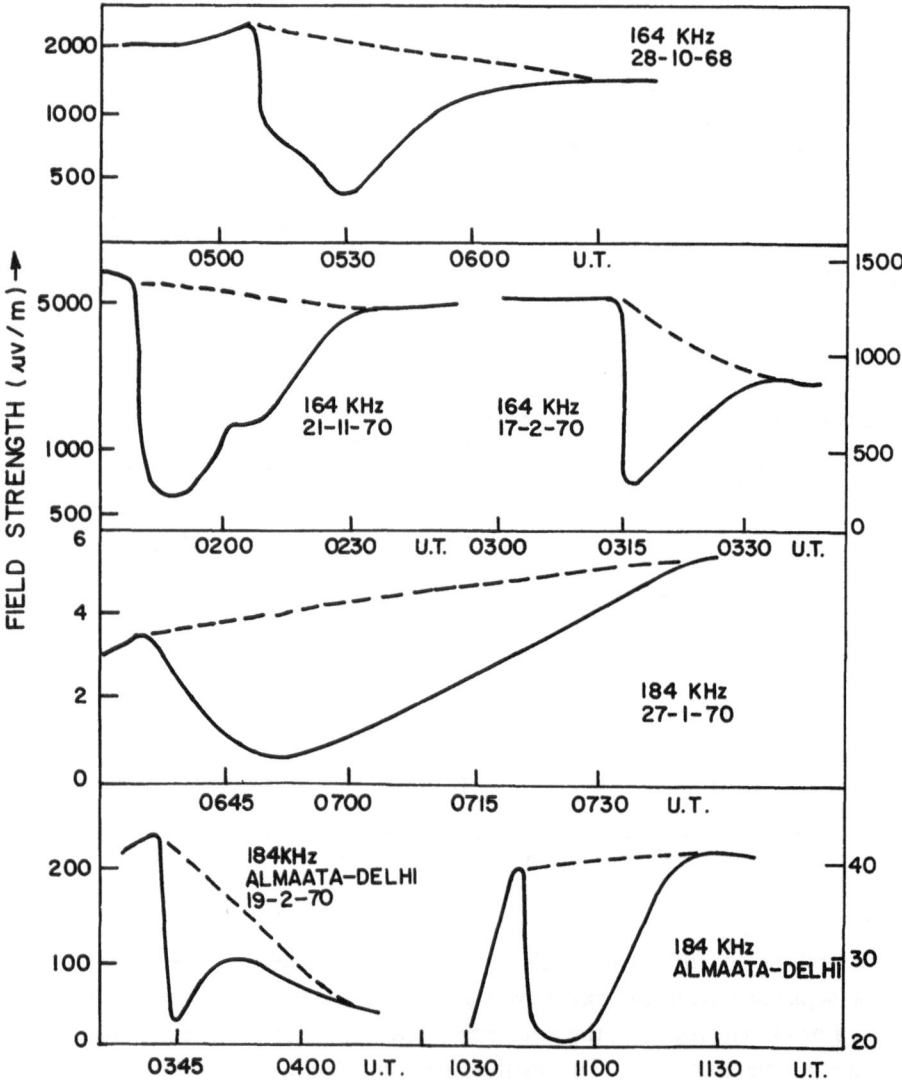

Fig. 17.   Negative SES observed at Delhi with Tashkent radio transmission at 164 kHz. Different subtypes are indicated.

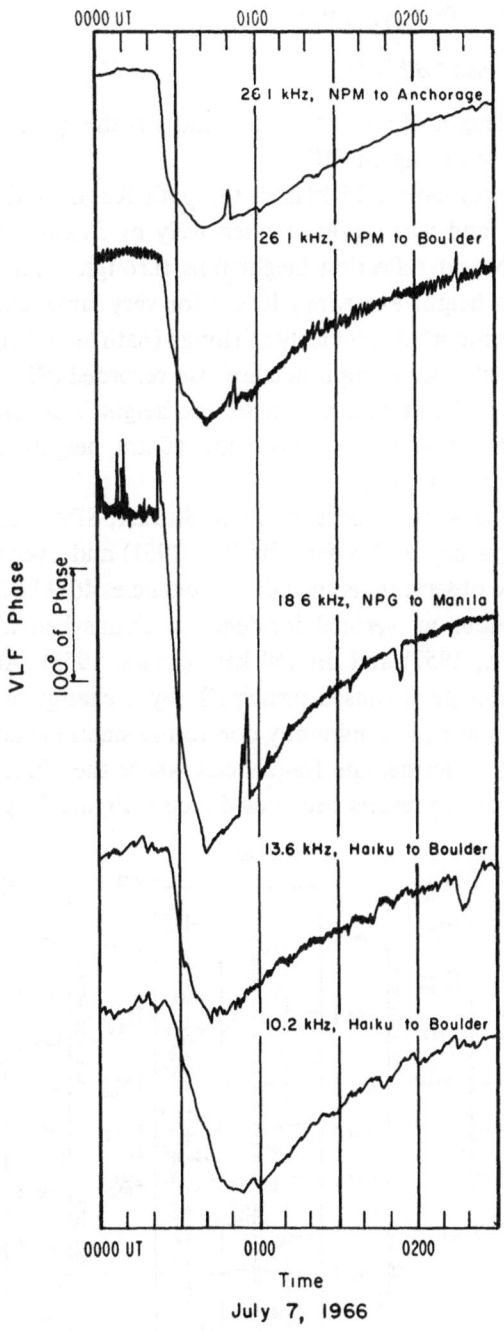

Fig. 18.   SPAs recorded for the event of July 7, 1966 for different paths and at different frequencies (from Donnelly, 1968c).

fictitious mirror-like surface given by:

$$\Delta h = - \lambda \sec I \, \Delta\phi / 720, \tag{3}$$

where λ is the wavelength of the radio wave and I is the mean angle of incidence of the downcoming wave during an SID.

SPAs were first recorded on 16 kHz in the path Rugby – Cambridge by Budden and Ratcliffe (1937) and was reported extensively by Bracewell and Straker (1949). At 16 kHz, the apparent reflection height falls through a few kilometers in about 5–10 min, the fall in height exceeding 10 km for very large solar events. Such measurements are now being made regularly at Slough (path length 108 km). Simultaneous measurements of the sky wave amplitude are also recorded (SFA). Since for this path and at this frequency, the undisturbed reflection height is around 70 km, for a large event the reflection height drops to levels below 60 km, heights not normally sounded by most ionospheric techniques.

Since the pioneering work of Bracewell and Straker, SPA observations have been extended upto a frequency of 2 MHz (Findlay, 1951) and over various path lengths. The early Cambridge observations included frequencies 40–113 kHz; soon afterwards there were observations at vertical incidence at Pennsylvania State University on 75 kHz (Houston Jr., 1957) and on 150 kHz (Jones, 1955). At frequencies around 16 kHz the phase change is caused principally by a change in reflection height; at 2 MHz such changes are predominantly due to the small deviation of the refractive index from unity. At intermediate frequencies where the effect is a combination of both, interpretation of the results becomes difficult. Figure 18 gives the observations

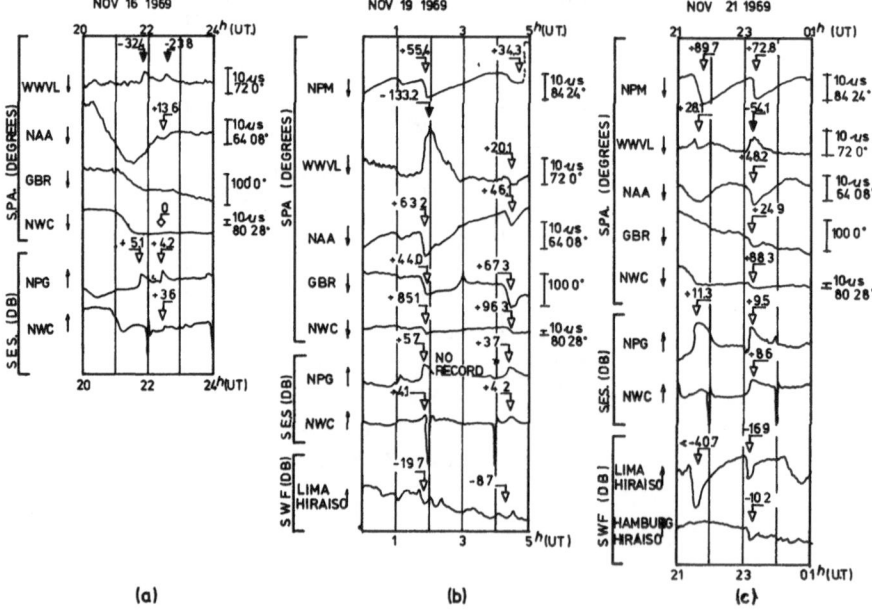

Fig. 19. Negative SPA events and related records of SESs and SWFs (after Ohshio, 1971). The black triangle, white triangle and white rectangle show a negative, positive, and a null SPA respectively.

recorded at Boulder for the flare of July 7, 1966 for the frequencies ranging from 10.2 kHz to 26.1 kHz over different paths.

The Japanese workers (Ohshio, 1968, 1971) have pointed out different types of SPAs: *regular* SPAs, *null* SPAs and *negative* SPAs. Occasionally no sudden phase anomaly is observed on VLF propagation circuits even though the path is sunlit and the solar X-ray flux is enhanced. This is referred to as a 'no SPA effect' or 'null SPA effect'. The null effect occurs when: (a) the enhancement of the intensity of the solar soft X-ray flux is small and (b) the effective solar zenith angle on the VLF propagation circuit is large.

Several negative SPAs have now been identified at Inubo Radio Wave Observatory from records of the circuit from Fort Collins (WWVL, 20.0 kHz) to Inubo for November 16, 19 and 21, 1969. These negative SPA phenomena are shown in Figure 19 together with records of positive (or normal) SPAs, null SPAs and sudden enhancements of signal strength (SESs) on other circuits. Two positive SPA events contiguous to a negative SPA event are shown in Figure 19 for reference. The negative SPA events, each of which includes one negative SPA phenomenon at least, are named event Nos. 1, 2, 3 and 4, and each of the positive SPA events is named event Nos. 1[+] or 2[+] in the order of time of occurrences. Several atlas of SPAs are available (see for example Chilton *et al.*, 1964).

2.4. Short Wave Fadeouts (SWF)

On HF radiowaves flares cause an abrupt decrease, sometimes a complete fadeout, known as Short Wave Fadeouts (SWF). Originally called Mögel-Dellinger effect, SWF is the first ionospheric flare effect discovered, and is the easiest to observe. Care should be taken to distinguish between the flare-associated SWF and abnormal fades that are caused by changes in propagation mode, by the presence of interference, or

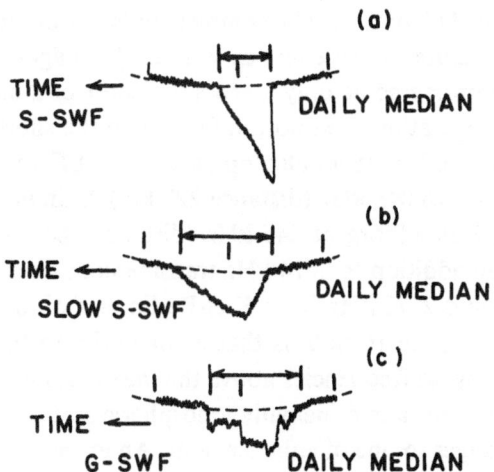

Fig. 20a–c. Representative shapes of different types of Short Wave Fadeouts (SWFs). Curve (a) shows a typical S–SWF; curve (b), a slow S–SWF; curve (c), a G–SWF.

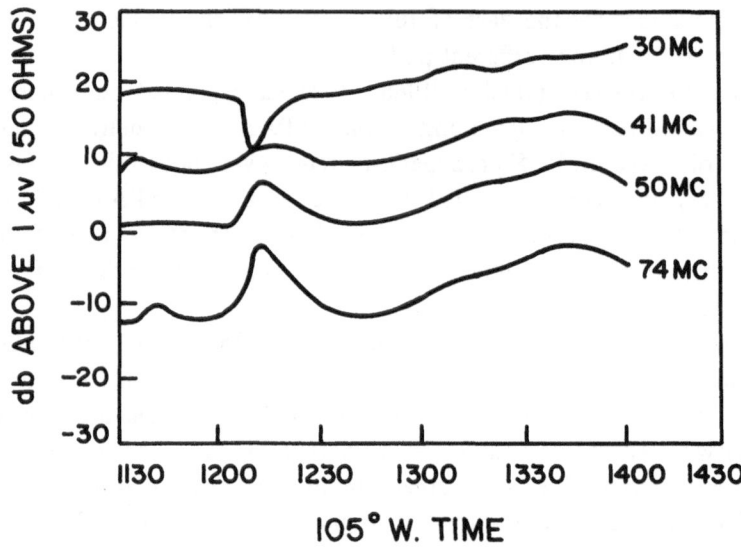

Fig. 20d. Enhancement in VHF signals recorded by Blair (1959) during the flare of July 19, 1958 over a 1295 km path (After Blair, 1959).

by disturbed ionospheric conditions associated with geomagnetic or dynamical effects.

SWFs are classified as:

S-SWF: Sudden dropout of the order of 1–5 min, depending on the recorder time constant and gradual recovery (Figure 20a).

Slow S-SWF: Drop-out takes from 5–15 min but decrease is steady and recovery gradual (Figure 20b).

G-SWF: Gradual disturbance; fading irregular in both drop-out and recovery (Figure 20c).

For most flares, the fadeouts may be complete in HF. Consequently, for such cases, the time of the maximum of the effect as well as the magnitude become uncertain. SWF is thus to be considered as only a gross indicator of a flare effect.

The frequency of operation need not, in fact, be in the shortwaves. Kühlungsborn (in German Democratic Republic) also operate in the LF and MF range with radio transmitters Decca at 128.048 kHz (distance 220 km), 'Stimme der DDR' at 185 kHz (distance 182 km), Kalundborg at 245 kHz (distance 180 km), Hörby at 1178 kHz (distance 225 km), in addition to 2614 kHz transmissions from Norddeich-Radio at a distance of 395 km and 2775 kHz from Kiel-Radio at a distance of 220 km.

Another important point to note is that as one goes to the VHF to frequencies above 30 MHz, that is, to frequencies above the maximum usable frequencies, where the propagation mechanism is principally ionospheric scattering, the absorption condition gradually changes to one of enhancement. An example is shown in Figure 20d. Here signal intensities measured simultaneously on 30, 41, 50 and 74 MHz over a 1295 km path between the Long Branch transmitting station at Long Branch, Illinois, and the receiving site at Boulder, Colorado during the solar flare of July 19, 1958 are given. The measurements have been reported by Blair (1959). The signals were transmitted and received on rhombic antennas and values were adjusted to 2 kW transmitter power. The changeover from an absorption effect at 30 MHz to enhancement effects on 41, 50 and 74 MHz is clearly evident.

2.5. Sudden Cosmic Noise Absorption (SCNA)

Like the SWF this is also an absorption effect, but observed with cosmic radio noise which is almost entirely due to a widely spread and diffuse source. Measurement is made usually in the frequencies 20–30 MHz, somewhat above the critical frequencies of the F-layer. The effect was first discovered by Shain and Mitra (1954). One of the

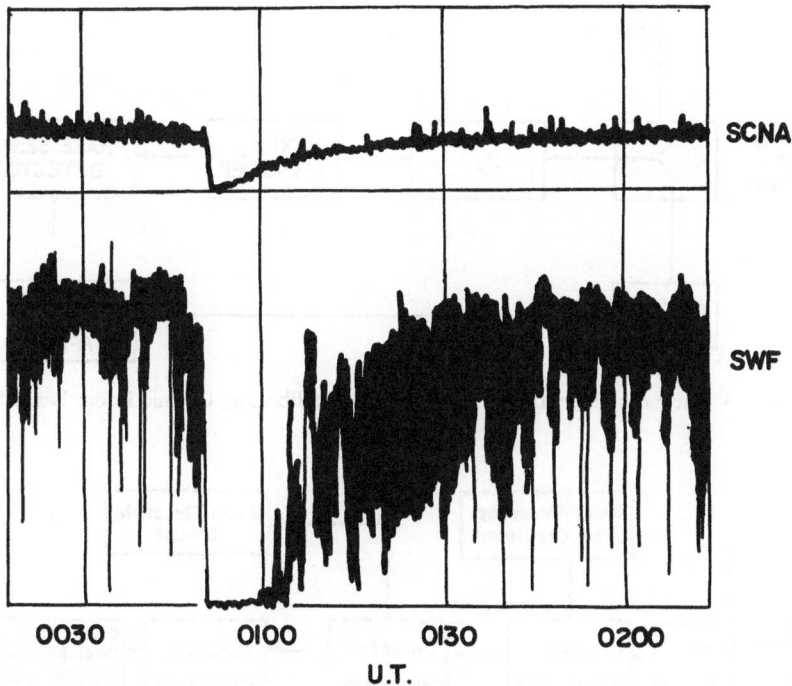

Fig. 21. One of the earliest records of Sudden Cosmic Noise Absorption (SCNA) (after Shain and Mitra, 1954).

earliest effects measured is shown in Figure 21. The method was used extensively during the IGY and the IQSY, and ever since. Although similar to SWF, the entire time-curve is almost always available (except for really outstanding events), making it thus a considerably more valuable tool than the SWF for the study of the flare-time changes in the ionosphere. In the original method of Shain and Mitra the total cosmic radio noise power was measured. In the improvement introduced by Little and Leinbach (1959) a servo-control unit continuously compares the noise received from the antenna with that generated by a local noise diode and adjusts the latter to maintain equality. In most antenna arrangements, the antenna beams are too wide to allow any reliable study of normal absorption, but are suitable for the study of SCNAs (and PCAs), although special antenna systems have sometimes been used for specific jobs.

The continually self-calibrating system developed by Little and Leinbach is called the Riometer (Relative Ionospheric Opacity Meter). It continually compares the received signal with that from a local noise standard at a switching frequency of many hertzs (340 Hz in the riometers used in NPL, New Delhi). The mode of operation of a typical riometer is shown in the block diagrams of Figures 22a, b. The operation is as below:

(i) The input of a superheterodyne receiver is switched alternately between the antenna and a local noise diode at a switching rate of 340 Hz. The receiver output is a

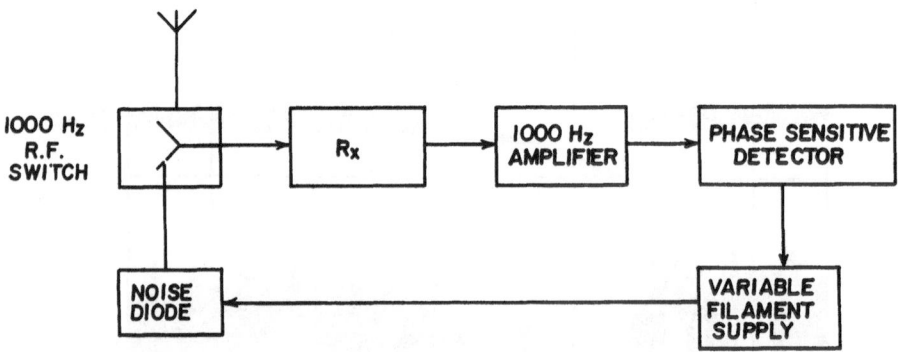

Fig. 22a.   A schematic diagram of a continually self-calibrating Cosmic Radio Noise System.

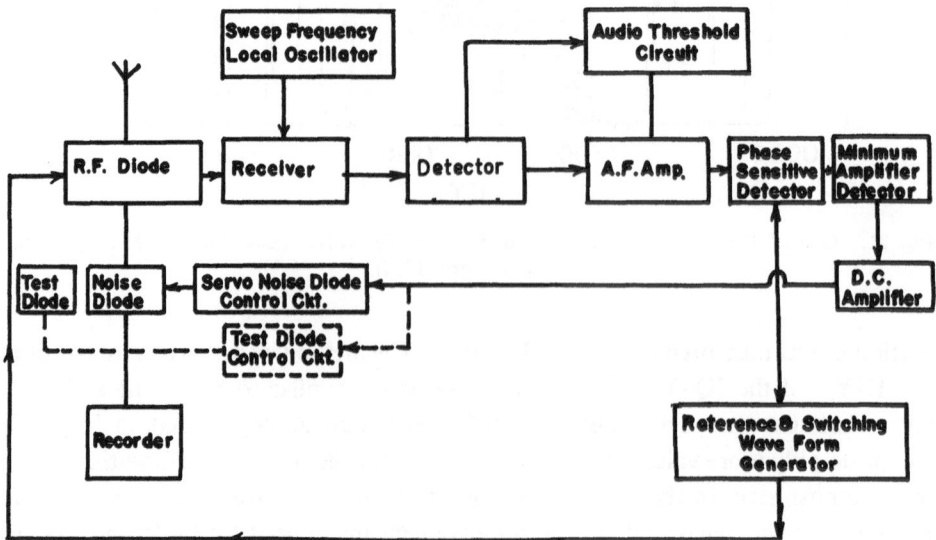

Fig. 22b.   A functional block diagram of a typical riometer.

square wave at the switching frequency with an amplitude proportional to the difference between the antenna noise and local noise generator output.

(ii) The square wave is amplified in an audio frequency amplifier and then fed to a
synchronous detector. A reference wave from the switching generator is also applied
to the synchronous detector. As a result, the d.c. signal appears at the output of the
synchronous detector; the amplitude of this d.c. output voltage is proportional to the
difference between the antenna signal and the local noise diode signal and its polarity
is determined by which of the two signals is stronger.

(iii) This d.c. signal from synchronous detector is then used to control the temperature of the noise diode filament (and hence the reference noise level) in such a way as
to make the two input signals at the receiver equal in strength.

As the antenna signal strength varies, the noise diode power is automatically and
continuously adjusted to bring it into equality. Since the noise power from the noise
diode is proportional to the d.c. plate current flowing through it, the d.c. current
through the noise diode is a measure of the received cosmic noise power. This noise
diode current is recorded on a recording milliammeter. The advantage of such a system
is that the receiver is used as a null detector rather than as an amplifier and consequently the receiver gain does not affect the final measurement.

There is a provision for reducing the maximum bandwidth of the LF strip from
100 kHz to 30, 15 or 3.5 kHz by switching in one of the three crystal filters. Normally
the riometers are operated with a bandwidth of 30 kHz. For the 20 MHz riometer this
bandwidth is usually reduced to 3.5 kHz to keep the interference to a minimum.

Absorption is estimated from the expression

$$A = 10 \log_{10} (P_0/P) \, \mathrm{d}B, \tag{4}$$

where P is the signal power received and P_0 is the signal power that would have been
received in the absence of the ionosphere at the same sidereal time.

The concept of equivalent temperature is generally used in discussing noise powers,
through the expression:

$$P = kT B,$$

where k is Boltzmann's constant, B is the effective bandwidth and T, the equivalent
temperature of the noise source. The equivalent temperature of the Galaxy varies
approximately as $f^{-2.7}$ from about 30000 K at 30 MHz to about 200 K at 200 MHz.
In understanding what we actually measure one must include: (1) an efficiency factor,
c (<1), for the antenna which represents the ohmic losses in a fictitious resistor at the
ambient temperature T_a, (2) a transmission factor, b, which represents the loss in the
cable (at a temperature T_c) feeding the receiver, (3) and a transmission factor, a (<1)
in the ionosphere at a temperature T_e (which is simply the electron temperature in the
absorbing region). Then we have the following expressions in terms of the galactic
equivalent temperature T_g:

True Ionospheric Absorption

$$A \, (\mathrm{dB}) = 10 \log_{10} (T_g/aT_g). \tag{5a}$$

Absorption Measured by an Ideal System with no losses

$$A' \text{ (dB)} = 10 \log_{10} \left[\frac{T_g}{aT_g + (1-a) T_e} \right].$$ (5b)

Absorption Measured by a Practical System

$$A'' \text{ (dB)} = 10 \log_{10} \frac{bcT_g + b(1-c) T_a + (1-b) T_c}{abcT_g + (1-a) bcT_e + b(1-c) T_a + (1-b) T_c}.$$

(5c)

In most observations where frequencies below 30 MHz are used, $T_g \gg T_a, T_c, T_e$ and Equation (5c) approaches (5a). At higher frequencies where the galactic temperature is comparable to the other temperatures and during PCAs or SCNAs when a becomes very small, the complete expression in (5c) should be used. An interesting consequence occurs when the ionospheric temperature exceeds the galactic temperature. In this case the difference between the noise powers received under quiet and disturbed conditions is given by:

$$\Delta T = bc (1-a) (T_g - T_e)$$

and consequently there is an enhancement in antenna temperature during the disturbed condition; and the enhancement increases as the absorption in the ionosphere increases.

Choice of frequency is equally important. The normal absorption is only a few tenths of a dB on 30 MHz, and it is consequently desirable to operate on as low a frequency as practicable in the presence of local and propagated interference. On the other hand, PCAs are sometimes so intense as to cause complete fadeout on 30 MHz and would require a second riometer operating at higher frequencies. At non-polar latitudes, it is advantageous to operate on the guardbands of the frequencies allotted for standard time transmissions; these guardbands are available to radio astronomy secondary allocations. Appropriate frequencies are 10 MHz, 15 MHz, 20 MHz and 25 MHz.

Different types of antenna arrangements have been used. For flare studies, relatively simple antenna systems can be used, although these may not be suitable if simultaneously the nature and magnitude of *normal* absorption are sought to be determined. A simple and useful arrangement is to use three-element Yagi antennas. Each Yagi consists of a driven element and a director, which are mounted on an artificial ground screen. That acts as the reflector element. The advantage of using the ground screen is that the antenna impedance is maintained constant for a wide variation of ground constants. If a more directive system is desired, then one may use a corner reflector, such as the one used by Sarada and Mitra (1962) which consisted of four half-wave dipoles separated by 0.3 λ and $\lambda/4$ above the ground, the angle between the corner reflectors being 90°. Such an antenna has a total gain of 17.2 dB and a beamwidth of 37° in N–S plane and 44° in E–W plane. A more directive antenna such as that used by Shain and Mitra (1954) in the original work establishing SCNA as a flare patrol

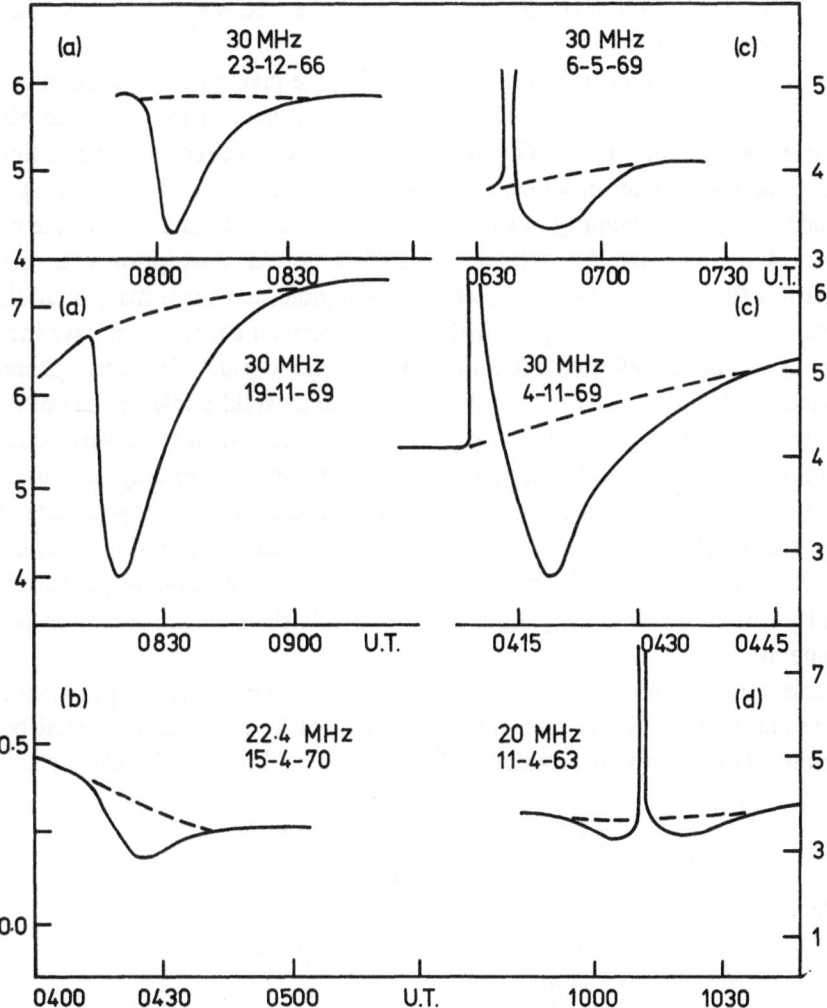

Fig. 23. Types of SCNAs (from records of NPL, New Delhi). Record-*a* shows a typical SCNA decreasing rapidly but with slow recovery. Record-*b* an *U*-type SCNA; record-*c*, one accompanied by solar burst at the beginning of SCNA; and Record-*d*, with a solar radio burst in the middle of the event.

technique, involving a broadside array of 5×6 dipoles, is normally unnecessary for flare studies.

As in the SEA and SES effects, SCNAs observed are of different types. In Figure 23 are shown the different types of SCNAs that are frequently observed on cosmic noise records. These are redrawn and represent the smoothed variations of the original SCNA records. In the most common type of SCNA observed (Figure 23a) the cosmic noise intensity decreases rapidly but the recovery is slow. Figure 23b represents the *U*-type SCNA, where the intensity falls slowly. On many occasions, particularly during the period of high solar activity, SCNAs are often accompanied by solar bursts. In Figure 23c the burst occurred at the beginning of the SCNA whereas in Figure 23d, it occurred in the middle of the SCNA. The size of an SCNA effect is reported as the excess absorption ΔA.

The ideal absorption measurement would be one giving the absorption of a radio signal passing vertically through an infinitely small area of the overhead absorbing region. However, for any practical antenna the signals are received over a finite solid angle. This solid angle in which the cosmic noise arrives at the receiving antenna depends on: (a) the obliquity of the rays arriving at the antenna and (b) the reception pattern of the antenna. An oblique ray suffers greater absorption over the longer path it has to traverse through the absorbing region, and there may, in addition, be a refraction effect in the F-region with its high electron density. The rays incident at large angles may be reflected back upward from the topside of the ionosphere, a case of 'window absorption'. During a flare, the cone or solid angle of the cosmic noise arriving at the antenna may be narrowed because the rays at large vertical angles of incidence which at normal times pass through the ionosphere may now be reflected back due to an increase in electron density at F-region and the rays at large vertical angles still reaching the antenna will now suffer greater absorption. This change in the solid angle in which the cosmic noise arrives at the antenna produced by the modification of absorbing region in its turn affects the effective reception pattern of the antenna.

A correction to the apparent absorption (A) registered with any practical antenna can be made for obliquity and antenna beam width. The zenithal pencil beam absorption ($\bar{A}$) can be then obtained. The expression involved is (Chivers and Hargreaves, 1965):

$$A = 10 \log_{10}\left[\frac{\int 10^{-(\bar{A}\sec\theta/10)} D(\alpha)\, D(\beta)\, d\omega}{\int D(\alpha)\, D(\beta)\, d\omega}\right]$$

in which

$$\cos\theta = \cos\alpha_0 (1 - \sin^2\alpha - \sin^2\beta)^{1/2} - \sin\alpha_0 \sin\alpha$$

$$d\omega = \frac{d\alpha\, d\beta}{(1 - \tan^2\alpha \tan^2\beta)^{1/2}}$$

α, β are the angles measured orthogonally from the centre of the beam, the α-plane passing through the zenith,

α_0 is the angle of the centre of the beam with zenith,

$D(\alpha)$ and $D(\beta)$ are the polar diagrams of antenna in H-plane and E-plane respectively.

The results of numerical computations based on the suitable modification of these relations can be approximated by a more simple expression:

$$\bar{A} = a(A)^b \text{ dB}$$

For an antenna system consisting of four collinear half wave dipoles connected in broad side and backed by 90° corner reflector used for 22.4 MHz cosmic noise equipment in Delhi, $a \approx 0.633$, $b \approx 1.17$. For the Yagi antenna used with 20 and 30 MHz riometers, $a \approx 0.564$, $b \approx 1.04$.

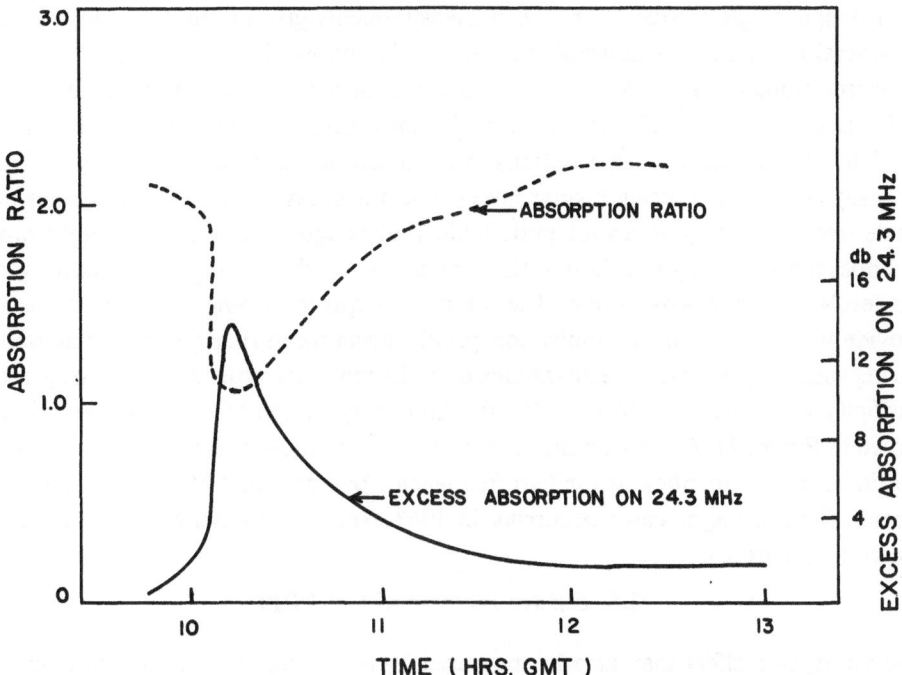

Fig. 24. SCNA measurements at two frequencies and evidence of flare time change in the frequency law (after Morriss, 1960).

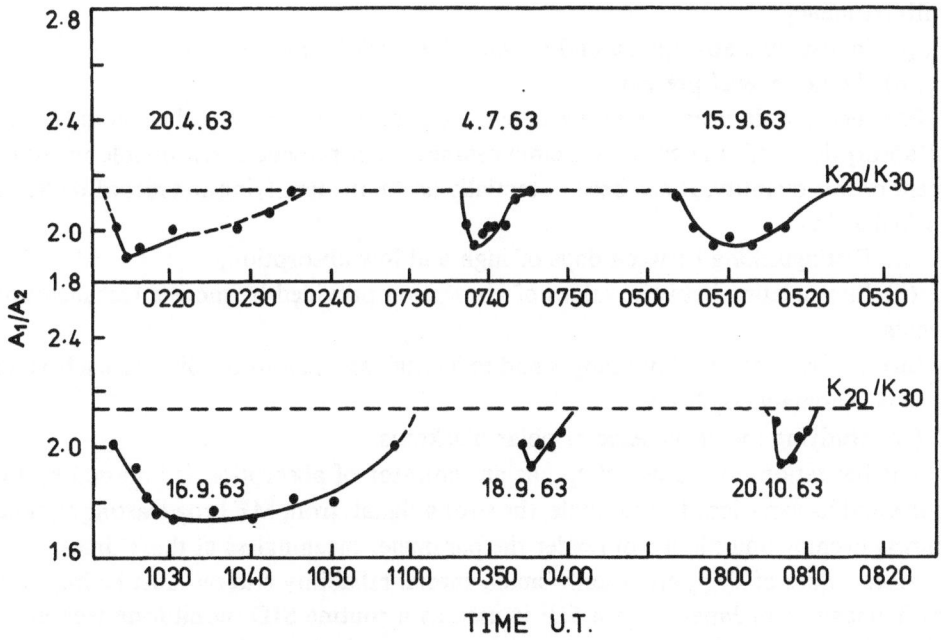

Fig. 25. Selected flare events for which the ratio of SCNA absorption at 20 MHz to that at 30 MHz decreases appreciably during a flare (after Mitra *et al.*, 1966).

Although single frequency riometer measurements give useful data, observations on several frequencies is desirable because of the following:

Observations on a single frequency give a measure of only the total absorption which is an integrated effect over a wide height range, ranging from 60 km to about 500 km, and the relative contributions from various levels in this height range cannot be easily separated. During a solar flare event, the excess absorption measured on a single frequency is again an integrated effect, although over a smaller height range, and does not allow any insight into the ionization distribution during the flare, nor of the levels of excess absorption. Use of multifrequency riometer observations can provide information on the ionization profile structure during quiet days as well as during solar flare events. The advantage of multifrequency observations during a flare was first demonstrated by Morriss (1960). An example from Morriss's observations is shown in Figure 24. An important point to note is the sudden change in the frequency relationship in absorption at the two frequencies 16.5 and 24.3 MHz as the flare progresses. Several major cases occurring in 1963 examined by Mitra *et al.* (1966) are shown in Figure 25.

2.6. Sudden Increase of $f_{\min}$ (SIF)

The absorption effect may also be measured by recording at close time intervals the changes in $f_{\min}$ with an ionosonde during a solar flare. The parameter $f_{\min}$ represents the minimum frequency for which the echo signal falls below the minimum recording level. The factors determining the minimum frequency are:

(i) The amplification of the recording system and the variation of its sensitivity with frequency;

(ii) ionospheric absorption and its variation with frequency and

(iii) the noise level present.

Provided certain precautions are taken, the parameter can be used as an indication of absorption and, under certain circumstances, can provide a reasonable measurement of this quantity. It has been potentially useful in identifying and recording SIDs, as well as in:

(i) Distinguishing between days of high and low absorption,

(ii) Interpolating between values of absorption provided by more direct measurements.

(iii) Studying day-to-day changes and to identify anomalous conditions, such as the Winter Anomaly condition.

(iv) Studying the occurrence of polar blackouts.

The limitation on the use of $f_{\min}$ as an indicator of absorption is imposed by the noise and its variation. For example, the strong signals from MF broadcasting stations during evening and night can render the parameter meaningless at these times.

Since values of $\Delta f_{\min}$ are usually small, special reliability criteria must be imposed. At Yamagawa in Japan, where SIF is used as a routine SID monitoring technique, SIF value is adopted on the basis that simultaneous increases in $f_{\min}$ occur in at least 3 of the four stations and that, furthermore, one of the values satisfied $\Delta f_{\min} \geqslant 1.0 \, \text{MHz}$

during the peak of the event. Absorption is connected with f_{min} through the equation:

$$A = \log \varrho = K(f_{min} + f_L)^2,\tag{6}$$

where K is a constant of proportionality, and f_L is the longitudinal component of the gyrofrequency. If $A_n (= \log \varrho_n)$ is the absorption under quiet conditions, and $(f_{min})_f$ and $(f_{min})_n$ are the flare and quiet time values of f_{min}, then the excess absorption $\Delta A (= \Delta \log \varrho)$ produced during a flare is given by:

$$\Delta A = K[\{(f_{min})_f + f_L\}^2 - \{(f_{min})_n + f_L\}^2].\tag{7}$$

2.7. Flare Effects Observed with Pulsed Transmitters

Flare effects can also be monitored with ionospheric absorption measurement techniques using vertical incidence pulsed transmission (A1 technique). One of the earliest absorption effect was reported by Appleton and Piggott (1954). A large number of flare effects were also recorded by Ganguly (1972) in Calcutta from measurements of absorption with pulsed transmitters at a number of frequencies. Such work is, however, very limited.

2.8. Sudden Frequency Deviations

Whereas all the SID effects discussed so far arise as a result of the excess ionization in the lower ionosphere, sudden changes in the *frequency* of HF radio signals propagated

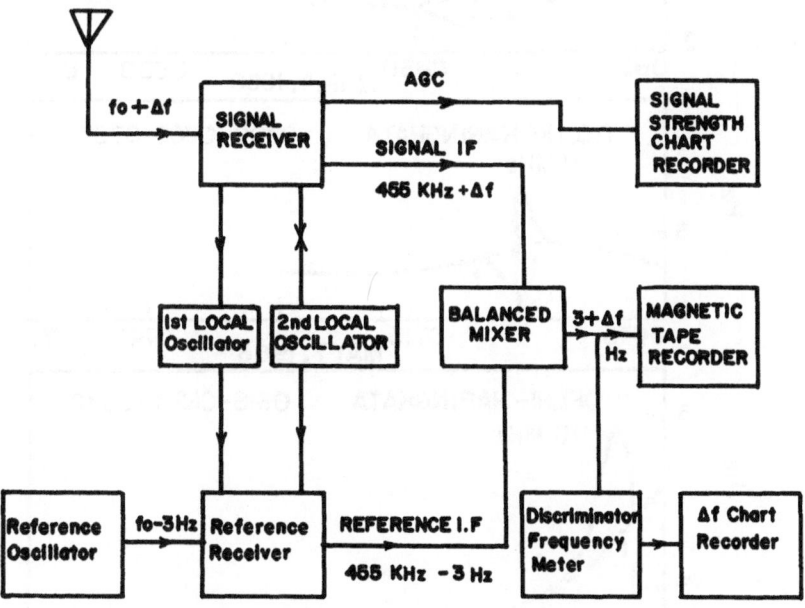

fo = Stable Transmitted Frequency
Δf = Frequency Deviation imposed by Ionosphere

Fig. 26.   Block diagram of a typical dopplometer.

via ionosphere originate in the ionization enhancements at heights above 100 km in the E and F-regions. In its occurrence as well as in the time development, an SFD effect is remarkably different from the other types of SIDs, firstly lasting for a much shorter period, and secondly exhibiting a rapid excursion from positive to negative effects, passing through a zero point at the time when the ionization enhancement is maximum. The frequency changes are usually of the order of a few hertzs. These small changes are registered by a dopplometer receiver of the type developed by Watts and Davies (1960), which compares the instantaneous received frequency of stable frequency transmissions with a local standard frequency usually stable to a few parts in 10^9 day^{-1}. The dopplometer equipment (Figure 26) consists of two identical, transistorized fixed frequency receivers operating on any of the standard frequency

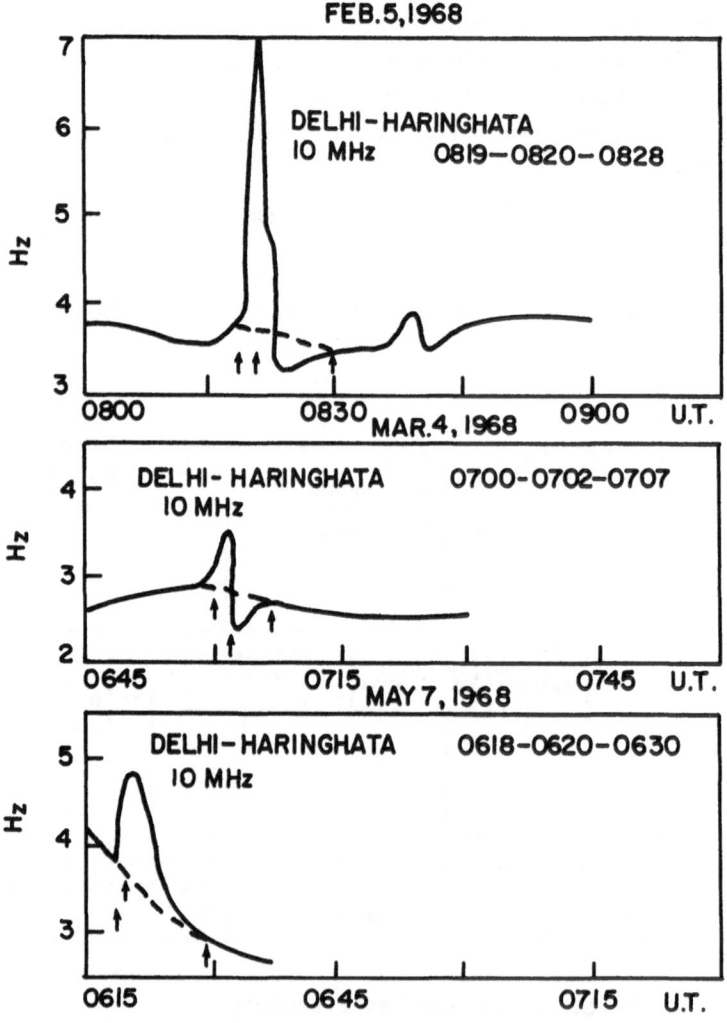

Fig. 27.   Selected Sudden Frequency Deviation Effects (SFDs) recorded at Calcutta with the standard frequency transmission ATA from Delhi at 10 MHz.

transmissions at 5, 10, 15, and 20 MHz. A stable frequency transmission is received through one receiver and the other receiver receives the reference signal from a standard frequency oscillator of high stability. The IF output from each receiver is coupled to a balanced mixer. Ionospheric changes causing frequency shift of the order of a few Hz cause the output of the balanced mixer to develop a difference beat frequency proportional to the phase difference. The difference frequency is then coupled to a discriminator which produces a voltage proportional to the difference frequency. The typical aerial consists of 3-element dipole array. In recent versions, instead of a reference receiver one uses frequency synthesizers to take f_0–3 Hz and generate (a) an input frequency replacing the 1st local oscillator, (b) an input frequency replacing the 2nd local oscillator, and (c) the 455 kHz–3 Hz (Baker et al., 1968).

A typical SFD record consists of a positive change in the frequency (Δf) at the beginning almost always followed by a longer recovery period of negative Δf. Some SFDs recorded at the field station in Calcutta with NPL transmission ATA from Delhi are shown in Figure 27.

The frequency deviation Δf from the operating frequency f is proportional to the rate of change of phase path P of the signal and is given by (Bennett, 1967):

$$\Delta f = -\frac{f}{c} \int_s \frac{d\mu}{dt} \cos \alpha \, ds, \tag{8a}$$

where μ is phase refractive index and α is the angle between the wave normal and the ray direction. If the ionization changes take place in the part of the path well below the height of reflection, then for vertical incidence:

$$\Delta f = \frac{R}{fc} \int \frac{dN_e}{dt} \, dh$$

with $R = f_N^2/N_e$, f_N being the plasma frequency.

Employing an equivalence theorem for Doppler shifts (Agy et al., 1965) for non-deviative case,

$$\Delta f = \frac{R}{f_v c} \int \frac{dN_e}{dt} \, dh, \tag{8b}$$

where $f_v = f \cos i$ is the equivalent vertical incidence frequency, i being the angle of incidence. Equation (8b) can be used to estimate electron density changes in a vertical column of unit cross-section (cm^2) needed to cause the observed Doppler shift in Hz. Thus,

$$\int \frac{dN_e}{dt} \, dh = \frac{c}{R} \Delta f \cdot f_v = 3.72 \times 10^8 f_v \Delta f$$

$$\Delta N_e = \frac{c f_v}{R} \int_0^t \Delta f \, dt. \tag{8c}$$

If there are several dopplometers operating on different frequencies, the frequency deviation for the same path is proportional to the reciprocal of the frequency i.e.,

$$\Delta f \propto \frac{1}{f}$$

if the change in phase path is brought about by formation of additional ionization in the non-deviative region below the reflecting level. If, however, the change in phase path is brought about by a change in the reflecting layer for mirror like reflection, the frequency dependence expected is of the form;

$$\Delta f \propto f.$$

This additional information can be obtained from the multifrequency measurements alone. It must be noted, however, that even when $\Delta f \propto 1/f$ in the SFD observations, the contribution to Δf from the deviative regions is often dominant with respect to the contribution to Δf from the nondeviative region.

2.9. Wave Interaction Technique

We have so far discussed relatively simple systems that can be, and often are, used on a synoptic basis in a large number of observatories around the world. We now discuss,

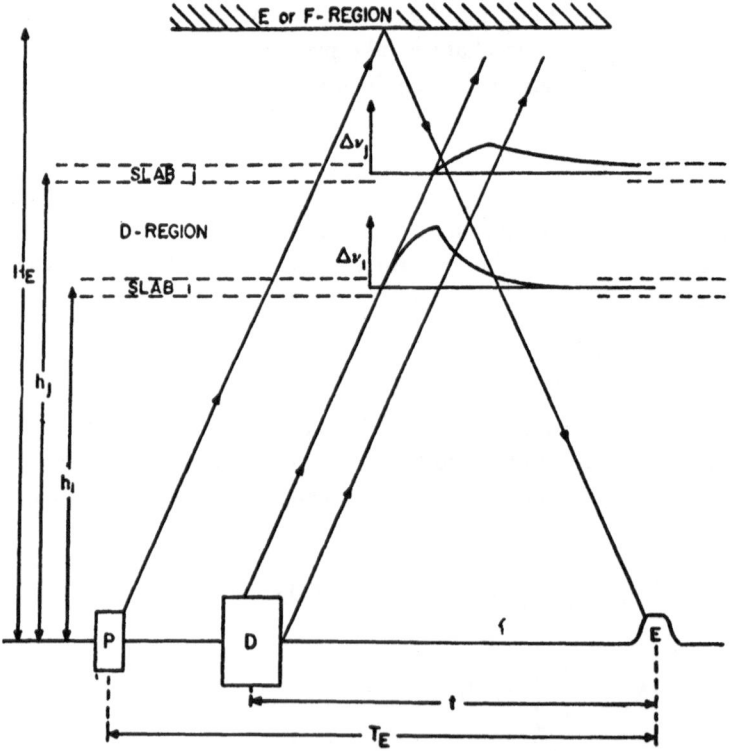

Fig. 28.   Basic principle of a Wave Interaction Experiment (after Ferraro and Lee, 1967).

in this and the following sections, techniques which are considerably more sophisticated and are currently in use in only a few places.

In the wave interaction technique, first reported by Tellegan (1933), a wanted wave is modulated in the ionosphere by radiation from a powerful disturbing wave. The basic principle of operation is outlined in Figure 28. Currently, almost all investigations use a pulse technique, originally introduced by Fejer (1955). In this, a long train of short disturbing pulses is transmitted at exactly half the reception frequency f of the train of wanted pulses. The time f^{-1} is made long compared to the duration of the transient electron heating caused by the disturbing pulse. However, every second wanted pulse travels through the transiently modified D-region either along its entire passage or only beyond a certain point of time along its passage through the D-region. The height discrimination is obtained by adjusting the relative time delay between the pulse trains. The best experimental data are being obtained by two high power installations, one of which is at University Park, Pennsylvania, U.S.A., and the other at Armidale, Australia. With the Pennsylvania equipment both phase as well as amplitude changes of the wanted wave are recorded providing additional information on collision frequency (Ferraro and Lee, 1967, 1968). A disturbing radiation of about 2MW on a frequency of 4.5 MHz is used with a $\sin^2$ shaped pulse. The wanted frequency is 2.2 MHz, with a rectangular pulse. Reliable electron density profiles can be obtained in about a minute compared to the typical sounding time of 30 min in previous installations.

Flare time soundings are, however, being done only by the Pennsylvania State University, and even in this place, only since the present high power equipment was established in November, 1967. Soundings have been made for more than a dozen flares. The experiment produces profiles of N_e, over the range 55–90 km, at the rate of about one profile every 3 or 4 minutes. Figure 29 shows the profiles obtained during

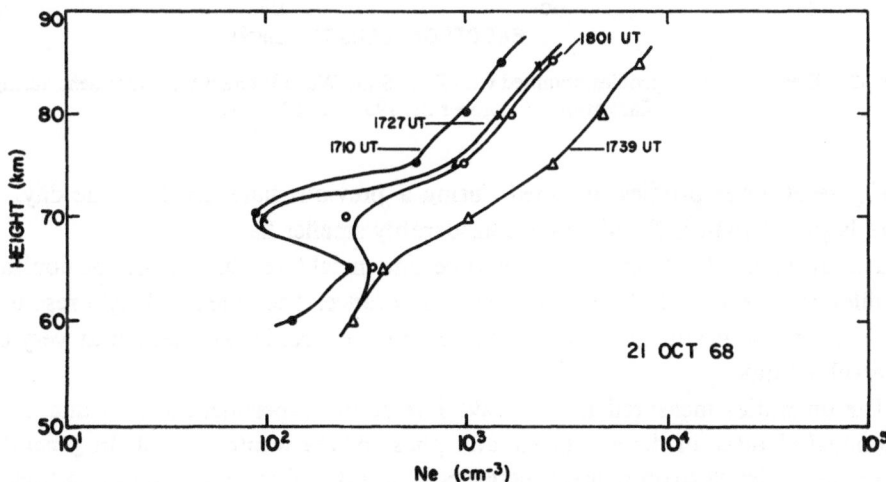

Fig. 29.   Electron density profiles obtained at State College, Penn State, with the High Power Wave Interaction Facility during the solar flare event of October 21, 1968 (after Rowe *et al.*, 1970).

the flare of October 21, 1968, occurring at about 1739 UT. This was a class 2B flare with a strong X-ray enhancement (about a factor of 50 below 8 Å). The normal profile is one obtained at about 1710 UT, just before the flare began. As the flare progressed, N_e at heights above 70 km began to increase without any appreciable change at and below 70 km (1727 UT profile). The peak effect was observed at 1739 UT. The profile at this time was quite different in shape from the normal one suggesting different production sources during normal and disturbed times. During the decay phase the profile at 1801 UT was identical with that of 1727 UT at levels above 70 km, but was considerably different below.

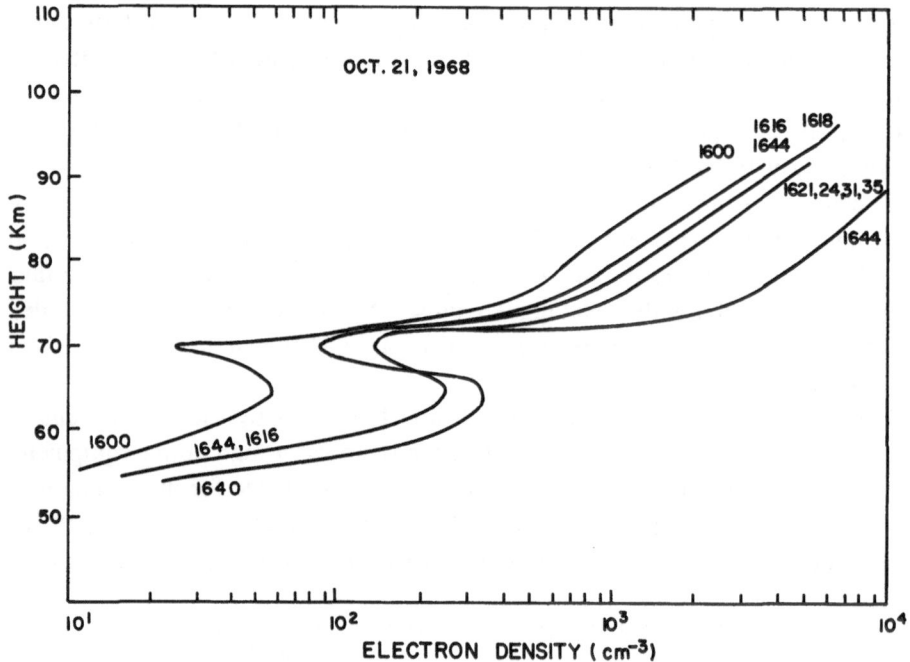

Fig. 30.   Electron density profiles obtained with Penn State Wave Interaction Equipment during the flare event of October 21, 1968, at 1614 UT.

Figure 30 gives profiles obtained during a previous flare on the same day. This flare began at 1614 UT and was a considerably smaller flare.

The question of reliability and of time and height resolutions of the ionization profiles obtainable with this technique is of relevance. The time and height resolutions are related; the height resolution becomes poor if profiles are needed at very close intervals of time.

The quantities measured in the wave interaction experiment are T_A and T_p, the fractional changes in the amplitude and phase of the wanted signal. In general, T_A increases as the electron density increases if all other factors are kept constant, but there is no simple relation between the interaction value at a certain altitude and the electron density value at that altitude. The T_A profile has both positive and negative regions with a cross-over typically between 67 and 69 km, and a negative peak of $7-10 \times 10^{-4}$ at about 63 km; this altitude increases with increasing collision frequency.

The reduction of T_A and T_p values to electron density and collision frequency is mathematically quite complex. For this reason the reduction of wave interaction data have in the past used the 'model' concept in which a postulated electron density profile is adjusted until agreement is obtained between the measured and calculated inter-action coefficients (Lee and Ferraro, 1969).

The Penn State Group uses a catalogue of some 400 electron density profiles of various shapes and magnitudes (including profiles with and without valleys) and corresponding interaction profiles which can be searched automatically. To find an electron density profile for a measured T profile, an initial search is made with large tolerances on the experimental profile (ranging from 300% at low altitudes to 30% at high altitudes). If some electron density profiles are found which will match the data within this tolerance, the tolerance is reduced 25% at all altitudes and the search is repeated. This procedure is continued, in steps of 25%, until one or no profile is found. If one profile is found at the smallest tolerance it is used; if no profile is found, the best-fitting profile at the previous step is selected by hand. In constructing this catalogue the electron energy loss coefficient G must be known. A constant value of 5.5×10^{-3} is used in Penn State catalogue, although there is some indication of a seasonal variation of this parameter.

In addition to the above 'Catalogue Search' method, the Penn State Group have recently also introduced an 'Overdeterministic Profile Inversion' method. This is based on a matrix technique for inverting integral equations, and is capable of arriving at electron density values at m number of heights by solving n number of simultaneous equations set up with experimental data available at n heights. For $n > m$ this leads to a solution of overdetermined simultaneous equations (see for example, Ferraro et al., 1974).

The profile inversion method is believed to be superior to the catalogue search method. Ferraro et al. (1974) find from numerous test results that the technique is capable of matching the data values within 5% and reproduces electron density profiles within 20% accuracy averaging over all height ranges. It is important, however, that the integration time be larger at lower heights than at higher. Ferraro et al. used an integration time of 1.25 min for heights 55–69 km, and of 5 s for higher heights. This ensures reliability of interaction data to within 10%.

2.10. Partial Reflection Observations

Belrose and his colleagues in Ottawa have recently been able to use partial reflection of radiowaves generally in the frequency range 1.6 to 6.0 MHz from heights of ap-proximately 60–90 km for monitoring a number of flare events with considerable success (Belrose and Cetiner, 1962; Montbriand and Belrose, 1972). An outline of the experimental arrangement is given in Figure 31. Since reflection occurs from irreg-ularity, the signal strength is very weak. The basic experimental measurement concerns

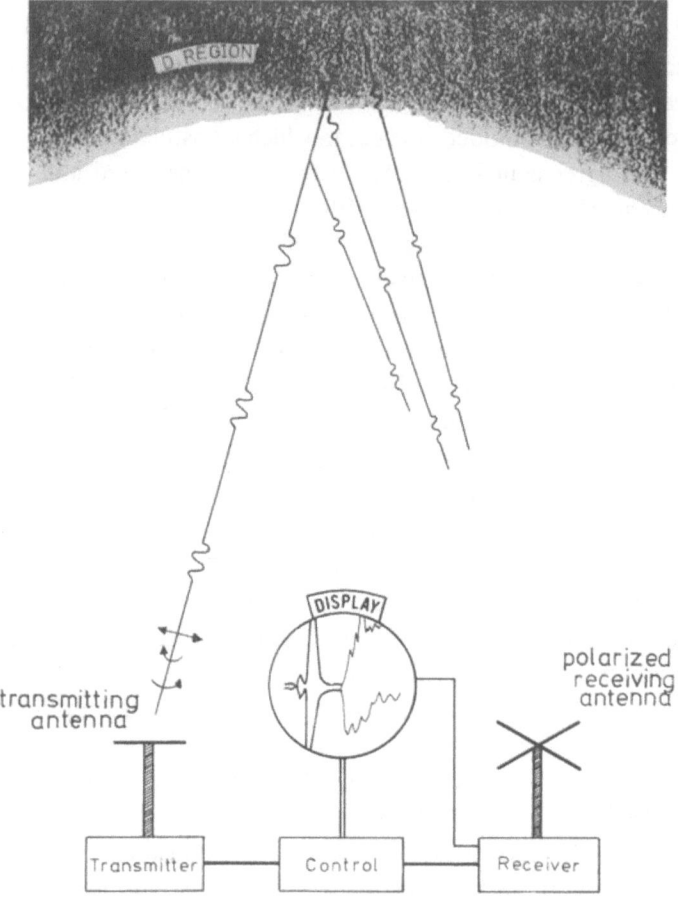

Fig. 31.   Basic principle of a Partial Reflection Experiment (after Belrose *et al.*, 1967).

the measurement of ratio A_x/A_o, (the relative amplitudes of the extraordinary and ordinary wave) as a function of height with a powerful pulsed transmitter. It is assumed that reflection is due to ionization irregularities $(\overline{\varDelta N_e}/\bar{N}_e)$. A_x/A_o is principally a function of collision frequency and of electron density through the equation:

$$\frac{A_x}{A_0} = \frac{R_x}{R_0} \exp\left[-2 \int_0^h (k_x - k_0)\, dh \right],\tag{9}$$

where $R_{x,o}$ are the Fresnel reflection coefficients for the extraordinary and the ordinary components and $k_{x,o}$ are the corresponding absorption indices. The latter are proportional to electron density, and R_x/R_o depends only on collision frequency. Thus Equation (9) gives a direct measure of electron density if collision frequency is known. The experiment also provides an independent measurement of v. At low heights where N_e

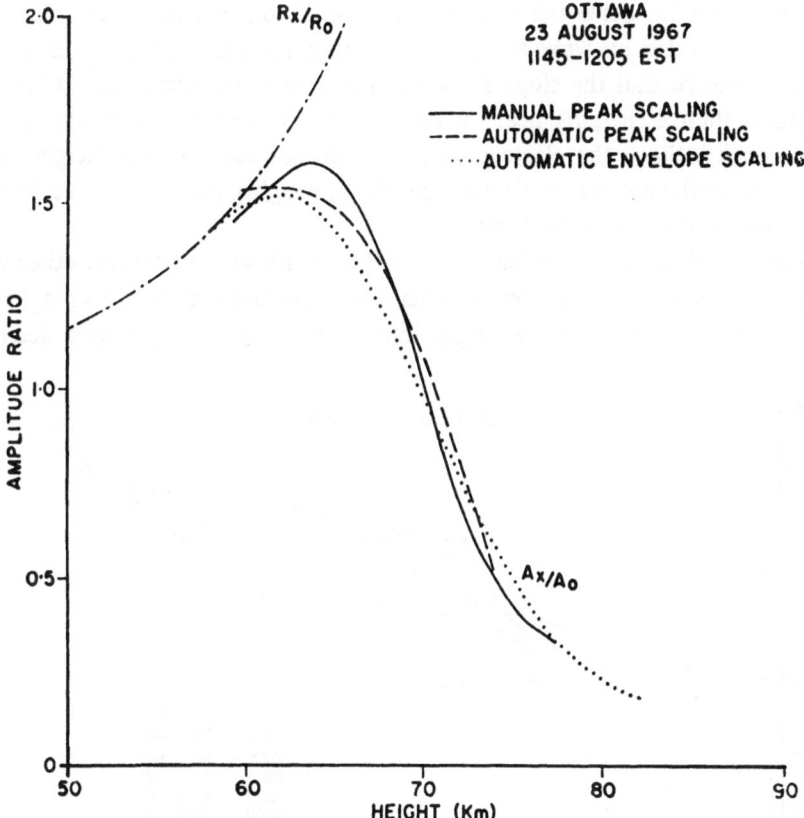

Fig. 32.   Relationships between A_x/A_0 and R_x/R_0 in a Partial Reflection Experiment
(after Belrose, 1970).

is small, the integrated differential absorption is small and

$$\frac{A_x}{A_0} \approx \frac{R_x}{R_0} \tag{10}$$

At these heights, therefore, A_x/A_o provides a direct measure of collision frequency. A typical example of measured A_x/A_o value is shown in Figure 32, with R_x/R_o fitted with A_x/A_o at low altitudes.

By employing two or more frequencies, and relatively large antenna arrays, electron density can be measured in the height range 50–90 km. At Ottawa, frequencies used range between 1.78 to 6.275 MHz.

The accuracy with which the electron density can be determined (especially at heights of and below 70 km) is of major concern, since large differences have been observed between profiles obtained at Penn State and at Ottawa.

Since the accuracy at which $N_e(h)$ is measured depends on the slope of $\log A_x/A_o$ vs height curve, we are concerned with the accuracy with which the height distribution of A_x/A_o can be measured. However, for a given precision in the measurement of A_x/A_o, the accuracy with which the electron density can be determined depends markedly on height, since the uncertainty depends on collision frequency and on electron density (also on wave frequency and on gyrofrequency). Belrose and Burke (1964) have shown that the slope must be measured more accurately at low and at high heights than at heights near 70 km, to obtain the same accuracy in a given electron density. Since the electron density increases rapidly with height, the percentage error in the computed electron density is roughly constant above 75 km, but increases very rapidly at lower levels.

Gardner and Pawsey (1953), Belrose and Burke (1964), and several other workers have deduced electron density from a measure of the slope of the $\log A_x/A_o(h)$ curve; the mean electron density in the height interval $h-1$ to $h+1$ (2 km) is determined,

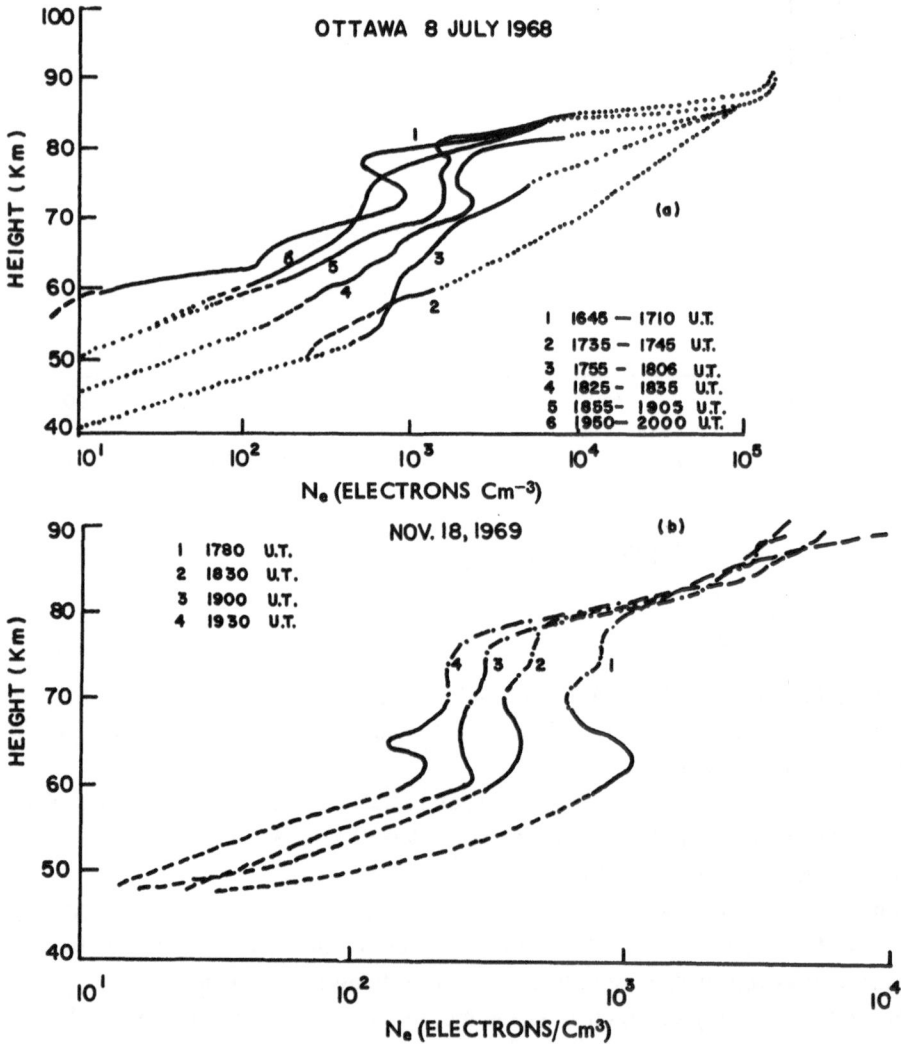

Fig. 33. Electron density profiles obtained with Partial Reflection Technique at Ottawa for (a) the large flare event of July 8, 1968 and (b) flare event of November 18, 1969 (after Montbriand and Belrose, 1972).

independent of the electron density below this height. The absolute magnitude of the ratio A_x/A_o is, however, inversely proportional to the average electron density below the reference height, and, therefore, the ratio A_x/A_o at the low height and $(A_x/A_o)_{max}$ give some information on the degree of ionization at heights for which no reflections are observed; but the distribution of the electrons is not determined.

Belrose believes that the $A_x/A_o(h)$ curve is a very sensitive and accurate measure of the electron density, in which any uncertainties can be reduced by smoothing the data and averaging over a longer interval of time. Smoothing, however, may eliminate fine structures and averaging over long intervals of time removes short-time ionospheric variations. Belrose *et al.* (1967) have considered errors in great detail, and have concluded that the electron density can be determined with an accuracy comparable to or exceeding that deduced by other groundbased (and even rocket) techniques.

A few examples of flare profiles deduced by Montbriand and Belrose (1972) are shown in Figures 33a, b. Figure 33a shows the time variations during the large X-ray event of July 8, 1968. The solidline curves were obtained from the analysis of 2.66 and 6.275 MHz differential absorption/partial reflection data; the dashed line extensions were obtained by normalising the amplitude of the O-ray (A_o) at 2.66 MHz to these curves; the dotted lines were based on riometer absorption and on a real-height analysis of ionogram recordings (corrected for underlying ionization) made before the flare. Profile 1 was taken immediately before the flare began, and profiles 2–6 were taken at various times during the flare.

Another example is given in Figure 33b. This was for the flare event of November 18, 1969. Here the solid line curves were from 2.66 and 6.275 MHz data and the lower dashed lines from the normalised $A_o(h)$ amplitudes.

Montbriand and Belrose have also given the profiles for the flare events of March 1, 1970; January 31, 1970; July 3, 1969 and May 2, 1969.

2.11. Satellite Radio Beacon Experiment – Changes in Total Electron Content (SITEC)

F-region changes during solar flares also appear in the measurements of total electron content with satellite radio beacon transmission when continuous monitoring is possible, as with geostationary satellites. The beacon satellites used exclusively for ionospheric research, transmit linearly polarized unmodulated waves of frequencies somewhat greater than the F-layer critical frequency of the ionosphere. Ground stations deduce information about the ionosphere by analysing received signals from the satellite. The ionosphere has three principal effects on a radio wave passing through it: (a) absorption, (b) refraction of the beam with subsequent attenuation due to divergence and (c) changes in the state of polarization (magnetoionic effect). The radio signals from orbiting satellites also exhibit characteristic ionospheric influences due to the motion of the satellite. These are: (a) dispersive Doppler frequency shift, (b) regular fading of the amplitude due to the Faraday rotation of the plane of polarization, and (c) irregular fading due to scintillation.

One of the techniques of investigating the ionosphere by means of beacon satellites involves the deduction of the integrated electron density upto the satellite height from the measurements of the dispersive Doppler frequency shift resulting from the change of phase path length between the satellite and the receiver due to the ionosphere (Ross, 1960). This can be done by measuring the departure from a harmonic relationship of the received signals from harmonically related transmitters. The reduction of phasepath length is dependent upon the electron content along the raypath.

A second technique for the determination of the total electron content involves the observation of the most easily recognizable effect, the Faraday rotation. Travelling through a magneto-ionic medium like the ionosphere, a linearly polarized electromagnetic wave experiences a rotation of its plane of polarization about the axis of propagation, the amount of rotation depending upon the integrated electron density and on the longitudinal component of the Earth's magnetic field. Thus the electron content upto the height of the satellite can be calculated.

A limitation of total electron content measurements is that there is no way of determining the distribution of electrons between the satellite and observer. Using a mean magnetic field factor $\bar{M}$ (equal to the value at 400 km), the observed Faraday rotation can be converted to an apparent electron content of the ionosphere N_T given by:

$$\Omega = \frac{K}{f^2} \bar{M} N_T \tag{11}$$

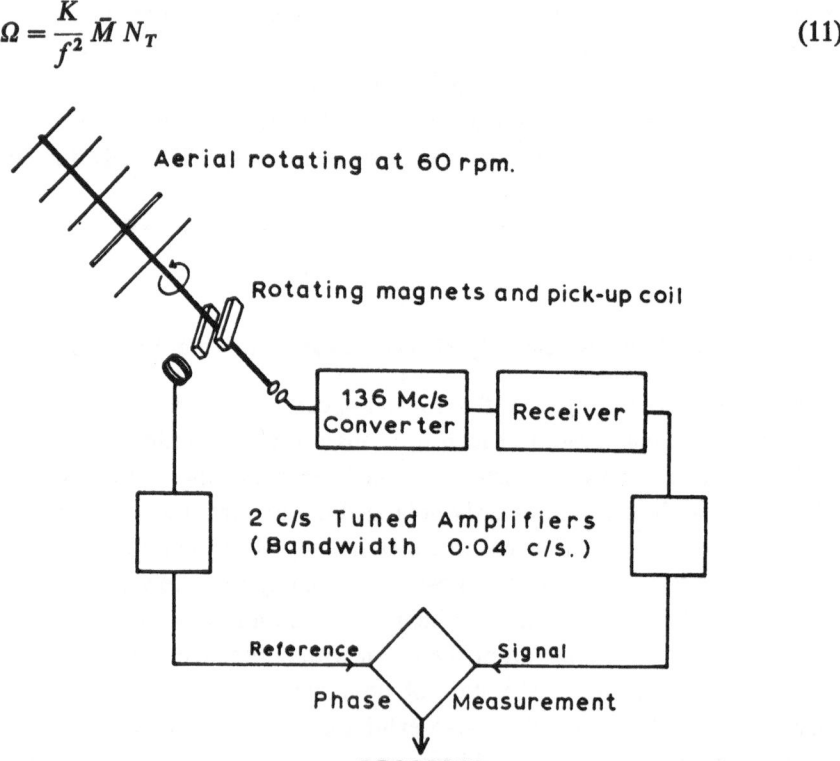

Fig. 34. Typical experimental arrangement of a beacon experiment with a geostationary satellite (after Titheridge, 1966).

with

$$N_T = \int N_e \, dh,$$

where K is a constant.

Titheridge (1972) has shown that what is measured is *not* the total electron content

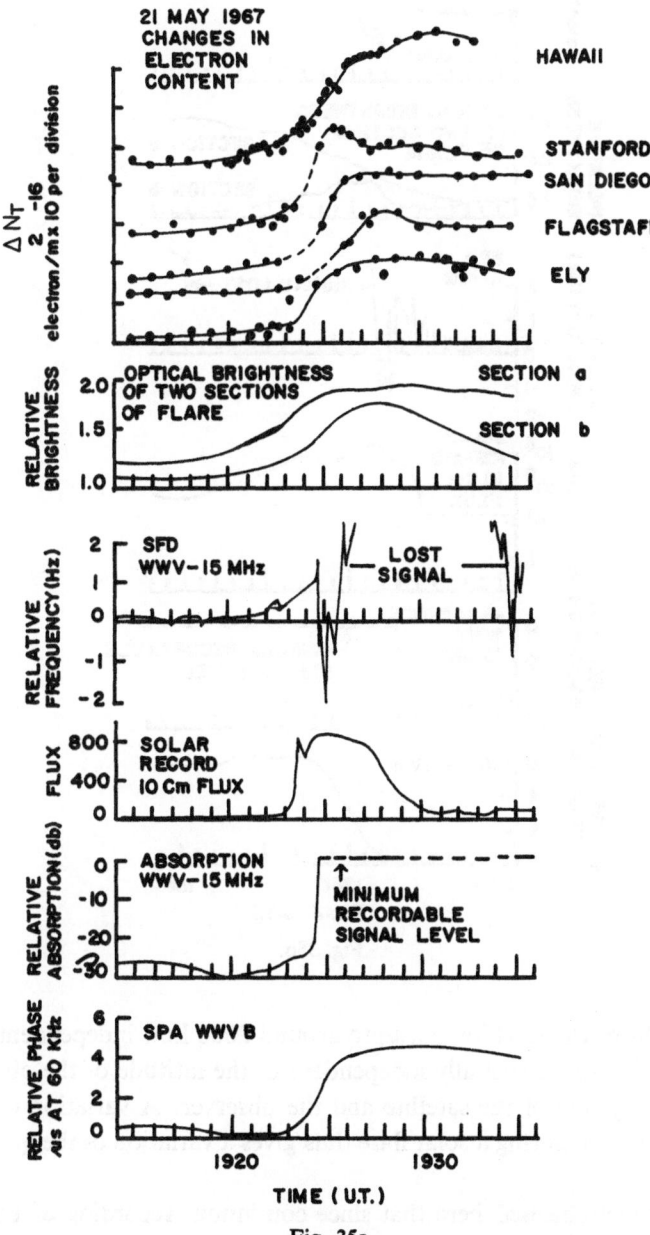

Fig. 35a.

Fig. 35a–b. Sudden Increase in Total Electron Content (SITEC) recorded at Hawaii, Stanford, San Diego, Flagstaff and Ely for the large flare events of May 21 and 23, 1967 (after Garriott *et al.*, 1967), (a) May 21 effects and (b) May 23 effects.

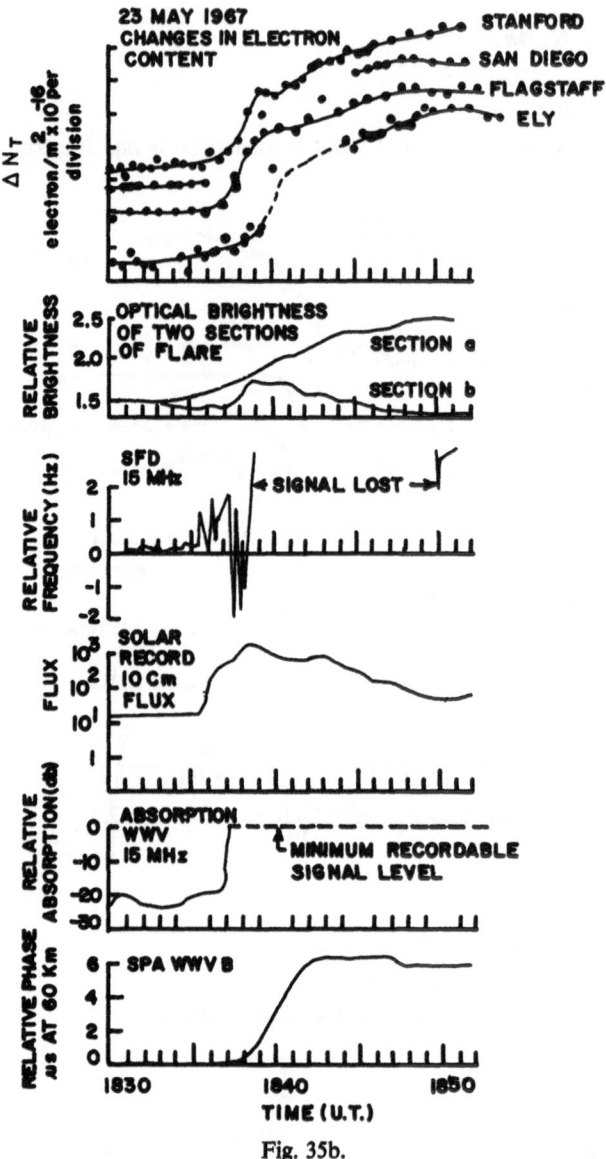

Fig. 35b.

upto the height of the satellite, but upto around 2000 km, independent of changes in the $N_e - h$ profile, and essentially independent of the latitude of the observing site, or the relative longitude of the satellite and the observer. A variation of the observed rotation Ω with time during a solar flare thus gives a variation of the quantity $\left(\int N_e \, dh \right)$ with time.

It should be emphasised here that since continuous recording of the Faraday rotation is required for the entire duration of the flare, transmissions from low orbiting satellites cannot be used, and one must use geostationary satellites. Since the angle between the ray path and the magnetic field is fixed, $\Omega - t$ is a direct measure of $N_e - t$.

The equipment most commonly used follows one first developed by Titheridge (1966) for communication satellite Syncom-3 which had a telemetry transmitter operating on a frequency of 137 MHz. A block diagram of this recorder is given in Figure 34. The signal from the rotating aerial is brought out through two coils, one mounted on the end of the aerial and one stationary. These coils are tuned to the beacon frequency and slightly overcoupled. Although this gives a slight loss of signal, it is much quieter and more reliable than the use of slip-rings. The signal then passes through a temperature-stabilised converter and a receiver with a bandwidth of a few hundred hertzs. The receiver operates without A.G.C., and the rectified output contains a 2 Hz component whose phase depends on the polarisation angle of the satellite signal. Four high-pass and four low-pass filter stages remove most of the noise from this output, and the 2 Hz component is then extracted by a tuned amplifier with a total bandwidth (at -3 dB) of 0.04 Hz. This results in a clean, noise-free signal suitable for accurate phase measurements. The amplifier is allowed to overload slightly on strong signals to allow the output to rise to its full value in a few seconds. If the satellite signal disappears, the 2 Hz output persists at a measurable level for nearly 1 min.

The long-term stability of the recorder is important. Titheridge claimed that in his

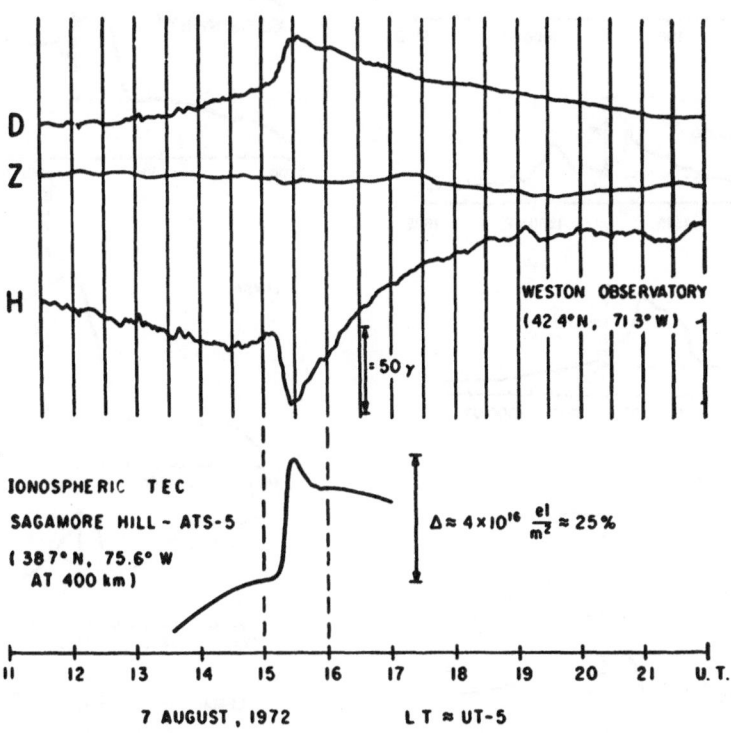

FLARE INDUCED CHANGES IN THE GEOMAGNETIC FIELD (SFE)
AND THE IONOSPHERIC TOTAL ELECTRON CONTENT (SITEC)

Fig. 36.   Total electron content effect during the flare of August 7, 1972 records at Sagamore Hill with ATS-5 satellite along with changes in the geomagnetic field (after Mendillo *et al.*, 1974).

equipment the overall drift during 10 months of operation at Auckland corresponded to a change of only 4° in the polarisation angle. Weekly or monthly calibrations are, therefore, adequate, and can be performed by recording a test signal transmitted from a fixed vertical dipole.

The satellite beacon experiment now provides the chief means of routinely monitoring F-region flare effects. Several flare effects have so far been reported with satellite radio beacon measurements. Two of the best documented cases are the flare series during the period May 21–23, 1967 (Figures 35 a, b) (Garriott et al., 1967) and the flare of August 7, 1972 (Figure 36). The sequence of five major flares that occurred in the first case produced a wide variety of ionospheric effects, many of them published and examined: the flares occurred at 19 23 UT on May 21, 1967, at 0005 UT on May 22, 1967, and at 1805, 1835 and 19 35 UT on May 23, 1967. The effects were recorded with the geostationary satellites ATS-1 at 137.35 MHz at Stanford University and San Diego, California; at Flagstaff, Arizona; at Ely, Nevada; and at the

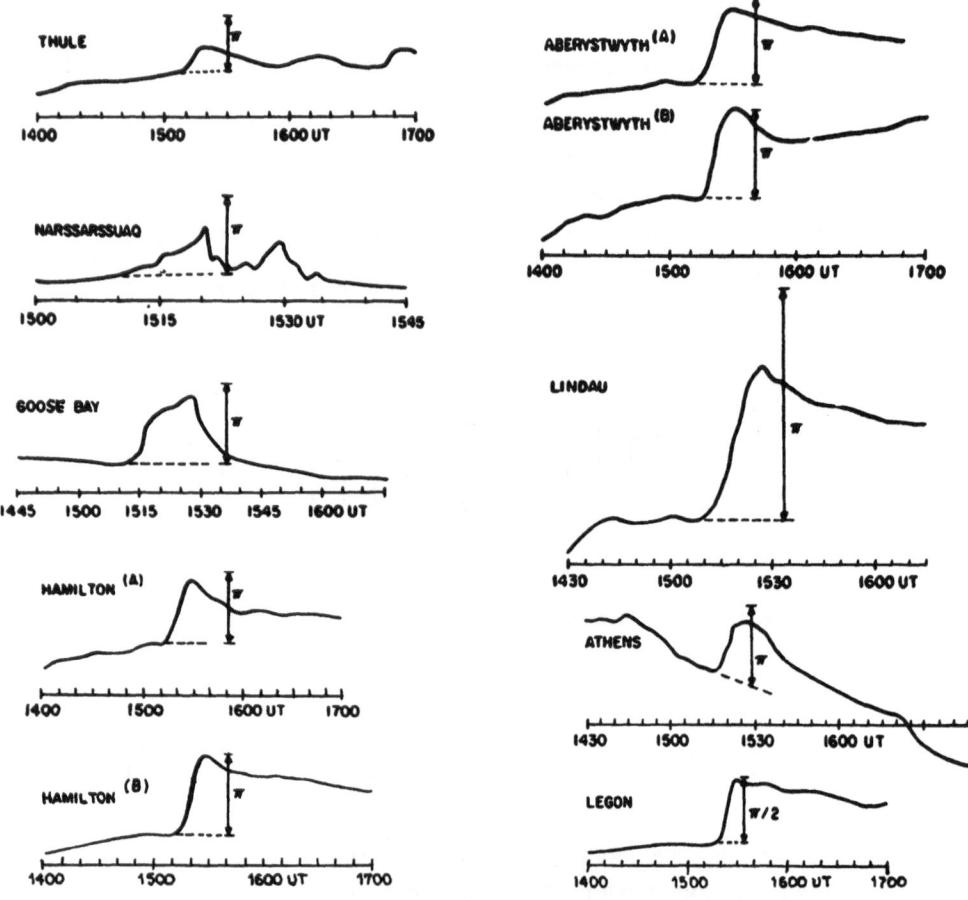

Fig. 37.   Total electron content changes observed with satellite beacon experiments globally in 17 stations for the outstanding flare event of August 7, 1972 (after Mendillo et al., 1974).

University of Hawaii. The events of May 21 and 23 are shown in Figures 35 a, b, along with the SWF observed simultaneously with WWV on 15 MHz and SPA with WWVB transmissions from Ft. Collins on 60 kHz. The electron content is seen to have increased by about 2×10^{16} el m^{-2} in a time interval of less than 2 minutes (about 5% of the total ionospheric content).

In the second case (Mendillo *et al.*, 1974) the first truly global coverage was achieved with 17 stations in North America, Europe and Africa, spanning over ten hours in local time and over 70 deg in latitude. The stations ranged from Thule, Greenland to Legon, Ghana, and ΔN_T ranged from 1.8 to 8.6×10^{16} el m^{-2}. On a percentage basis, all the SITECs fell within 15 to 30% range. The observed effects reproduced from the paper of Mendillo *et al.* are shown in Figures 36 and 37.

In addition to these, several other cases have been reported. The Stanford Group have reported a number of SITEC events obtained at the AFCRL Radio Observatory at Sagamore Hill with transmissions from the synchronous satellites ATS-3 and ATS-5 on 137.35 and 136.47 MHz during the period August 1968 to August 1970. A total of 19 cases were identified.

A disadvantage of the technique is the fact that the observed effect is an integrated one of effects occurring over the entire height ranges of 100–2000 km. Nevertheless since *F*-region detection techniques are few, this provides one of the most useful routine techniques for the *F*-region currently available.

2.12. Incoherent-Scatter Technique

The technique makes use of the scattering of radio waves from the random thermal fluctuations in electron density existing in a plasma in quasi-equilibrium, and measures the power, frequency spectrum (or autocorrelation function), and polarization of the scattered signal returned from the ionosphere. From this one may obtain the following parameters of the ionospheric plasma:

(1) Electron density (N_e), (2) electron temperature (T_e), (3) ion temperature (T_i), (4) ionic composition (molecular ions of similar masses, such as N_2^+, O_2^+, NO^+ cannot however be distinguished), (5) ion-neutral particle collision frequency (ν_{in}), (6) ion-ion collision frequency (ν_{ii}), (7) magnetic field, (8) photoelectron velocity distribution, (9) mean plasma drift velocity and (10) drift velocity of electrons relative to ions (current density).

The basic requirements for the experiment are: (i) a powerful transmitter (a few megawatts peak power) and (ii) a large transmitting/receiving antenna (e.g. a parabola of a few hundred meters in diameter) for a monostatic system, or two large antennas where the transmitter and the receiver are not located in the same place. It is possible to get a nearly complete picture of the state of the ionosphere as a function of time and altitude, with height resolutions smaller than a scale height and time resolutions of a few minutes or less. In the *F*-region, where the signal-to-noise ratio is highest, a time resolution of, or better than, 5–10 min is achieved by most observatories. A height resolution of, or better than, 10–20 km is common for power measurements,

TABLE II

Incoherent-scatter facilities

Facilities	Coordinates	Operating frequency (MHz)	Antenna	Peak power (MW)	Altitude coverage	
					Power (km)	Spectrum (km)
Jicamarca, Peru	11.95°S, 76.87°W (dip = 2.0°)	49.92	290 × 290 M	4.0	150–8000	200–1200
Arecibo, Puerto Rico	18.3°N, 66.75°W (51.7°)	I 40.12 II 430	300 M Spherical reflector	1.3 2.0	50–2000	90–1400
Millstone Hill, U.S.A.	42.6°N, 71.5°W (73.0°)	I 440 II 1295	68 M Parabola 25 M Parabola	3.0 4.0	100–1000	200–900
St. Santin, France	44.65°N, 2.19°E (61.0°) (Bistatic)	935	20 × 100 M reflector (vertical) 40 × 200 M reflector (oblique)	0.075	95–500	95–500
Malvern, U.K.	52.09°N, 2.14°W	400.5	42.7 M Parabola	8.0	90–1000	200–750
Chatanika, Alaska	65.1°N, 147.5°W (76.5°)	1290	26.4 M Parabola	4.0	70–1000	150–500

but the resolution is poorer for spectrum measurements. In the St. Santin station only one altitude can be observed at a time. Each observation (of both power and spectrum) requires 5–6 min, when single height is monitored. Complete height profiles require 1–2 h. This is a major disadvantage for flare studies.

The incoherent-scatter technique is one of the most versatile systems available for ionospheric studies. It is, however, an extremely sophisticated and expensive setup. It is, available only in a few places in the world, but fortunately the few that exist include different geographical areas. The equatorial area is covered by the Jicamarca observatory in Hauncayo; there is a low latitude station in Arecibo, Puerto Rico; there are two midlatitude observatories, one in North America (Millstone Hill, Mass.) and the other in France (St. Santin); and recently an auroral station has been established in Alaska at Chatanika. Major features of these installations are summarised in Table II.

For flare studies there are two major difficulties with this arrangement. One is that in a sophisticated setup of this kind, the scientific interests are very diverse, and flare work must compete with many other interests. These include not only a time sharing between radioastronomical and ionospheric programmes, where both programmes are undertaken at the same station, but a question of rescheduling of the operational arrangement, since the central point in a flare study is continuous time coverage. The other inadequacy is that the incoherent-scatter facilities are currently confined within the American and European Zones, and cannot, therefore, follow the Sun all the time.

Another insufficiency concerns the rather large time necessary for each observation. Since it is now well established that solar flare radiation (10–1030 Å) affecting the E

and F regions often have rise and decay times less than the 5–10 minute resolution of most incoherent scatter measurements, and since the time constants for electron loss in the E and $F1$ regions are less than 5–10 min, it is believed (Donnelly, private communication) that most incoherent scatter measurements lack sufficient time resolution for studies of the ionospheric effects of flares. Also, only 1 or 2 flares per year for a given geographic location are large enough to be observable above 100 km.

As noted before, the incoherent-scatter sounding is generally effective for the E-region and above. Nevertheless recent attempts to probe down to lower heights have, in some cases, extended the observations to well within the D-region. Although their reliability is yet to be established, an encouraging feature is that there is general agreement in electron density at heights *above* the electron density ledge ($\sim$85 km) with rocket borne experiments. It seems, therefore, that the equipment is capable of monitoring electron density changes from 85 km upward all through the F-region and upto several thousand kilometers in some cases.

Taylor (1974) has discussed the relative accuracies of flaretime changes in the incoherent scatter measurements at different levels in the ionosphere. He points out that the quality of the result depends partly on the statistical measurement error of N_e, and partly on several systematic uncertainties in the baseline from which ΔN_e must be determined. In the D-region, both types of errors contribute since N_e is small and the signals are contaminated with clutter. The only saving feature is that ΔN_e is

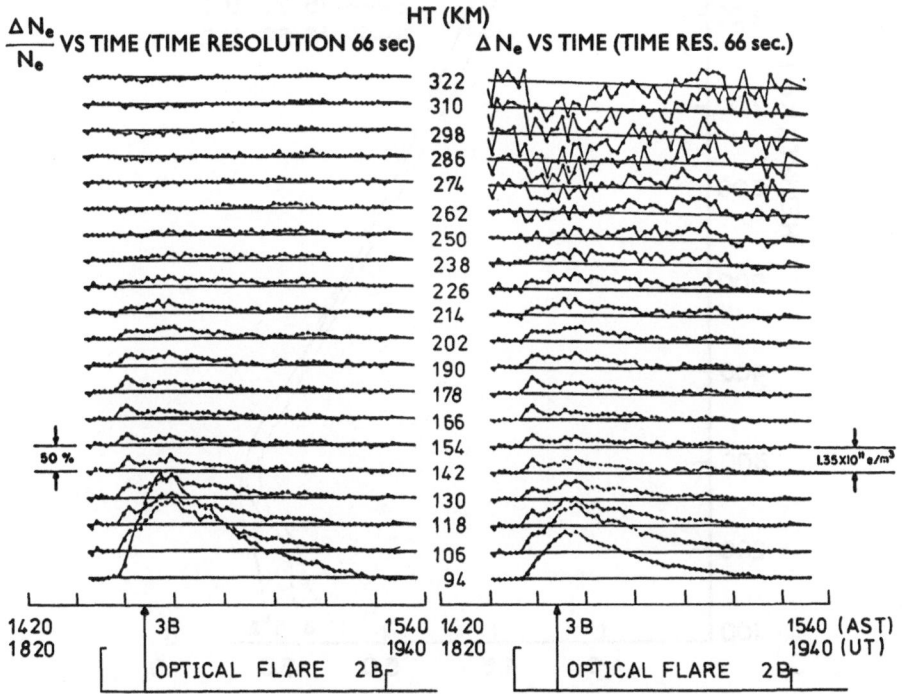

Fig. 38.   Enhancements in electron densities (ΔN_e) and the relative changes ($\Delta N_e/N_e$) obtained with the incoherent scatter facility at Arecibo during the flare event of May 21, 1967 (after Thome and Wagner, 1971).

exceedingly large. In the F-region, while there is the important advantage of large N_e, which can be measured with relatively small errors, ΔN_e is small and is of the same order as the irregular fluctuations (in both amplitudes and periods). There is an additional complicating feature: as we shall see later, the electron temperature T_e is considerably enhanced during a flare, although the ion temperature T_i remains relatively unchanged. The scattering cross-section in the F-region is reduced if T_e/T_i is enhanced. One must make appropriate correction for this. In the E-region, the flaretime results are the most accurate, since both N_e and ΔN_e are large enough to be measured accurately. Taylor feels that the incoherent scatter gives its 'best' flare results in the height range from about 90 to 130 km.

Several flare effects have been reported with incoherent-scatter facility giving detailed variations in the electron density with time during the course of the flare for

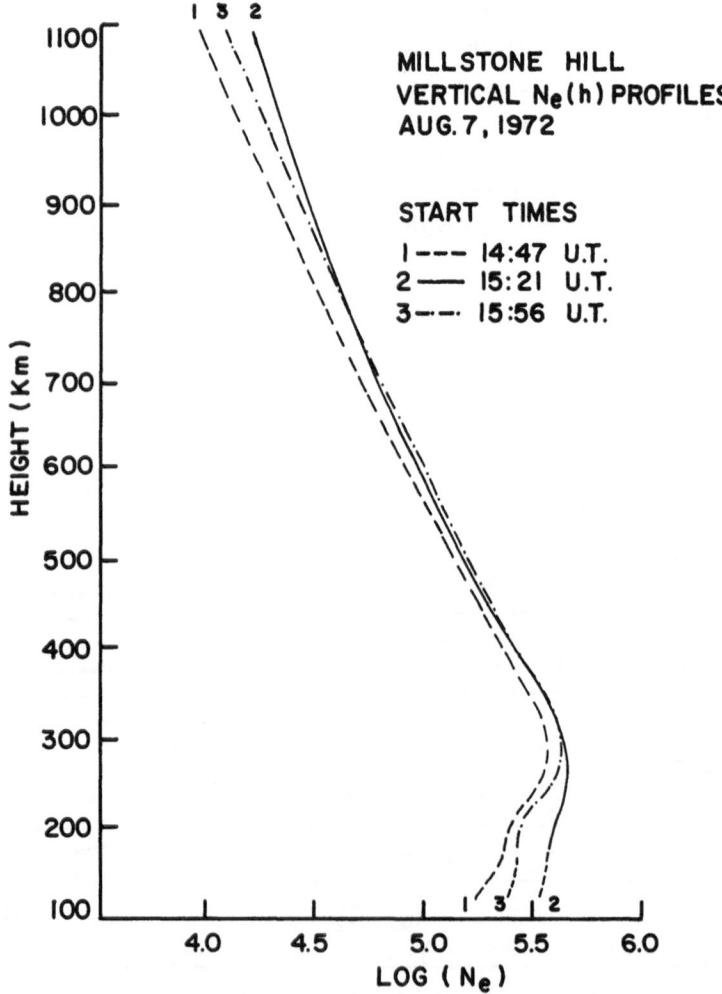

Fig. 39.   Changes in electron densities from 125 to 1200 km for the event of August 7, 1972 observed with the Incoherent Scatter Facility at Millstone Hill, USA (after Mendillo and Evans, 1974).

heights above 90 km and covering the E and the F-regions, including the topside ionosphere. Effects reported so far are, however, few. Nevertheless these are very rich in information content. Major effects were reported by Thome and Wagner (1971) for the flares of May 21 and 23, 1967. Figure 38 gives the time profiles of enhanced electron densities obtained from back-scatter power fluctuations at selected heights for the flare of May 21, 1967. These were obtained at Arecibo from simultaneous observations of the following: total power in the ion component of the backscattered spectrum as a function of height and time on 430 MHz; plasma line echoes observed in 0.5 MHz steps from 3.5 to 9.5 MHz below the transmitted frequencies, phase-path length variations and absorption on two frequencies in the F-region. The results indicate:

(1) Fairly large enhancements in the E-region, decreasing with height in the F-region, but with measurable enhancements upto 300 km.

(2) Delayed time of onset of the enhanced ionization at levels above the E layer, becoming most pronounced at F layer heights.

(3) A negative fluctuation at F-region heights over the flare affected region.

Another case reported (Mendillo and Evans, 1974) is the flare of August 7, 1972, observed with the incoherent-scatter facility at Millstone Hill. In this case measurements were made from 125 to 1200 km considerably beyond any height previously reported. The electron density profiles are shown in Figure 39 for three consecutive

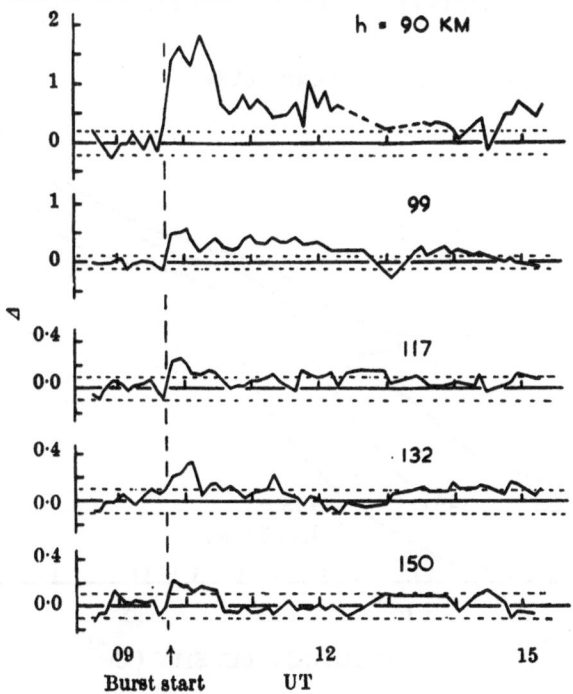

Fig. 40.   Flaretime ionization enhancements in the E-region for the flare of February 20, 1970 obtained with the Incoherent Scatter Facility at Malvern (after Taylor and Watkins, 1970).

runs before, during and after the maximum flare activity. The electron density enhancement was 100% at 125 km and 60% at 200 km.

There have also been some attempts to determine flaretime ionization enhancements down to 90 km by using incoherent-scatter radar. One such case is shown in Figure 40 giving $\Delta = (N_e - N_{eo})/N_{eo}$ as a function of time for five heights 90, 99, 117, 132 and 150 km for the large solar flare that occurred on February 20, 1970. The recordings were made with the incoherent-scatter radar at Malvern, and have been reported by Taylor and Watkins (1970). At 90 km, the enhancement was more than 100% at its peak and lasted for at least 3 hours, whereas at 132 and 150 km there were only marginally significant changes, lasting for about an hour. In addition to the flare effect of February 20, 1970, two of the distinctive events have been identified at Malvern during about 300 h of observations: these occurred on 11 December, 1968 at 1144 UT and 19 August, 1970 at 1113 UT.

2.13. Rocket Soundings

The final method that we will discuss is that of direct rocket soundings. While its importance is obvious, the practical problems of proper timing of the launchings make this an exceedingly difficult experiment to perform. The only rocket firings

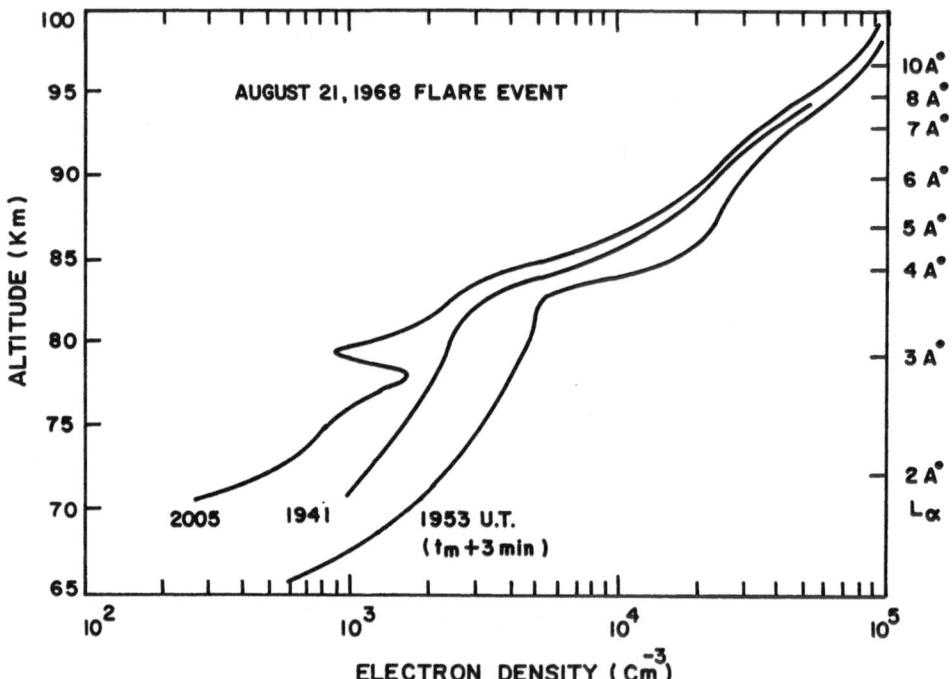

Fig. 41.   Results of three rocket measurements of electron density profiles during the flare of August 21, 1968. Flights were at 1935 UT (3 min after peak X-ray flux), 1941 UT (during decay phase) and 2005 UT near the end of flare, but still not normal (after Somayajulu and Aikin, 1969).

made so far are by Somayajulu and Aikin (1969, 1970) who have successfully obtained electron density profiles from two series of flare time launchings. The first series consisted of two rocket launchings during the decay phase of a class 1 flare that occurred on January 16, 1968. The second series were flown into a class 1B flare in August 21, 1968, the first flight being approximately 3 min after the peak of the flare, the second 12 min later during the decay phase and the third near the end of the flare (but not entirely a normal condition). The payloads were instrumented for the simultaneous determination of ionizing Lα and X-rays along with the electron density. The development and decay of ionization are given in Figure 41.

2.14. A Comparative Evaluation of Electron Density Profile Errors

If we leave out the purely monitoring type of techniques which yield electron density profiles only indirectly and the errors are likely to be quite large (e.g. SES, SFA, SCNA, SPA), the errors in electron density and height in the direct techniques can vary quite substantially from one method to another. Thrane (1973) has given a

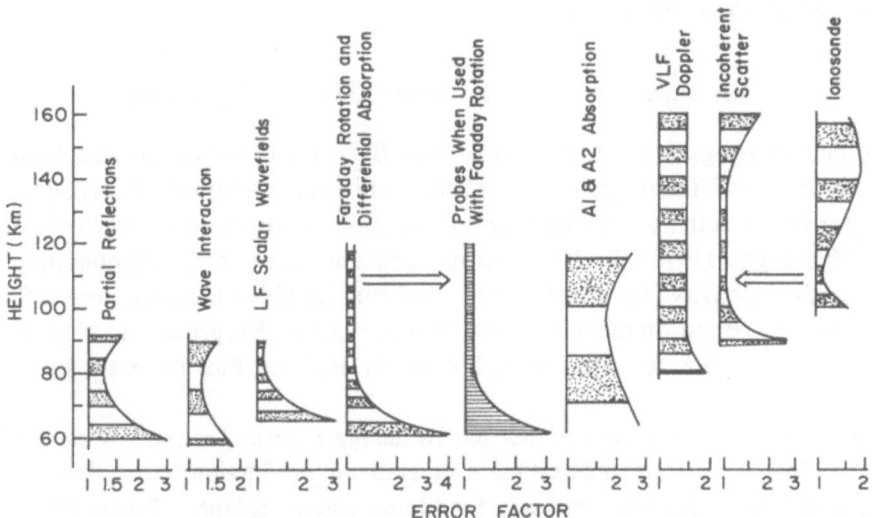

Fig. 42. A comparative study of errors in electron density and height measurements in different SID techniques (after Thrane, 1973)

comparative study of these errors. While there is some difference of opinion about these errors, and also in deciding to what extent these are to be modified during flare conditions when suddenly $(N_e \times v)$ weight factor descends downwards, this is still a very valuable comparison. This is shown in Figure 42.

CHAPTER 3

FLARE RADIATIONS RESPONSIBLE FOR
IONOSPHERIC EFFECTS

The major wavelength areas of interest to upper atmospheric physics are:

(i) The broad Schumann continuum from 1350–1750 Å, the far-ultraviolet responsible for the dissociation of O_2.

(ii) The whole Spectrum from 1350–100 Å, which contributes to the ionization of the upper ionosphere. This region includes:

(a) Lα at 1216 Å responsible for the production of quiet day *D*-region,

(b) He I (584 Å) and He II (304 Å) lines, which are quite prominent, and contribute substantially to *F*-region ionization.

(iii) *X-rays above 8 Å*, contributing to the lower *E* region.

(iv) *X-rays below 8 Å*, negligible under quiet conditions, but the main source of flare time *D*-region ionization.

3.1. Quiet-Day Solar Flux in the EUV and X-Ray Regions

The flux values generally used in the usual EUV photo-ionization calculations are mostly those of Hinteregger given for solar minimum conditions (Hinteregger, 1969) and for higher activity some adjustments are made using the measurements of Hall and Hinteregger (1970). The latter measurements however, cover only conditions upto intermediate activity ($F_{10.7} = 142$ units), and considerable uncertainty exists for high solar activity period. In the region 260–1300 Å, the AFCRL group find a flux increase by a factor of 1.5 for the period for which 10.7 cm flux increases by a factor of 2.

There has been some feeling that the Hinteregger fluxes for the low solar activity are too low, giving an underestimate of electron production rates. Swartz (1972) found 40% more heating necessary for global energy balance. Taylor (1972), in a rocket flight launched immediately following the solar eclipse on 7 March 1970 at 1930 UT over White Sands, New Mexico (10.7 cm flux = 171 units) observed in the region 1–800 Å flux between 4–6 times larger than the medium Sun values of Hinteregger (corresponding to values based on 11 March 1967 flight – with 10.7 cm flux having a value of 141.6 units). Higgins (private communication) finds, from an apogee scan of a rocketborne experiment launched on 21 November 1968, an increase in flux over the Hinteregger values of the order of 1.5 to 2.0. In Table III are given Hinteregger's minimum solar activity and midsolar-activity flux values, along with the new flux values of Taylor.

It should also be mentioned that estimates made from sunrise measurements of

TABLE III

Flux distribution in different XUV bands

Wavelength or range (Å)	Identification	Hinteregger's results [a] (10⁹ photons cm⁻² s⁻¹)		March 7, 1970 [b] Taylor's estimates (10⁹ photons cm⁻² s⁻¹)	Ratio = $\dfrac{7/3/70}{\text{Hinteregger}}$
		Medium Sun	Min. Sun		
800–770	H Cont	0.39		1.8	4.6
790.2, 790.1	O IV	0.26		1.2	4.6
787.7	O IV	0.13		0.61	4.7
786.5	S V	0.08		0.37	4.6
780.3	Ne VIII	0.12		0.56	4.7
770.4	Ne VIII	0.23		1.1	4.3
770–740	H Cont	0.19		0.89	4.7
765.1	N IV	0.18		0.84	4.7
760	O V	0.08		0.37	4.6
740–710	H Cont.	0.09		0.42	4.7
710–680	H Cont.	0.04		0.18	4.5
703 group	O III	0.23		1.1	4.8
800–630	unresolved	0.39		1.8	4.6
629.7	O V	0.92		5.0	5.4
625.3	Mg X	0.25		1.5	6.0
609.8	Mg X	0.5		3.0	6.0
599.6	O III	0.08		0.37	4.6
584.3	He I	0.89		5.3	5.9
554 group	O IV	0.31		1.7	5.5
521.0	Si XII	0.19		1.2	6.3
508 group	O III	0.08		0.37	4.6
504	He I Cont.	0.50		2.9	5.8
499.3	Si XII	0.38		2.4	6.3
465.2	Ne VII	0.16		0.88	5.5
630–460	unresolved	0.44		2.4	5.5
460–370	integral	0.63		3.5	5.6
368.1	Mg IX	0.56		3.1	5.6
364.8		0.17		0.95	5.6
360.7	Fe XVI	0.36		2.7	7.5
335.4	Fe XVI	0.72		5.5	7.6
303.8	He II	5.4		28.7	5.3
284.1	Fe XV	1.1		9.2	8.4
370–280	unresolved	2.03		11.9	5.8
280–260			0.5	2.7	5.4
256			0.3	1.6	5.3
260–240			0.4	1.9	4.7
240–220			0.4	2.2	5.5
220–205			0.3	1.7	5.6
205–180			3.8	13.7	3.6
180–165			3.2	18.6	5.8
165–60			0.75	13.7	18.2
60–1			0.12	8.9	74.1

[a] Medium Sun values based on 11 March 1967 flight (10.7 cm flux 141.6 units). Minimum Sun values based on Hinteregger's quiet Sun model (Hinteregger, 1969).

[b] Launched immediately following the solar eclipse on 7 March 1970 at 19 30 UT over White Sands, New Mexico (10.7 cm flux 171 units) (Taylor, 1972).

ionospheric total electron content using satellite radio beacon transmissions give a total EUV flux increase by a factor of 2 from sunspot minimum to maximum.

1300–100 Å. Hinteregger (1969) has given the flux values in the continuum as well as in the principal lines. Variations with solar activity have also been reported by Hinteregger and his colleagues for Lα (1216 Å), Lβ (1026 Å), He I (584 Å), Fe XVI (335 Å), He II (303.8 Å), Fe XV (284 Å). In the region 260–1300 Å, Hinteregger and Hall (1969) find a flux increase by a factor of 1.5 for the period for which 10.7 cm flux increase by a factor of 2. Estimates from sunrise effects in satellite beacon experiment give EUV flux increase of a factor of 2 from sunspot minimum to maximum.

Lα (1216 Å). Measurements of Lα have been made from 1949 onwards over the entire solar cycle. The early measurements by the U.S. Naval Research Laboratory were made by photon counters, and the later measurements by ion chamber technique. The values quoted so far range between 0.1 erg cm^{-2} s^{-1} during the first low solar activity observations to some 6 erg cm^{-2} s^{-1} during IGY–IGC, a period of sunspot maximum. Values below 1 erg cm^{-2} s^{-1} are, however, not considered reliable. Nicolet and Aikin (1960) accepted a value of 3 erg cm^{-2} s^{-1} as the best average. Hinteregger (1961) obtained a value of 3.3 erg cm^{-2} s^{-1} during a 1960 flight. Russian observations with a radiation detector mounted in Spaceship II yielded flux values of 5–6 erg cm^{-2} s^{-1} during August 9, 1960. Kreplin *et al.* (1962) from observations on satellite 1960 Eta 2 during the period July 13–August 3, 1960, found that the day-to-day variation did not exceed 18%.

The nature of solar cycle variation of Lα is not known. It is not even known if any variation exists at all. Comparison with absorption measurements indicate that a variation of a factor of 2 is not unlikely.

He 304. The He II (304 Å) emission, which alone accounts for some 10% of the total flux in the region 60–850 Å, has a flux density of about 6×10^9 cm^{-2} s^{-1} (or about 0.40 erg cm^{-2} s^{-1}), and varies with solar activity. Observations by OSO-1 satellite showed that the emission increases by about 33% during a period when the sunspot number increased from 0 to 94 and the 2800 MHz flux increased from 76 to 125×10^{-22} Wm^{-2} Hz^{-1} (March 11–22, 1962).

X-rays (100–1 Å). The radiation is highly variable with solar activity (Acton *et al.*, 1963).

Kreplin *et al.* (1962) estimate that the flux of X-radiation between 1953 to 1959 increased by factors of 600–1000, 60 and 7, in the ranges 2–8, 8–20 and 44–60 Å, respectively.

A more reliable estimate of the solar cycle variation in the 2–12 Å region by Wende (1969) using Anton Type 213 GM tubes from Injuns I and III, flown in 1961–62 and 1962–63, respectively, and Explorer-33 and 35, flown in 1966 and 1967 shows that the fluxes observed by Injun I and Explorer-33 and 35 are about an order of magnitude higher than those measured by Injun III. We note that the maximum variation in efficiency between these GM tubes is less than a factor of 2, so that the observed difference is not instrumental.

In the region 20–100 Å the information available until recently was either insufficient or inaccurate. The most abundant source of information was the NRL observations in the band 44–60 Å, but it was recently realised that the flux measured includes a component from 1–20 Å flux which must be removed. This reduces the published flux values by about $\frac{1}{3}$. A major information came from Manson's measurements (Manson, 1972) made on August 8, 1967 for an active, but non-flaring conditions. There is only weak line emission below 60 Å and several lines of small intensity between 60 to 110 Å. The flux values are given in Table IV in Model 1. Table IV also gives two other models, chosen by Rowe *et al.* (1974) in a recent work. A

TABLE IV

Characteristics of Solar XUV flux

Spectral range (Å)	Model 1	Model 2	Model 3	Remarks
	(erg cm^{-2} s^{-1})			
1–20	Spectral distribution constructed from published flux in 1–8 Å and 8–20 Å ranges based on a spectrum including free-free (bremsstrahlung) free-bound, and line emission.			Same for all three models
30–40	0.012	0.012	0.012	Model 1 (30–100 Å) are from Manson
40–50	0.017	0.021	0.021	(1972) based on a rocket flight made on
50–60	0.029	0.030	0.030	August 8, 1967 for an active, but non-
60–70	0.029	0.032	0.032	flaring condition. Model 2 and 3
70–80	0.024	0.025	0.025	(30–100 Å) are from Hinteregger for
80–90	0.028	0.030	0.030	11 March 1967 ($F_{10.7}=144$)
90–100	0.021	0.021	0.021	
100–1025.7	$H_{AS} \times 2$	H_{AS}	$H_{AS} \times 2$	H_{AS} refers to Hinteregger flux for 11 March 1967. ($H_{AS} \times 2$) refers to fluxes in all bands raised by a factor 2.

matter of considerable concern is that the Manson flux values coupled with Hinteregger's fluxes above 100 Å or the Hinteregger flux over the entire wavelength region, give electron densities considerably lower than the experimental values in the region between 90 km and the height of maximum ionization in the E-region. The discrepancy is already noticeable at 90 km where the theoretical value of N_e is lower by a factor of 2. There is need to increase the Hinteregger XUV flux by at least a factor of 2 in the region 100–1027 Å. Model 3 in Table IV takes this into account. The solar activity variation in this region is not well known, but must exist. An unpublished and preliminary reduction of 44–60 Å fluxes from data obtained with NRL ionization chamber experiments aboard Solrad's 7B, 8 and 10, OGO-4 and OSO-5 by Horan and Kreplin (private communication) indicates that the flux varies substantially with solar activity, from about 0.05 erg cm^{-2} s^{-1} during low solar activity ($70 < F < 100$) to 0.15 erg cm^{-2} s^{-1} increase in flux in this region.

3.2. Flaretime Enhancements in XUV Fluxes

Since ionospheric ionization can occur in the entire wavelength region below 1216 Å, any enhancements in any of the wavelengths at and below 1216 Å are of interest. Over this region, large enhancements have been observed in the bands 0.5–3 Å, 1–8 Å and 8–20 Å and in many of the EUV lines. At the lower end of the spectrum, the very large spectral hardening of the radiation below 10 Å and the increasing ionization efficiency of the radiation with decreasing wavelength, allows wavelengths down to 0.1 Å to contribute at the lowest D-region levels. The peak ionization level for the entire XUV region is shown in Figure 43. Levels for selected EUV and X-ray lines are also shown.

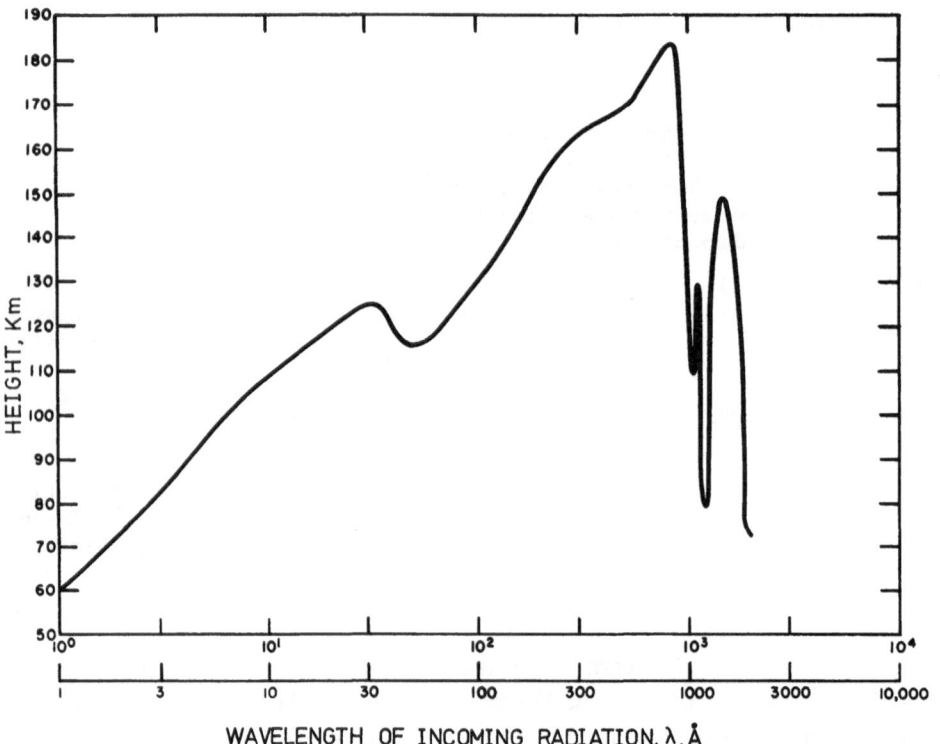

WAVELENGTH OF INCOMING RADIATION, λ, Å

Fig. 43. Level of 50% absorption as a function of wavelength of incoming radiation.

The spectral changes which occur in the X-ray emissions of a flare are best dealt with in two regions; a low energy one (soft X-rays, with wavelengths longer than about 1 Å and energies less than 12 keV of primary interest for SID effects), and a high energy one (hard X-rays, with wavelengths shorter than about 1 Å and energies greater than 12 keV see, for example, Arnoldy *et al.*, 1968).

The gross changes in the spectrum of the soft X-ray flares were first observed by Culhane *et al.* (1964) by using proportional counters. During the course of the flare,

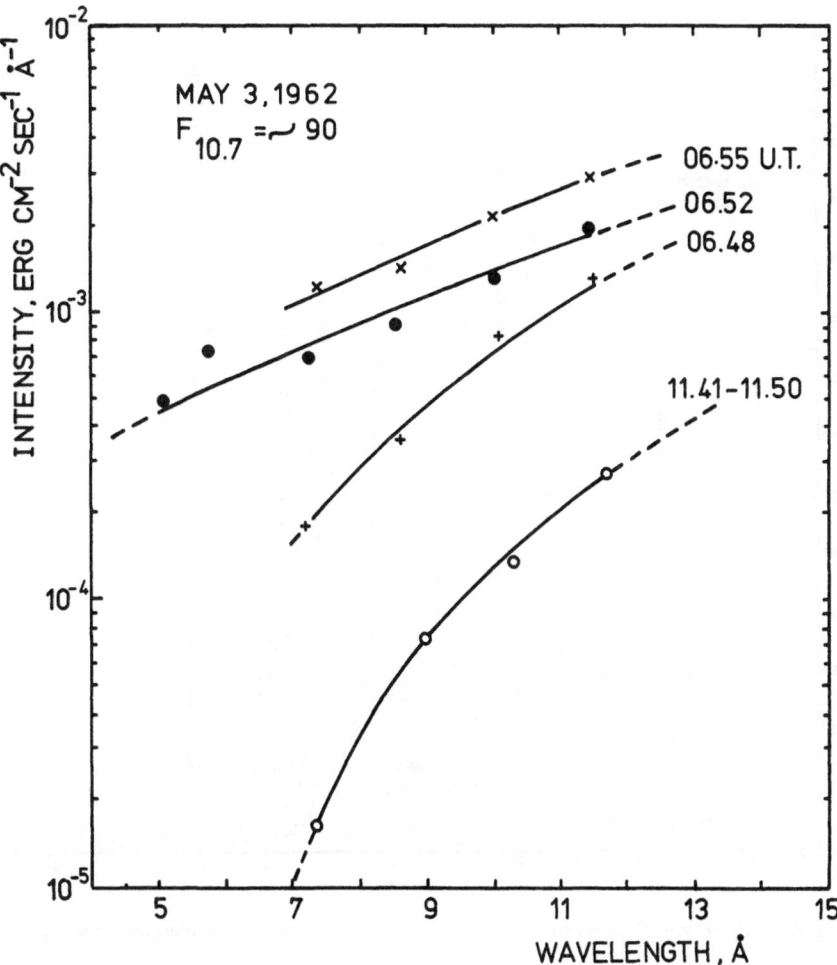

Fig. 44.   Hardening of solar X-ray spectrum during the flare of May 3, 1962
(after Culhane *et al.*, 1964).

the spectrum clearly hardens with respect to the pre-flare spectrum. An example is given in Figure 44. This gives the spectral development during the flare of such spectral hardening which allows radiations to penetrate deeply into the ionosphere (the absorption coefficient rapidly decreases with decreasing wavelength) and produce ionization at low levels. The difficulties and errors involved in deriving this spectrum from such measurements with proportional counters have later been discussed at length by Lepine and Hall (1972) along with examples for quiet conditions.

Changes in the emission line structure for wavelengths longer than 1 Å have been observed through the use of satelliteborne Bragg spectrometers (Figure 45). Not only does the continuum flux increase appreciably, but lines appear at wavelengths shorter than in the preflare spectrum. These have been identified; ionization states as high as Fe xxv are seen. Such high ionization states are indicative of regions of very high temperatures, typically of the order of 10^7 to 10^8 K, presuming that the concept of

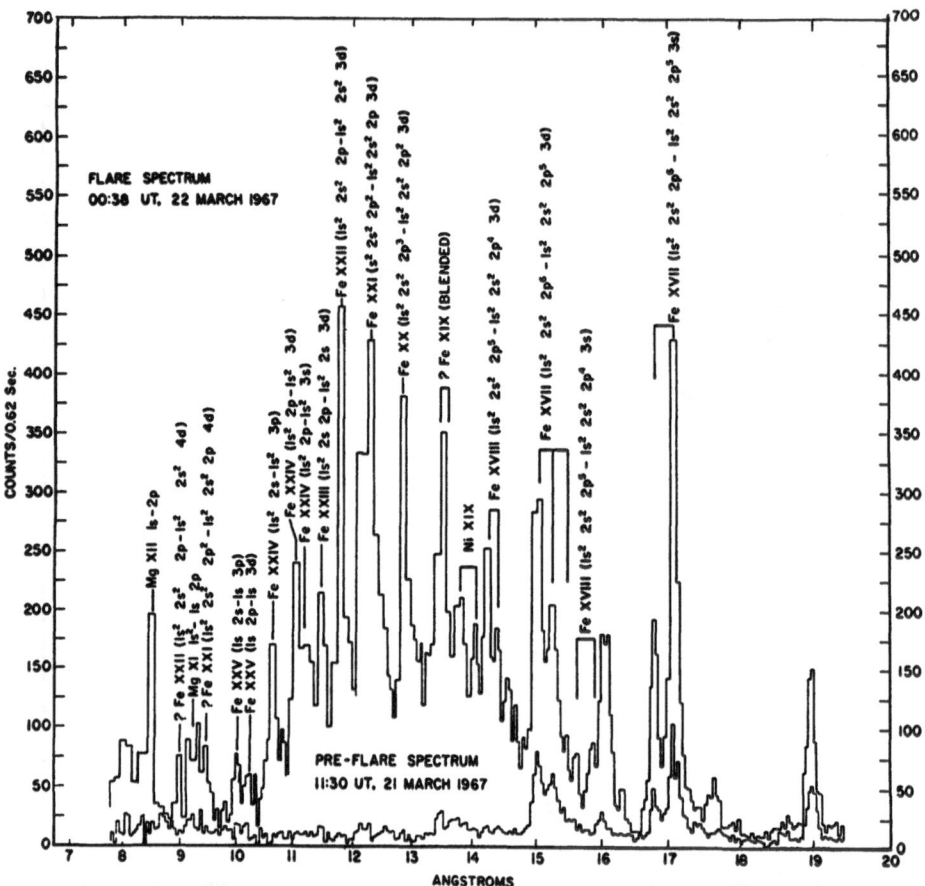

Fig. 45. Flaretime changes in solar emission line structures (after Neupert *et al.*, 1967).

temperature is still valid during the flaring process. Neupert *et al.* (1967) found that lines of Fe XXIV and Fe XXV are present during some flares, and that emissions from the highest stages increase rapidly during the flare onset, while the lower stages are observed later in the event. For ionization calculations it is important to have information on the relative contribution of fluxes in the lines and the continuum. All calculations assume that for $\lambda < 10$ Å, the electron production due to continuum predominates: a conclusion that needs further examination.

Few observations of solar extreme ultraviolet radiation (100–1030 Å) have been made during flares. Such measurements are not easy to make, because the EUV spectrum is a jungle of emission lines and the percentage enhancement during flares is much smaller at EUV wavelengths than for X-rays. Some of the early observations with OSO-1 were reported by Neupert (1964); the events reported were March 22, 1962 (2220 UT); March 13, 1962 (1440 UT); April 20, 1962 (1958 UT) and were observed on 304 Å, 284 Å and 335 Å. Hall and Hinteregger (1969) have reported some fascinating OSO-3 measurements of EUV flashes. They observed some lines that

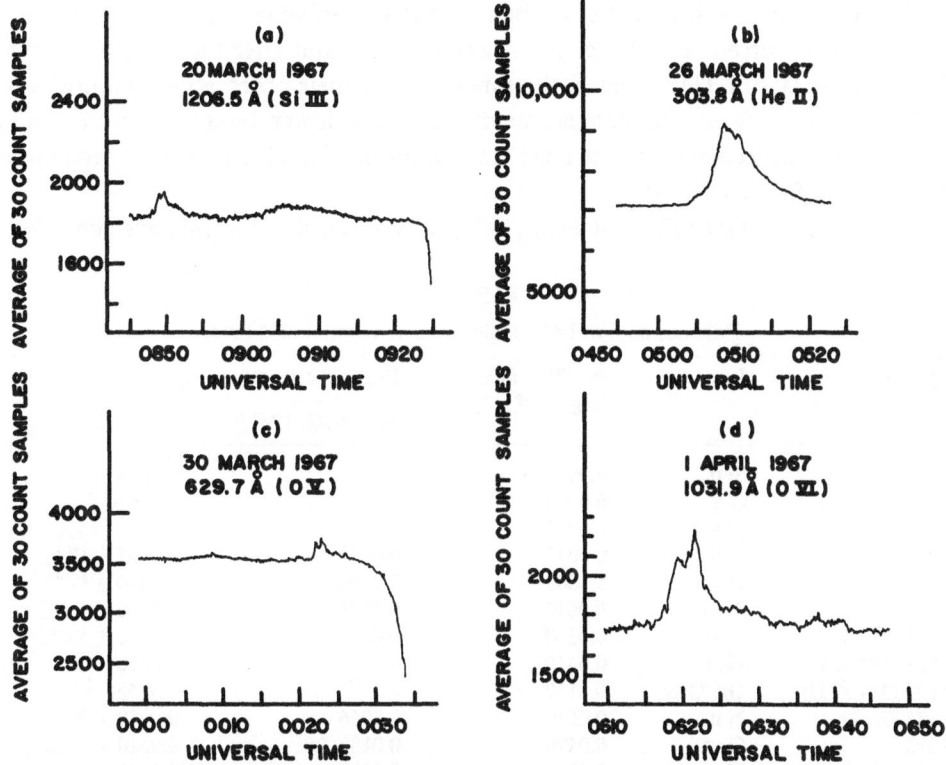

Fig. 46. Time dependence of EUV enhancements recorded with an extreme ultraviolet spectrometer aboard OSO-III (after Hall and Hinteregger, 1969). Parts (a) and (c) show a drop off in flux at the right side of the figure caused by atmospheric absorption.

increased by 200% during a flare. Some of the EUV lines observed showed appreciable fine-time structures (0.5 min minor peaks). The EUV enhancements usually peaked several minutes before the reported times of Hα intensity. Hall and Hinteregger report that the EUV lines normally considered to originate in the hot sparse corona of the Sun showed small slow enhancements while the lines of moderate ionization potential (from C, O, and N) exhibited much larger and faster enhancements.

The time dependence of the EUV enhancements takes various forms. Figure 46 shows a few examples. Table V gives the flux enhancements (in erg cm^{-2} s^{-1}) of several EUV lines observed by Hall and Hinteregger (1969) during a 3$^-$ flare starting at 0022 UT on 22 March, 1967. Since the time for a complete scan was 5.44 min, the enhancement record for an individual line had very coarse time-resolution, except for a few cases where the second order of the line was also present. However, the enhancement was prolonged; the main enhancement seemed to be from about 0028 UT to 0038 UT, peaking around 0031–0032 UT. The scan showing the largest enhancements began at the long wavelength end at 0030.1 UT and, therefore, the longer wavelength lines showed approximately their peak enhancements. By the time the scan reached the short wavelength, the enhancement was down to perhaps half its peak value. Table V shows the maximum observed enhancements of various lines (reduced from percentages of the flux observed before and after the flare) in the scan beginning at 0030.1 UT, listed in the time-order of their observation. According to Hall and Hinteregger, the enhancements listed are lower bounds; the maximum percentage enhancements for the lines at the bottom of the list may have been larger by at least a factor of two.

More recently, Donnelly and Hall (1973) have given the flux enhancements for a

TABLE V

Enhancement in EUV region near maximum of 2 flares

Wavelength Å	Ion	Pre-flare flux [a] (erg cm^{-2} s^{-1})	Flux enhancement (erg cm^{-2} s^{-1})	
			March 22, 1967 [b]	March 27, 1967 [c]
1265	Si II	0.0075	–	0.0024 ± 49%
1260 (630 Å II)	O V	0.0300	–	See 630 Å
1243	N V	0.0048	–	0.003 ± 48%
1239	N V	0.0072	0.0022	0.0057 ± 17%
1216 Lα	H I	5.2000	0.9360	0.4300 ± 12%
1207	Si III	0.0610	0.0671	0.0370 ± 4.3%
1176 Group	C III	0.0370	0.0740	0.0200 ± 5.6%
1169 (584 Å II)	He I	0.0310	–	See 584 Å
1109 (554 Å II)	Gr. O IV	0.0110	–	See 554 Å
1085 Group	N II	0.0110	0.0066	0.0030 ± 20%
1032	O VI	0.0470	0.0188	0.0110 ± 11%
1026 Lβ	H I	0.0700	0.0070	0.0025 ± 24%
991 Group	N III	0.0130	0.0130	0.0048 ± 12%
977	C III	0.0940	0.0658	0.0280 ± 8.4%
973 Lγ	H I	0.0170	0.0024	0.0017 ± 34%
950 Lδ	H I	0.0074	0.00096	0.00059 ± 58%
911 (304 Å III)	He II	0.3700	–	See 304 Å
680–912 Cont.	H I	0.2120	0.0424	0.0250 ± 31%
834 Group	O II, III	0.0128	0.01152	0.00090 ± 84%
790 Group	O IV	0.0068	0.00544	0.0024 ± 18%
770	Ne VIII	0.0061	–	0.00064 ± 45%
765 Blend	N IV, III	0.0049	0.00196	0.00073 ± 125%
760 Group	O V	0.0020	0.0016	0.00043 ± 48%
630	O V	0.0300	0.0030	0.0045 ± 31%
625	Mg X	0.0084	–	0.0015 ± 57%
608 (304 Å II)	He II	0.3700	–	See 304 Å
584	He I	0.0310	0.00217	0.0051 ± 12%
554 Group	O IV	0.0110	0.01320	0.0019 ± 25%
465	Ne VII	0.0070	0.00105	0.0013 ± 52%
304	He II	0.3700	0.0259	0.0310 ± 13%

[a] Preflare flux values derived by Donnelly and Hall (1973) from observations of March 11, 1967 made shortly after launch of the satellite with a 4% increase to correct for the slightly greater increase in solar activity on March 27, 1967.
[b] From Hall *et al.* (1969), based on OSO-3 scanning spectrometer measurements. Percent enhancements given by Hall *et al.* have been converted here to flux enhancements using preflare flux values given by Donnelly and Hall (1973). Results include time variations during the flare.
[c] From Donnelly and Hall (1973), based on OSO-3 scanning spectrometer measurements. The time dependence of the broadband 10–1030 Å flux derived from SFD observations was used as a self-consistent correction for the time dependence of the OSO-3 observations.

somewhat weaker flare of 2114 UT March 27, 1967 from the same satellite and measured with the scanning spectrometer. These are listed in the last column along with relative estimated errors in percentage. Donnelly and Hall also estimate the preflare flux values from OSO-3 measurements made on March 11, 1967 shortly after the launch of the satellite before there was any appreciable deterioration of the sensitivity; these were increased by 4% to correct for the somewhat larger solar activity occurring on the day of the flare. These preflare flux values are given in column 3 of Table V.

Observations with OSO-3 have been followed by observations with other satellites with increasing time and spatial resolution; of major significance are the spectroheliograph observations made with OSO-6 satellite. The Harvard college observatory experiment in OSO-6 either measured spectroheliograms at selected wavelengths or scanned over the spectral range 300–1400 Å at specific spatial locations. The spatial resolution was 35″ and the wavelength bandwidth was 3.2 Å. The raster size of the spectroheliogram was either the full disc of the Sun or a field of view of 6.8′ by 7.9′. The EUV flares observed with OSO-6 have been tabulated by Wood et al. (1972) and a comparison with SFDs have been made by Donnelly et al. (1973) (discussed in Section 10.2.2). The flare observations were mainly in C II 1335 Å line, C VI 1032 Å line and Mg X 625 Å line. This is particularly valuable, since under nonflaring conditions, the first comes from the chromosphere, the second from the chromosphere-corona transition region and the third from the lower corona. According to Hall (1971) who has examined the time dependence of flare EUV emissions with accompanying Hα observations, the lines from the chromosphere and the chromosphere-corona transition are generally more impulsive than those coming from the corona: the first precede the Hα maximum by 3–4 minutes (chromosphere lines such as the lines of hydrogen and Si III) and about 2 minutes (lines of He II, C III, O V and O VI normally coming from chromosphere-corona transition region); the second maximise after the Hα maximum. The OSO-6 EUV flares numbered 205 (Wood et al., 1972).

3.3. X-Ray Spectral Development

A major factor determining the magnitude, spatial distribution and temporal variation of the flare time ionization in the lower ionosphere is the spectral development of the X-ray energy emitted during the course of a flare. However, it is necessary to obtain such information separately for each case in view of the fact that the flare X-ray emission below 10 Å exhibits complex wavelength and time dependences that vary from flare to flare. Moreover, it is difficult to make useful experimental measurements with high wavelength and time resolution; only few such measurements are available (Neupert et al., 1967; Culhane et al., 1969).

In Figure 47 some results of flare time X-ray spectral developments have been shown for specific flare events (see also pp. 78–79).

The shift in spectrum towards the shorter wavelengths together with the enhancement in flux level during flare events is evident from this figure. Even for a weak flare of 15 June 1964 (Landini et al., 1966) the spectrum is appreciably hard.

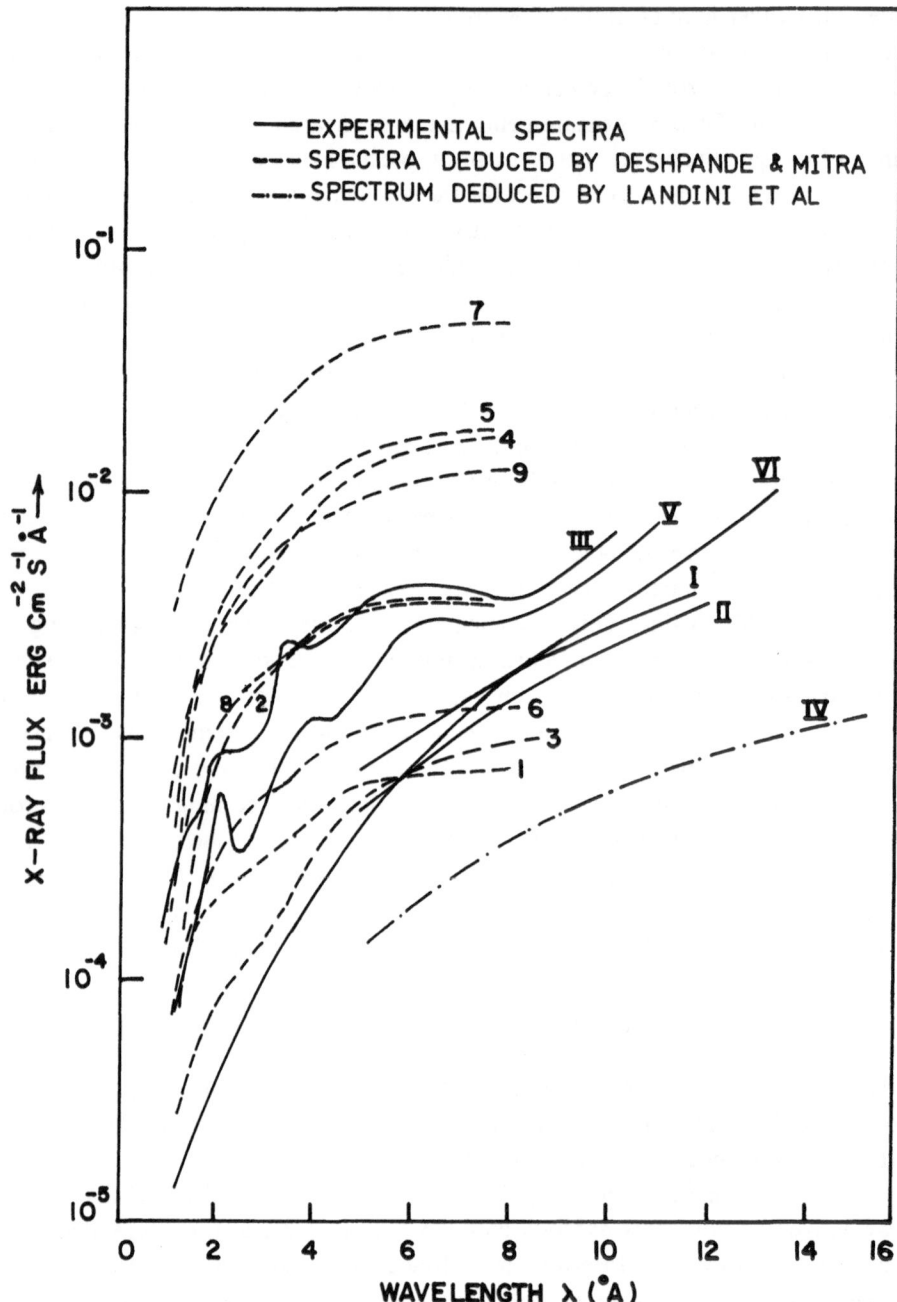

Fig. 47. Examples of X-ray spectral developments during solar flare. Dotted curves are theoretical ones deduced by Deshpande and Mitra (1972a).

The second important feature that the high resolution measurements have established, is the presence of numerous line emissions in this wavelength region to which we have already referred. The continuum emission is, however, dominant in 1–10 Å band, being an order of magnitude more intense than the total line emission (Meekins et al., 1968). The slope of the continuum spectrum gives electron temperatures well above 10^7 K.

Since the spectral developments vary from flare to flare, and flux measurements with high wavelength and time resolutions are scarce, it is important to consider how best and in what manner one may utilize the large volume of integrated X-ray flux measurements that are available for specific wavelength bands (0.5–3 Å, 1–8 Å, 8–20 Å) for obtaining the spectral distributions. Exponential distributions for a single temperature over the entire band 0.5–3 and 3–8 Å, as well as distributions given by power law functions have been considered. The entire time histories of X-ray spectra during flares are then deduced for the selected events.

The quantity that is measured by the X-ray detector is the ionization current $I(\lambda\lambda)$ or the count rate $C(\lambda\lambda)$ and is related to the differential energy spectrum $F(\lambda, T)$ in the wavelength range $\lambda\lambda$ over which the detector is sensitive, by the relations

$$I(\lambda\lambda) = D_i \int_0^\infty \varepsilon_{\lambda\lambda}(\lambda) F(\lambda, T) \, d\lambda$$

or

$$C(\lambda\lambda) = D_c \int_0^\infty \lambda\varepsilon_{\lambda\lambda}(\lambda) F(\lambda, T) \, d\lambda. \tag{12}$$

D_i and D_c are the detector constants and $\varepsilon_{\lambda\lambda}(\lambda)$, the efficiency of the detector as a function of wavelength. Information on $F(\lambda)$ can be obtained only if a multidetector response is registered. Even then, the appropriate spectrum can be developed only by searching for a distribution $F(\lambda)$ such that it will produce the observed response in the given set of detectors. One can, in principle, try any type of complex spectrum model including line emissions superimposed on the continuum. Another way is to express the wavelength dependence of the X-ray energy by an analytical function and derive the constants from the measured response. In this latter approach it necessarily implies that the continuum is dominant. This is believed to be the case in the region 1–10 Å, the main wavelength region of interest for D region flare effects.

Three distributions are generally considered. Planck's distribution for a black or a grey body, is suitable when the X-ray emission from the whole of the Sun is under consideration. Thus it is useful for the study of quiet day solar X-rays and the slowly-varying component. In case of flare X-rays when the emission is localized in the flaring region, the choice of an exponential function as employed by Mandel'štam (1965) and others (Sengupta, 1968) may be more appropriate. This is because the representation of X-ray spectrum given by:

$$F(v) \, dv = A \exp(- hv/kT) \, dv$$

or

$$F(\lambda) \, d\lambda = B\lambda^{-2} \exp(- C/\lambda T) \, d\lambda \tag{13}$$

is the same as expected for the wavelength dependence of X-ray energy emitted from a hot flare plasma by mechanisms such as free-free and free-bound transitions or thermal bremsstrahlung (Culhane, 1969). The full expressions for these are:

Free-Free Emission

$$F_{ff} = 7.15 \times 10^{-50} N_1 N_e \sum_z \frac{N_z}{N_1} Z^2 \exp\left[\frac{-143.89}{\lambda T}\right] \times$$

$$\times \frac{\bar{g}(Z, T, C/\lambda)}{T^{1/2}\lambda^2} \, d\lambda \text{ erg cm}^{-2} \text{ s}^{-1} \text{ Å}^{-1}, \tag{14}$$

where λ is in Å, T is electron temperature in units of 10^6 K, and $\bar{g}(Z, T, C/\lambda)$ is a temperature average of the free-free Gaunt factor. N_e and N_z are number of electrons and ions of nuclear charge Z, (N_z/N_1) is the element abundance relative to hydrogen.

Free-Bound

$$F_{fb} = 6.01 \times 10^{-53} N_1 N_e \exp\left[-\frac{143.89}{\lambda T}\right] \times$$

$$\times \sum_z \frac{N_z}{N_1} \sum_i \frac{N_{i+1}}{N_z} \frac{G \, d\lambda}{\lambda^2 T^{1.5}} \text{ erg cm}^{-2} \text{ s}^{-1} \text{ Å}^{-1}, \tag{15}$$

where

$$G = \sum_{n=n_0}^{\infty} \frac{\varrho_n X_{i,n}^2}{n} \exp\left[\frac{0.012 X_{i,n}}{T}\right] g(z, n, c/\lambda).$$

In this expression ϱ_n is the number of positions in the nth shell which are free to be occupied and $X_{i,n}$ is the ionization potential of an electron in the nth shell of an ion of ionization stage i.

The unit emission volume is

$$N_e^2 \, dv = \beta \, (\text{cm}^{-3}).$$

If we make an assumption that the Gaunt factor is slowly varying function of temperature, these expressions can be represented, in the simple manner, by Equation (13).

Another functional form for the spectrum is:

$$F(\lambda) = K\lambda^m, \tag{16}$$

where m can take only positive values. A power law form is expected in a situation where the flare associated hot plasma that emits X-rays has a temperature gradient.

The relation between the integrated X-ray energy flux $F(\lambda\lambda)$ in a given band $\lambda\lambda$ that produces a current $I(\lambda\lambda)$ in the detector sensitive to this band is given by:

$$F(\lambda\lambda) = I(\lambda\lambda) \left[\frac{\int_{\lambda_1}^{\lambda_2} F(\lambda) \, d\lambda}{D_i \int_0^{\infty} \varepsilon_{\lambda\lambda}(\lambda) F(\lambda) \, d\lambda} \right]. \tag{17}$$

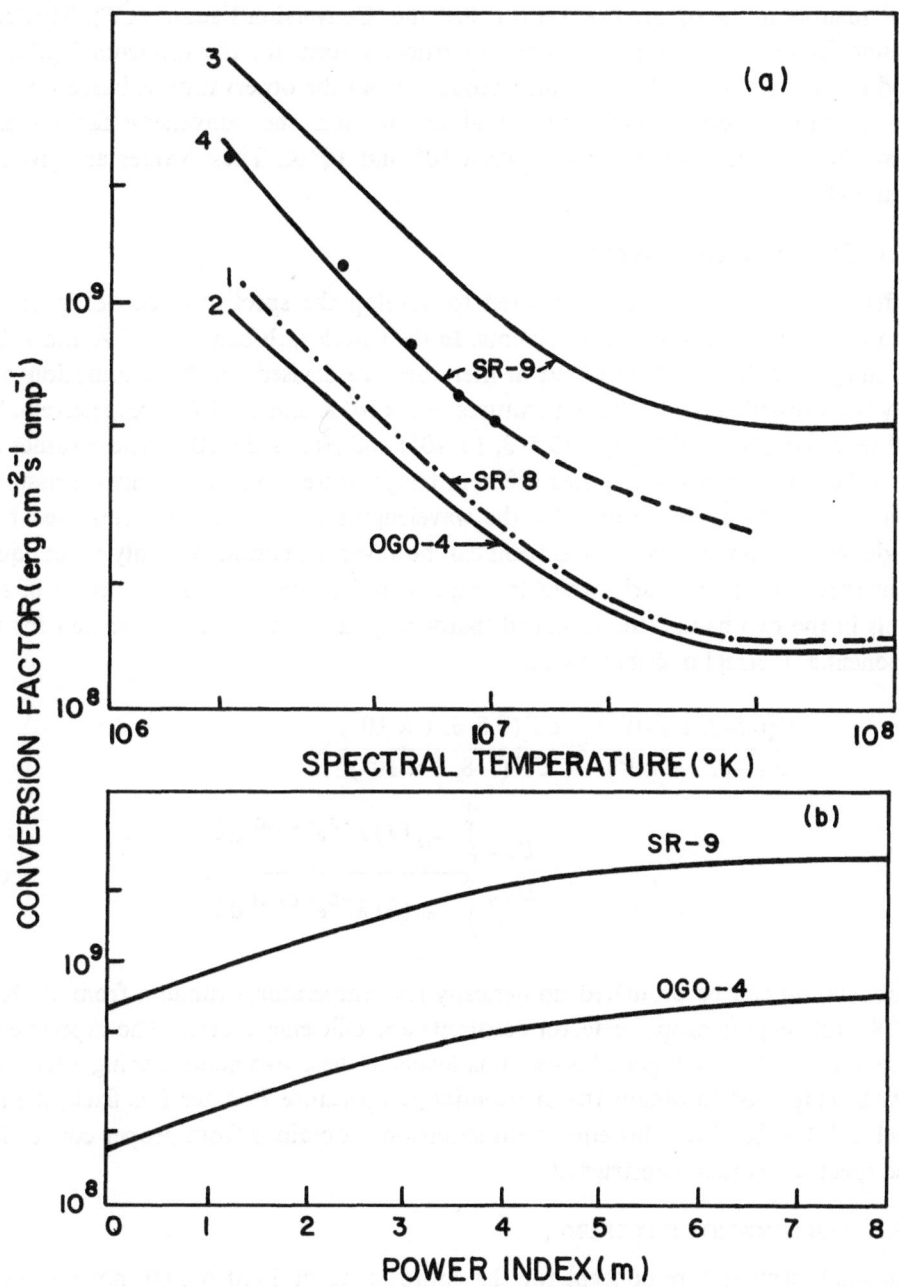

Fig. 48.   Conversion factors for 1–8 Å detectors aboard Satellites SOLRAD-8 and 9 and OGO-4 for (a) exponential distributions and (b) power law distributions (after Deshpande and Mitra, 1972a).

The quantity in the square bracket is called the 'Conversion Factor' (*CF*). This conversion factor can be evaluated when a particular form for the spectrum $F(\lambda)\,d\lambda$ is used and consequently the flux value reduced from the observation is based on such an assumption. Employing exponential distribution the conversion factors have been obtained for temperatures between 10^6 and 10^8 K. These values are given in Figure 48.

3.3.1. THE EQUIVALENT SPECTRUM

Different procedures must be adopted to develop the spectrum depending on the nature of available X-ray measurements. In the wavelength bands 0.5–3 Å and 1–8 Å the integrated flux levels reported in literature are devised on the assumption of a grey body distributions with temperatures of 1×10^7 K and 2×10^6 K respectively. We will represent these values by $F(0.5\text{–}3, 1 \times 10^7)$ and $F(1\text{–}8, 2 \times 10^6)$. These values are not valid if the spectrum requires different temperatures. We now make the assumption that the X-ray spectrum over the wavelengths 0–8 Å can be represented by a single temperature T. Such a spectrum can however, be considered only as an equivalent spectrum, particularly when the exponential model is used. The ratio of flux levels in the two bands (the so called 'hardening ratio') can then be written for the exponential thermal distribution as:

$$\frac{F(0.5\text{–}3, 1 \times 10^7)}{F(1\text{–}8, 2 \times 10^6)} = \frac{CF(0.5\text{–}3, 1 \times 10^7)}{CF(1\text{–}8, 2 \times 10^6)} \times$$

$$\times \frac{D_{i03}}{D_{i08}} \frac{\int_0^\infty \varepsilon_{03}(\lambda)\,\lambda^{-2} e^{-C_2/\lambda T}\,d\lambda}{\int_0^\infty \varepsilon_{08}(\lambda)\,\lambda^{-2} e^{-C_2/\lambda T}\,d\lambda}. \tag{18}$$

This relation can be evaluated numerically for temperatures ranging from 10^6 K to 10^8 K, utilizing the proper detector constants and efficiency factors. The experimental observations of the integrated X-ray flux levels in these two bands during a flare can now be employed to obtain the distribution temperature T. After T is fixed, the flux level in 1–8 Å band for this equivalent spectrum is obtained from proper conversion. The spectrum is then constructed.

3.3.2. THE COMPOSITE SPECTRUM

The single temperature concept for the entire range of 1–10 Å may not always be valid, especially during large flare events. The next step is to make the assumption that the spectra below and above 3 Å are represented by two distributions. With $F_1(\lambda)$ and $F_2(\lambda)$ as the two distributions, one can write:

$$F(0.5\text{–}3, 1 \times 10^7) = CF(0.5\text{–}3, 1 \times 10^7)\,D_{i03} \int_0^\infty \varepsilon_{03}(\lambda)\,F_1(\lambda)\,d\lambda \tag{19}$$

and

$$F\left(1\text{--}8, 2 \times 10^6\right) = CF\left(1\text{--}8, 2 \times 10^6\right) D_{i08} \left[\int\limits_0^3 \varepsilon_{08}\left(\lambda\right) F_1\left(\lambda\right) \mathrm{d}\lambda + \right.$$

$$\left. + \int\limits_3^\infty \varepsilon_{08}\left(\lambda\right) F_2\left(\lambda\right) \mathrm{d}\lambda \right]. \qquad (20)$$

Using the exponential function at temperatures T_1 and T_2 for the two spectra and introducing a condition that these two spectra should match at 3 Å, one arrives at the expression:

$$\frac{F\left(1\text{--}8, 2 \times 10^6\right)}{F\left(0.5\text{--}3, 1 \times 10^7\right)} = \frac{CF\left(1\text{--}8, 2 \times 10^6\right)}{CF\left(0.5\text{--}3, 1 \times 10^7\right)} \frac{D_{i08}}{D_{i03}} \left[\frac{E\left(\varepsilon_{08}, 0\text{--}3, T_1\right)}{E\left(\varepsilon_{03}, 0\text{--}3, T_1\right)} + \right.$$

$$\left. + \frac{E\left(\varepsilon_{08}, 3\text{--}8, T_2\right)}{E\left(\varepsilon_{03}, 0\text{--}3, T_1\right)} \exp\left\{-\frac{C_2 \times 10^8}{3}\left(\frac{1}{T_1} - \frac{1}{T_2}\right)\right\}\right]. \qquad (21)$$

The factors $E(\varepsilon_{\lambda\lambda}, \lambda_1 - \lambda_2, T)$ indicate the integrals evaluated for spectral distribution at temperature T, over a wavelength interval $\lambda_1 - \lambda_2$, and for efficiency factors of a detector sensitive to the wavelength band. This expression can be solved numerically for a pair of temperatures T_1 and T_2. The results of such computations for X-ray detectors on OGO-4 satellite are shown in Figure 49. In this figure the ratio $F(1\text{--}8, 2 \times 10^6)/F(0.5\text{--}3, 1 \times 10^7)$ is plotted against the temperature T_2 for a set of different values of T_1. The measurements of flux levels in the two bands alone can not provide the values of both T_1 and T_2. Estimate of either T_1 or T_2 need to be based on some other considerations. Once the two temperatures are obtained, the flux levels in the two bands are computed with the help of the appropriate conversion factors and the complete spectrum is built up.

An independent estimate of temperature T for any band may be obtained by considering the pre-flare or quiescent measurements of flux levels in any band, in addition to flare values in that band. The Relative Enhancement (R.E.) in the flux level in a given band is a ratio of flare time flux level to pre-flare or quiescent level (reduced for the same spectral temperature) and can be written as:

$$\text{R.E.} = \frac{\int\limits_0^\infty \varepsilon_{\lambda\lambda}\left(\lambda\right) F\left(\lambda\right) \mathrm{d}\lambda}{\int\limits_0^\infty \varepsilon_{\lambda\lambda}\left(\lambda\right) F_0\left(\lambda\right) \mathrm{d}\lambda} \qquad (22)$$

Thus some information on flare time spectrum $F(\lambda)$ can be obtained from relative enhancement in a given band $\lambda\lambda$, if the quiescent (or pre-flare) spectrum $F_0(\lambda)$ is known. It is sufficient to evaluate the integral $\int_0^\infty \varepsilon_{\lambda\lambda}(\lambda) F(\lambda) \mathrm{d}\lambda$ for a set of spectrum models for this purpose.

The quiet time level in the band 0.5–3 Å is below the detection threshold of the

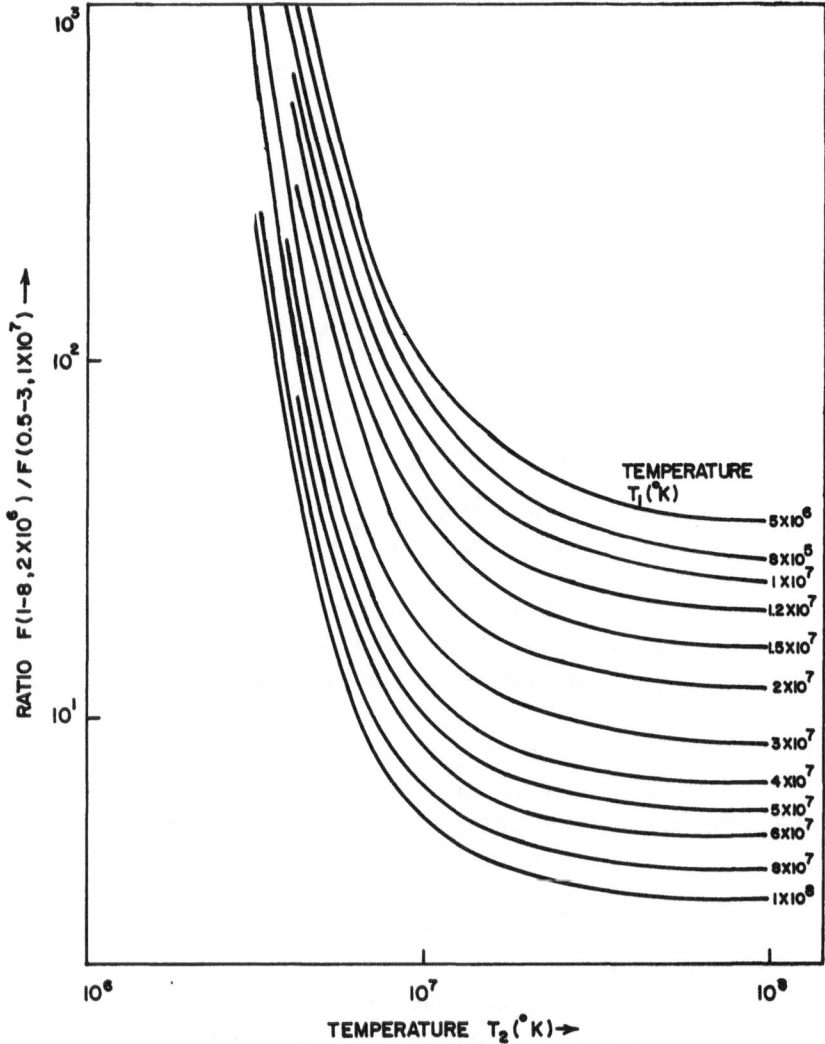

Fig. 49. Ratio of detector response $F(1-8, 2 \times 10^6 \mathrm{K})/F(0.5-3, 1 \times 10^7 \mathrm{K})$ against the equivalent flux temperature T_2 for 3–8 Å flux in the composite spectrum models for different values of 0.5–3 Å flux temperature T_1.

detector, and hence relative enhancement is generally not known. Nor is the spectral distribution in this wavelength region reliably known even for quiet conditions. When the spectrum is to be built from integrated flux values at 0.5–3 Å and 0–8 Å, the most convenient procedure is to obtain T_2 from the quiet and flare time observations of 0–8 Å flux. For this band quiet time spectrum is known to correspond to a temperature of 2–$3 \times 10^6 \mathrm{K}$ and the relative enhancement during a flare is experimentally available. However, during a flare the contribution to this band from 0–3 Å X-rays may become comparable with that from 3–8 Å. The necessary correction factor may be obtained by computing a theoretical spectrum in this range with $T = 2$–$3 \times 10^6 \mathrm{K}$,

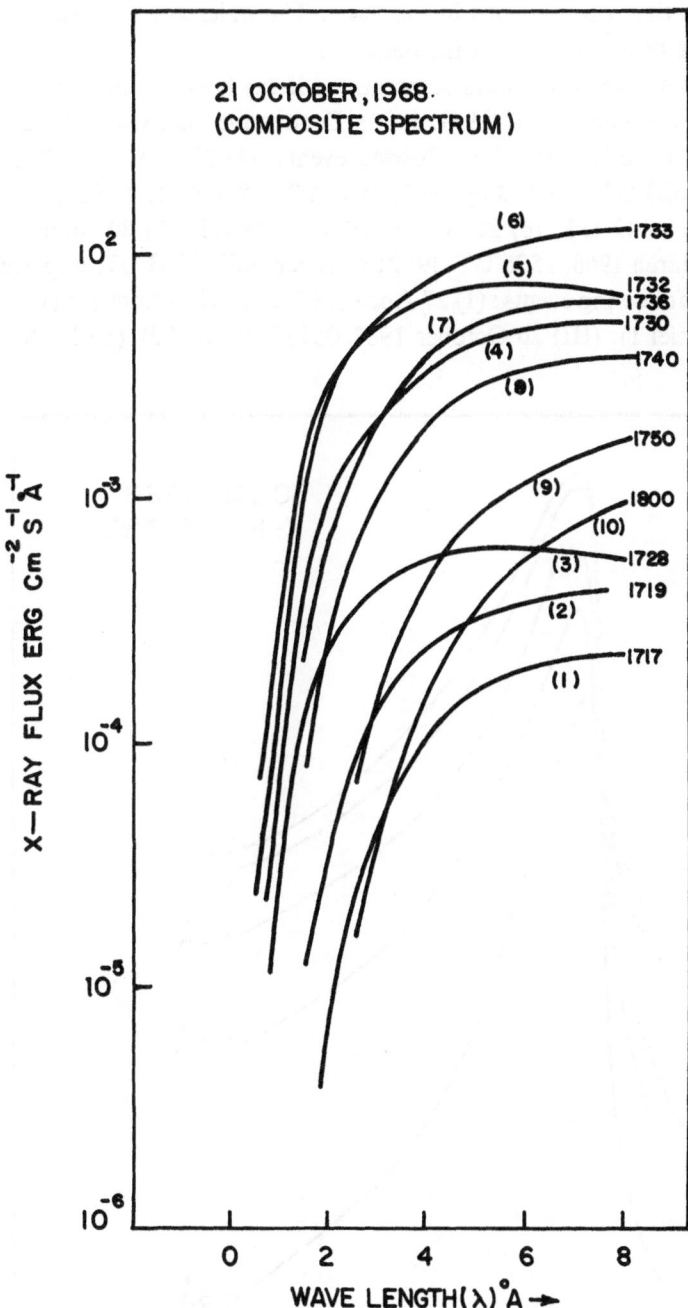

Fig. 50. Spectral development of solar X-rays during the flare of October 21, 1968, deduced from observations of 0.5–3 Å and 0–8 Å flux (after Deshpande and Mitra, 1972a).

and using the flare time observations in the band. The quiescent level in this band is mostly below the threshold level of the detector.

On the above basis, and using values of integrated X-ray fluxes in the bands 0.5–3 Å and 1–8 Å given by Kreplin *et al.* (1969, 1970), Deshpande and Mitra (1972a) have derived spectral distributions for the following events: (1) 29 July 1967, 2131 UT; (2) 30 July 1967, 0627 UT; (3) 30 July 1967, 0553 UT; (4) 1 August 1967, 1737 UT; (5) 2 August 1967, 0049 UT; (6) 28 August 1967, 2325 UT; (7) 30 January 1968, 0505 UT; (8) 25 March 1968, 1506 UT; (9) 21 October 1968, 1733 UT. Experimental profiles are for the following events: (I) 27 April 1962, 1412 UT (Ariel 1); (II) 3 May 1962, 0650 UT (Ariel 1); (III) 26 October 1967, 0613 UT (Ariel 2); (V) 21 December

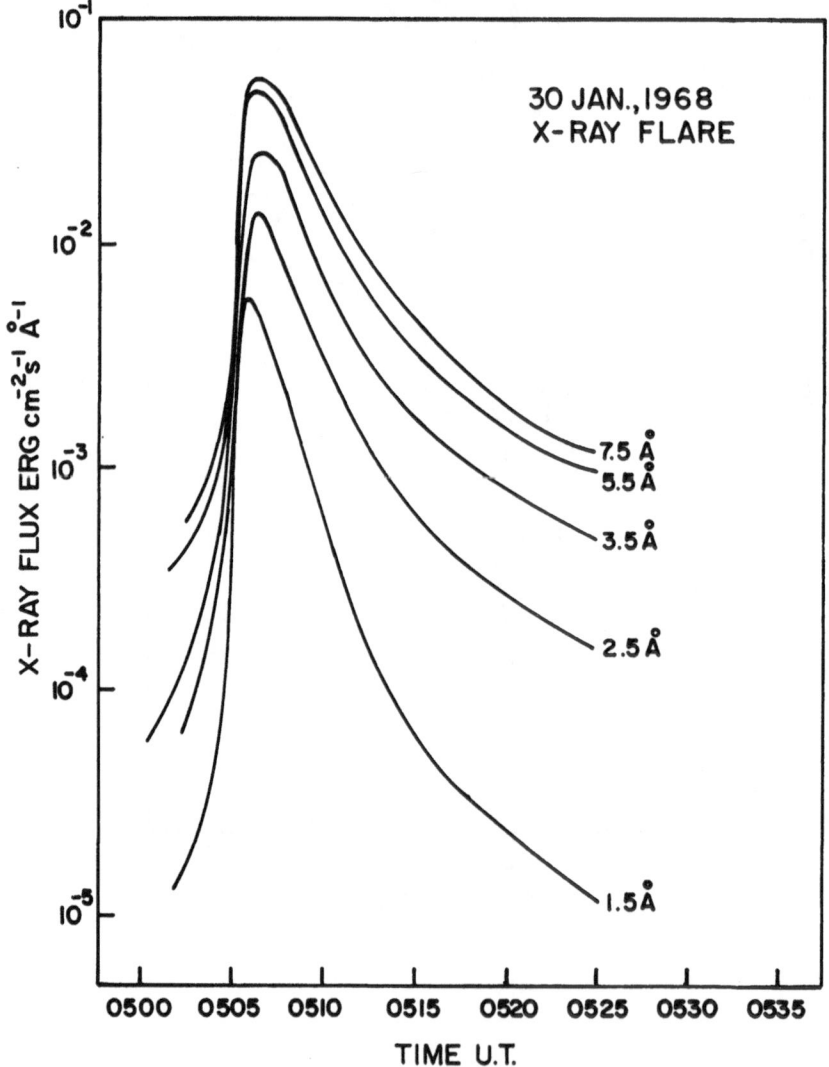

Fig. 51. Spectral development in the X-ray during the flare of January 30, 1968 deduced from observations of 0.5–3 Å and 0–8 Å flux (after Deshpande *et al.*, 1972b).

1967, 0154 UT (Ariel 2); (VI) 16 January 1968, 1914 UT (Rockets). Distribution (IV) is by Landini *et al.* for flare event of 15 June 1964, using $F(\lambda)=a+b\lambda^2+c\lambda^3$. It can be seen that the spectral distributions derived here are in close agreement with the experimental data. We further notice that the spectral temperature is between 1×10^7 and 2×10^7 K. This is true for the weak X-ray flare of 30 July 1967 (1946 UT) as well as for the strong event of 30 January 1968. When the proper weightage is given to the contribution from 0.5–3 Å band as in the composite spectrum method, the average temperature for T_1 is around 2×10^7 K and that for T_2 is around 9×10^6 K.

Figure 51 gives, after Deshpande *et al.* (1972b) the variation with time of the X-ray flux at different wavelengths during a solar flare that occurred on 30 January, 1968. This was a large solar flare for which the associated SIDs were large and were recorded over a wide frequency range in Delhi and Calcutta.

Figure 50 gives the X-ray spectra during growth and decay of a solar flare that occurred on 21 October 1968, and discussed previously by Rowe *et al.* (1970). Electron density profiles are available for this flare for different times during the growth and decay.

An interesting point to note is the way T_1 and T_2 vary during a flare. Two examples are shown in Figure 52; these refer to flare events of 30 January 1968 and 24 August 1967. While $T_1=T_2$ before the flare and at the end of the flare, T_1 increases much more rapidly than T_2 as the flare progresses, so that during the peak of the event T_1

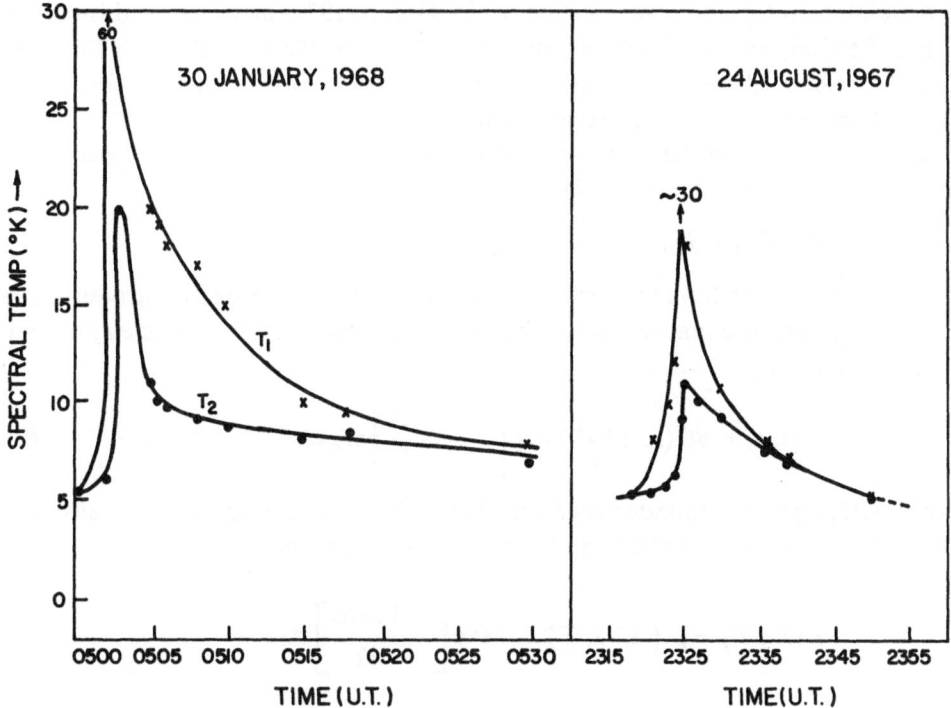

Fig. 52. Variation of spectral temperatures T_1 and T_2 during the flares of 30 January 1968 and 24 August 1967 deduced from observations of 0–3 Å and 0–8 Å flux (after Deshpande and Mitra, 1972a).

and T_2 are substantially different. It does seem, therefore, that the assumption of a single temperature is not appropriate during a solar flare. It is to be noted, however, that the choice of T_1 and T_2 is not unique and, therefore, the spectrum thus deduced may be called as a most probable one.

3.3.3. SPECTRAL DEVELOPMENT BASED ON FREE-FREE (BREMSSTRAHLUNG) AND RADIATIVE RECOMBINATION TRANSITIONS

The accuracy in the spectral development is crucial in any analysis of flaretime ionization in terms of D region chemistry. It is, therefore, important that these are constructed very carefully. The procedure of Deshpande and Mitra (1972a) is an improvement over the power law assumptions made earlier by Rowe et al. (1970) for the flare of October 21, 1968. Other improvements have been sought. One major approach, recently attempted by Rowe (unpublished information) and by Montbriand (1973) makes use of a table of factors to correct Solrad-9 grey body fluxes to fluxes and temperatures based on bremsstrahlung and radiative recombination transitions. The method is based on the theoretical work of Culhane (1969) which interprets the continuum in the 0.1–8 Å region as originating predominantly from a combination of free-free (bremsstrahlung) and free-bound (radiative recombination) transitions, defined by Equations (14) and (15) respectively, and the assumption that the energy contained in the continuum below 8 Å is at least an order of magnitude greater than that contained in the lines. In this approach the same temperature is used over the two wavelength intervals (see, however, Section 3.3.2). Horan (1970) made use of this concept quite effectively and obtained the relationship between the ratio $F(1-8)/F(0-3)$ and T_e, and calculated for a set of temperatures the coefficients needed to convert the above fluxes into values of corrected solar flux.

Since the solar continuum is made of a combination of free-free and free-bound transitions, the total continuum flux is given by:

$$F = F_{ff} + F_{fb}$$

in which both terms have the common factor $\lambda^{-2} \exp(-143.89/\lambda T)$. Introducing the equation for the emission volume β, one obtains for flux $F(\lambda, T)$ at wavelength λ and at temperature T,

$$F(\lambda, t) = \beta C(\lambda, T) \lambda^{-2} \exp\left[-\frac{143.89}{\lambda T}\right] \text{erg cm}^{-2} \text{s}^{-1} \text{Å}^{-1}, \qquad (23)$$

where $C(\lambda, T)$ is a complicated coefficient for the free-free and free-bound transitions. The total flux over the wavelength interval $(\lambda_1 - \lambda_2)$ is then

$$F(\lambda\lambda, T) = \beta \int_{\lambda_1}^{\lambda_2} C(\lambda, T) \lambda^{-2} \exp\left[-\frac{143.89}{\lambda T}\right] d\lambda. \qquad (24)$$

As before the ratio of the total fluxes in two wavelength regions can be used to determine the temperature of the source using Equation (18) or a slight modification

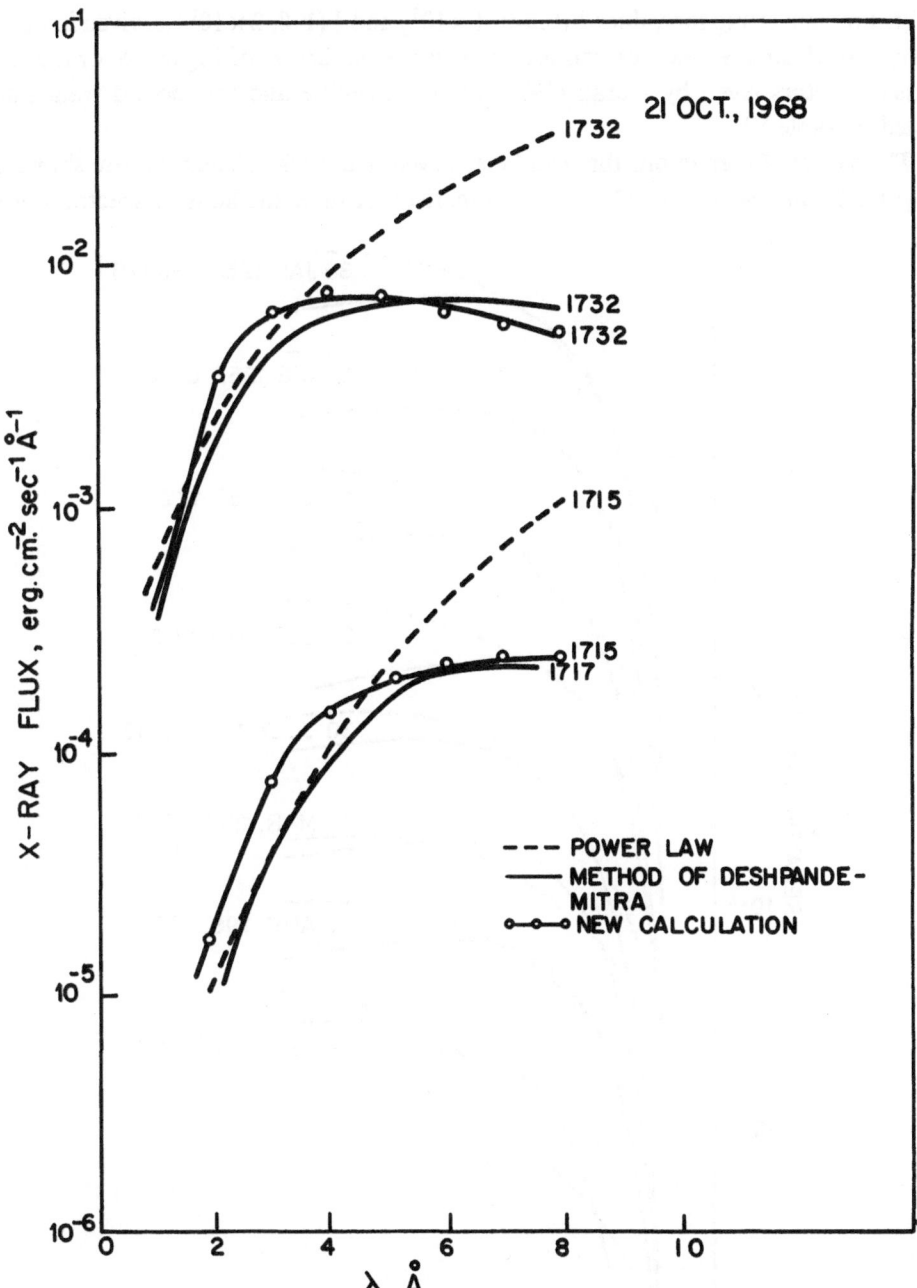

Fig. 53. Spectral distribution derived for the flare of October 21, 1968. Three different methods are indicated: use of power law, composite temperature concept, and use of correction factors to convert Solrad-9 greybody fluxes and temperatures based on bremsstrahlung and radiative recombination spectra (after Rowe, private communication).

of it. For converting the values $F(0.5\text{–}3, 1 \times 10^7)$ and $F(1\text{–}8, 2 \times 10^6)$ as given in *Solar-Geophysical Data Series*, one can use the conversion factors of Figure 48, or the conversion factors given by Horan (1970) for the free-free and free-bound transitions listed in Table VI.

The nature of corrections that can be achieved with such calculations are shown in Figures 53 and 54. Figure 53 gives a comparison of three methods of spectral devel-

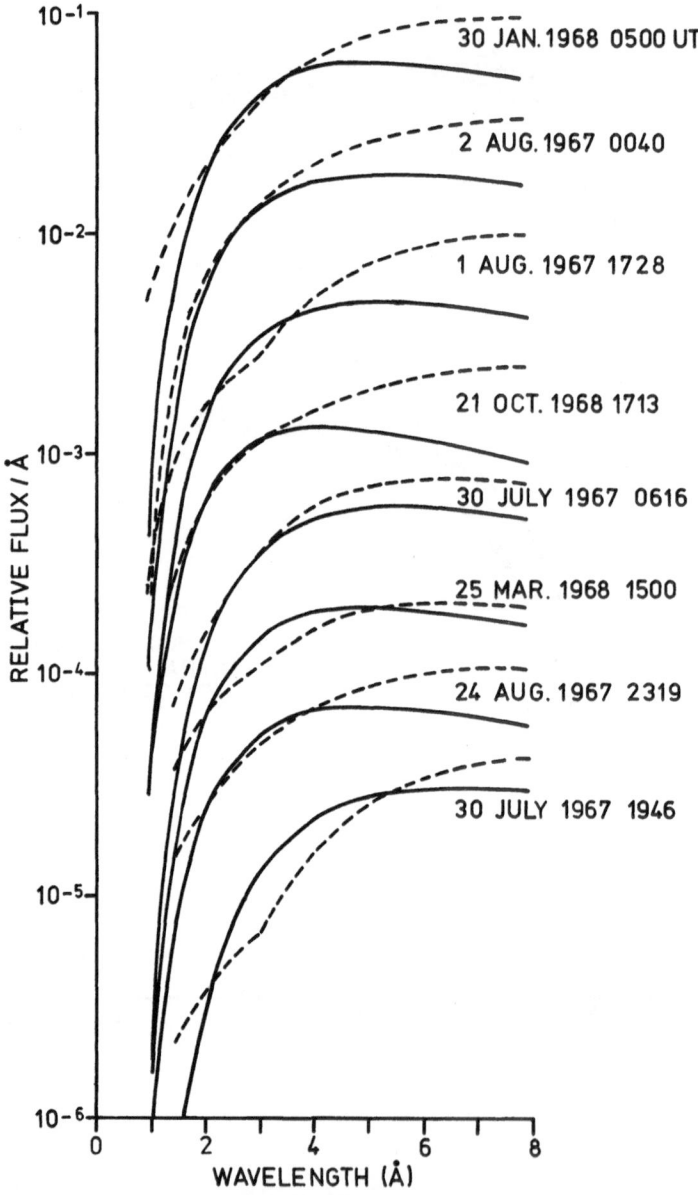

Fig. 54. Comparison of spectral distributions (solid curves) derived by Montbriand (1973) on the basis of free-free and free-bound transitions and a single source temperature, and derived by Deshpande and Mitra (1972a) (dotted curves) with two temperatures for a number of flare events (after Montbriand 1973).

TABLE VI

Solrad 9 conversion factors for the free-free
and free-bound transitions (Horan, 1970)

Temperature (10^6 K)	$k(0–3, T)$	$k(1–8, T)$
1	–	0.464
2	0.0209	0.968
3	0.188	0.878
4	0.439	0.753
5	0.652	0.652
6	0.797	0.574
7	0.886	0.513
8	0.936	0.465
9	0.962	0.428
10	0.972	0.397
11	0.973	0.372
12	0.969	0.352
13	0.962	0.335
14	0.952	0.321
15	0.942	0.308
16	0.932	0.298
17	0.922	0.289
18	0.912	0.282
19	0.908	0.275
20	0.894	0.270

opment for the flare of October 21, 1968. The dotted curve is the spectrum developed with a power law distribution – this is the one originally used by Rowe *et al.* (1970) for study of *D*-region aeronomy (see also Chapter 9). The solid line curve gives the distribution derived on the basis of the two temperature concept of Deshpande and Mitra (1972a) as given in Section 3.3.2. The third curve is a recent correction by Rowe (private communication) on the basis of free-free and free-bound transitions. In this particular example the last two calculations agree very well in the wavelength region 2–8 Å. In Montbriand's calculations (Figure 54) where a detailed comparison is made for a number of flare events, the agreement is good from 2–4 Å, but below 2 Å and above 5 Å, calculations from 3.3.3 give fluxes lower than in Deshpande and Mitra model.

The basic conceptual difference in the two (two temperatures for the ranges 0.5–3 Å and 1–8 Å in the case of Deshpande and Mitra and one single source temperature in Montbriand's calculation) can only be adequately resolved when high resolution flux values (in absolute values) are available below 2 Å.

3.4. Photoelectrons During Flares

Best *et al.* (1970) have recently reported a comparison of satellite Cosmos-261 measurement of EUV flux in the range $\lesssim 200$ Å and $\lesssim 100$ Å and the intensities of fresh

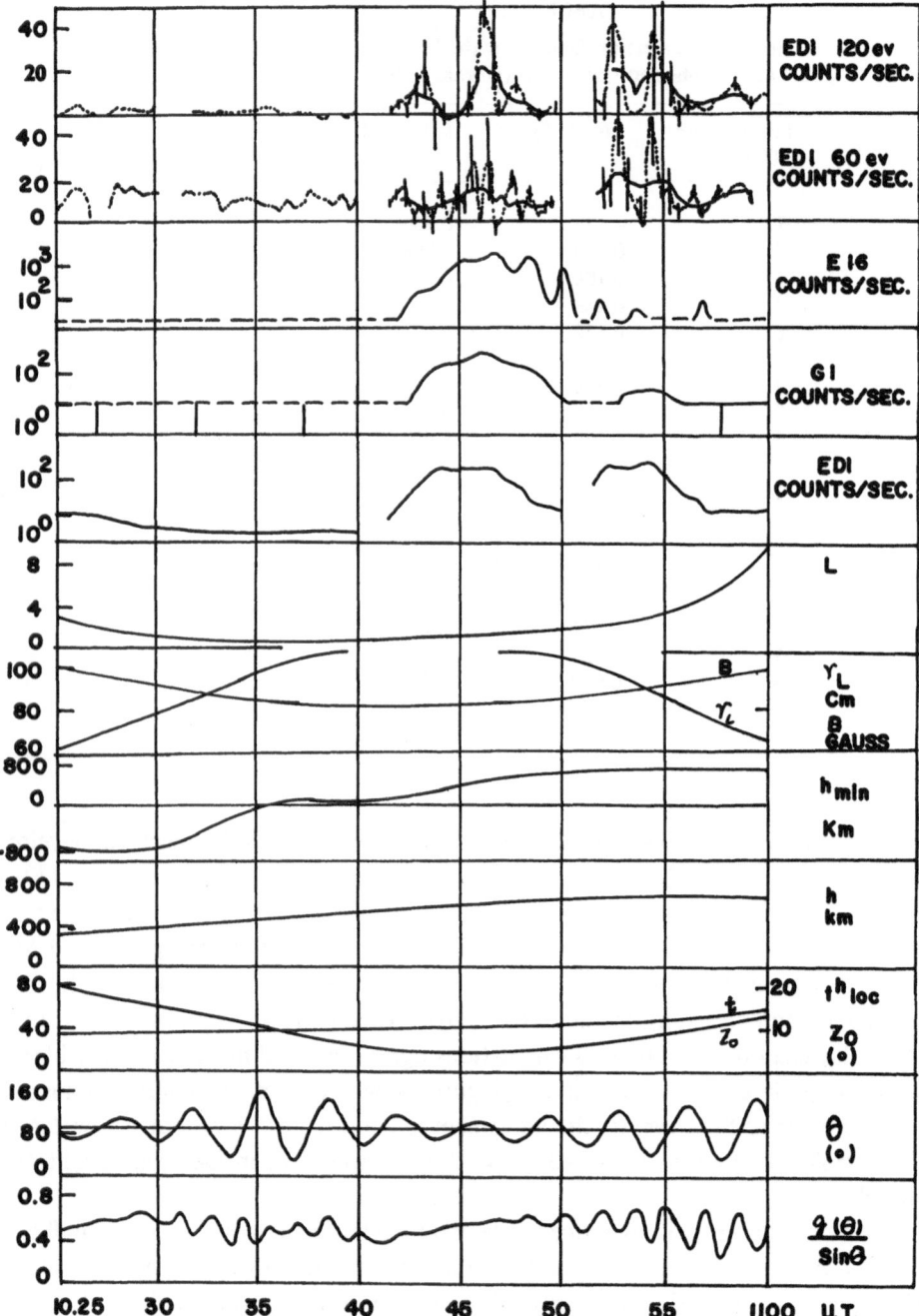

Fig. 55. Increase in photoelectron flux observed with Cosmos-261 satellite during the flare of December 27, 1968. The figure includes the photoelectron energy range of 60 and 120 eV, reflecting intensity variations of solar EUV radiations in the range $\lambda \leqslant 200$ Å and $\leqslant 100$ Å. EDI 120 eV and EDI 60 eV represent photoelectron flux intensities at 120 and 60 eV (after Best *et al.*, 1970). The different parameters in the diagram are shown opposite.

photoelectrons of 60 and 120 eV, as well as a number of groundbased SID measurements, including crochets for the solar flare of December 27, 1968. Photoelectrons of two specific energies 60 ± 10 eV and 120 ± 10 eV were chosen; these originate from absorption in the upper atmosphere of solar EUV radiations of $\lesssim 200$ Å and $\lesssim 100$ Å respectively. It was seen that 60 eV photoelectron flux density on December 27, 1968, was about 20% higher than in preceding days even before the flare started and increased by about 50% during the flare. For 120 eV photoelectrons, the time history was analogous (though more erratic) with the amplitude of enhancement during the flare time (Figure 55). EDI 60 and EDI 120 refer to photoelectron flux intensities with energies of 60 and 120 eV. To obtain absolute flux values, these are to be multiplied by the factor 1.3×10^5 el cm^{-2} s^{-1} ster^{-1} count^{-1} s. The measurements were made near the Brazilian magnetic anomaly, but the high background radiation was measured and subtracted for each spectrum. Large bursts of photoelectron intensities could be recognized approximately at 1042, 1046, 1052.4 for 120 eV and 1045.7, 1052.8, 1054.3 for 60 eV. During the first two events (at 1042 and 1046) the X-ray enhancements and SID effects have not yet begun, although Hα observations at Anacapri S-observatory showed the beginning of an effect at 1043 UT and at Haute Provence Observatory even at 1038 UT. This means that in the optical and UV regions flare activity assumed significance before X-rays ($\lesssim 10$ Å) – a result in agreement with those of Donnelly (1969a), and Nestorov-Krivsky (1967) who observed flare effects in the E-region to occur earlier than those in the D-region.

It is possible that the photoelectron flux enhancements were caused by some ionospheric electron acceleration upto ~ 100 eV and/or sharp generation of the parallel currents near the Sq current vortex (where Cosmos-261 was located at time of the flare). The crochet measurement, however, showed that the effect began only with the enhancement of X-rays; and consequently the photoelectron flux enhancement can only be ascribed to EUV flux changes.

EDI$_{120}$ eV	– Intensity of photoelectrons with $E = 120$ eV (counts s^{-1}) in the interval $\Delta E = 0.2E$. For transformation to electron cm^{-2} s^{-1} ster^{-1} multiply this value by 1.3×10^5 el cm^{-2} s^{-1} ster^{-1} counts $^{-1}$ s.
EDI$_{60}$ eV	– Intensity of photoelectrons with $E = 60$ eV (calibration factor the same as for 120 eV)
EI6	– Integral intensity of electrons with $E > 150$ keV
GI	– Counting rate of the Geiger counter (protons $E > 50$ MeV)
EDI$_b$	– the hard radiation background registered by the same spectrometer of soft electrons
L, B, h_{min}	– McIlwain invariant coordinates
r	– Larmor radius of 60 eV electrons
h	– satellite height
t_{loc}	– local time at the positions of the satellite
Ze	– solar zenith angle at the position of the satellite
θ	– the angle between the detector axis and local magnetic field
$g(\theta)/\sin\theta$	– anisotropy factor of photoelectron production by solar radiation.

CHAPTER 4

SID PHENOMENOLOGY

Routine SIDs of the types described in Sections 2.1–2.7 are now available over many years, and statistical relationship between them, and between any of the SID event and the solar ionizing radiations have been examined by many authors. Much of the early works relating to the statistical characteristics of the SIDs concerned the use of Hα emissions in the absence of information of X-ray fluxes. Amongst the major works in this line are the work of Bracewell and Straker (1949) giving the statistical properties of SCNAs observed at Hornsby on 18.3 MHz and that of Ellison (1953) giving the statistical properties of SEAs recorded on 27 kHz.

That the use of Hα emissions can be misleading was shown by Lindsay (1964), who showed, from a detailed comparison of some 60 X-ray flares detected by the satellite OSO-1 and Hα flares reported at these times, that (a) the flare-associated X-ray spectrum varied widely from flare to flare, sometimes having no counterpart in Hα, (b) the X-ray events were more frequent, and (c) residual X-ray enhancements existed after the end of the Hα flare on many occasions.

Use of X-ray and EUV observations for the study of statistical properties of SIDs is more recent. One of the earliest works to use the X-rays for SID studies was by Mitra (1966) who used the X-ray and EUV flux changes (and time histories, wherever available) of 15 selected flare events for detailed quantitative examination of the accompanying SIDs, which included principally SCNAs and SFDs. Another early work by Kreplin *et al.* (1962) led to the conclusion that SIDs occur only when the X-ray flux for $\lambda < 8$ Å exceeds a threshold value of 2×10^{-3} erg cm^{-2} s^{-1}.

More recently a number of exhaustive works have appeared. These include the work of Deshpande *et al.* (1972a) covering the SIDs observed during the 2-yr period January 1966 to December 1967 along with X-ray fluxes registered in 10–50 keV (OGO-I-III satellites), in 0–3 Å and 8–20 Å bands (Solrad-8) as well as in 2–12 Å (Explorer-33 and 35) and in 8–12 Å bands (OSO-3); the work of Richards (1971) which examined a number of SFDs with simultaneously recorded EUV flux and the work of Dasgupta *et al.* (1973) relating SIDs to microwave bursts. Table VII gives a list of selected statistical surveys.

4.1. Occurrence of SIDs with X-Ray, EUV Flares and Microwave Bursts

4.1.1. *D*-REGION SIDs AND X-RAY FLUXES AND MICROWAVE BURSTS

Not all X-ray flare enhancements below 20 Å cause SID effects. The percentage of X-ray flares producing *D*-region SIDs of different types is given separately for each

TABLE VII

Major works dealing with the statistics of SIDs

	Authors	SID type	Remarks
With Hα flares	Bracewell and Straker (1949)	SPA	The first statistical survey of SPA
	Ellison (1953)	SEA	
	Mitra and Shain (1953)	SCNA	The first statistical survey of SCNA
	Mitra (1954)	SEA, SPA, SCNA	
	Sarada and Mitra (1962)	SCNA	
	Mitra et al. (1964)	SPA, SEA, SCNA, SES	
	Dasgupta et al. (1973)	SWF, SCNA, SEA, SPA, SES, SFD	8230 cases of optical flares with dual-importance classification
	Albee and Bates (1965)	SPA	
	Bain and Hammond (1974)	SPA	2710 cases of GBR (16 kHz) observed at Slough (107 km)
	Donnelly (1970)	SFD	Ten years of Boulder observations (1961–69)
With X-rays	Mitra (1966)	SCNA, SEA, SES, SPA	
	Deshpande et al. (1972a)	SEA, SES, SWF, SPA, SFD, Magnetic crochets	
	Schwentek et al. (1970)	SWF, SCNA	34 SIDs correlated with OGO-IV and Solrad-9 flux measurements in 0.5–3 Å 1–8 Å and 40–60 Å bands
	Ohle et al. (1974)	SFA, SWF, SCNA, SPA	66 cases. Seasonal differences identified
With solar radio noise	Kaufmann and Mendes (1970)	SPAs received in Sao Paulo, Brazil from NSS (Annapolis 21 kHz) GBR (Rugby, 16 kHz) and Omega (Trinidad, 12 kHz)	207 SPA cases compared with 7 GHz solar bursts
	Dasgupta et al. (1973)	SWF, SCNA, SEA, SPA, SES, SFD	1670 microwave bursts used of types C-1 and 3, C-2 and 4, C-20, C-22, C-45 and 46
With EUV	Richards (1971)	SFD	78 EUV events obtained from OSO-3 satellite

band in Table VIII. The highest percentage of occurrence of about 90% exists for X-ray flares in 10–50 keV band. High percentage occurrence also exists apparently with SPA and SWF. VLF/LF effects are best correlated with 0–3 Å flux, and HF absorption with 8–20 Å flux. Although SFDs are caused principally by EUV radiations, there is strong correlation with X-ray events, the X-ray and EUV events being highly correlated, excepting that the latter occur first.

TABLE VIII

Percentage association of SIDs with X-ray flares and micro-wave bursts

Satellite and band	Total cases examined	Percentage for all SIDs	Percentage for different types of SIDs						
			SPA	SEA	SES	SWF	SCNA	SFD	Crochets
With X-rays (Deshpande et al., 1972a)									
OGO-I-III 10–50 keV	81	89	72	33	42	63	43	32	16
SR-80-8 Å, 8–20 Å, 0–3 Å	442	30	22	9	7	26	11	11	4
Explorer-33 2–12 Å	321	76	60	31	42	67	38	30	16
OSO-3 8–12 Å	134	63	49	13	18	43	31	19	7
With microwave bursts (Dasgupta et al., 1973)									
Burst types C-1 and 3	903		31	9	18	22	5	17	
Burst types C-2 and 4	188		55	19	35	43	14	19	
Burst type C-20	282		34	8	20	23	4	8	
Burst type C-22	117		44	15	31	37	6	8	
Burst types C-45 and 46	180		69	35	47	61	29	41	

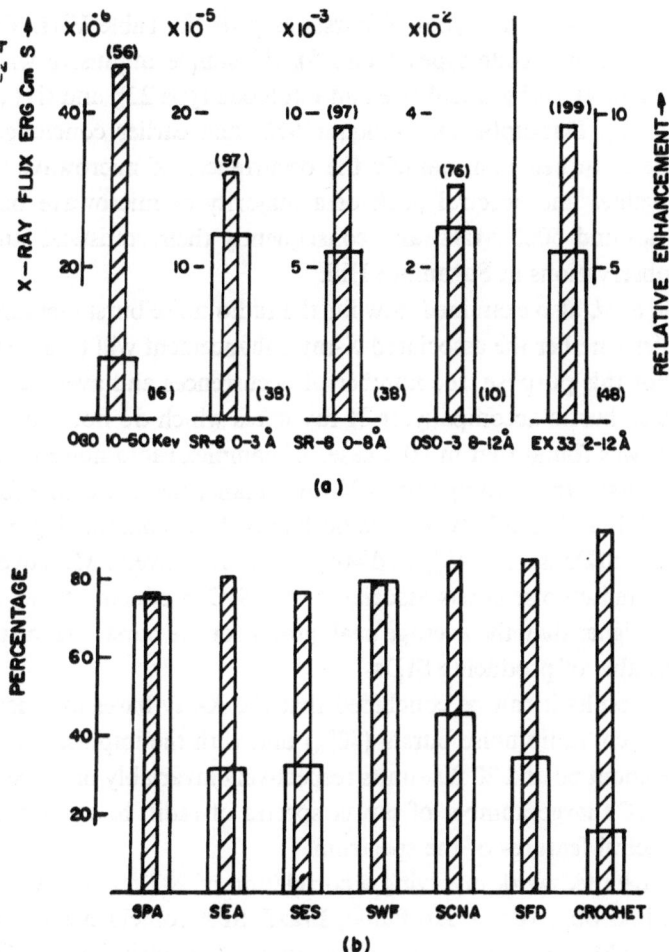

Fig. 56. (a) Size of SID-producing X-ray flares belonging to two groups (b) occurrence percentage of different types of SIDs with SID-producing X-ray flares belonging to two groups (after Deshpande *et al.*, 1972a). (Group I, hash-line bars; Group II, clear bars).

Deshpande *et al.* examined whether every X-ray enhancement capable of producing the SID, is a major solar event. Their analysis showed that in about 80% cases of SID-producing X-ray enhancements, Hα flares as well as radio noise bursts also occur concurrently. These are necessarily the major flare-SID events. In Figure 56a, b are shown the average size of X-ray fluxes and occurrence percentage of different types of SIDs for SID-producing X-ray flares of two categories. In the first (Group I), these X-ray flares are concurrent with Hα as well as radio noise bursts, while in the second (Group II) Hα or radio burst or both are absent. The X-ray events belonging to the first category show high flux enhancements. These also give rise to all types of SIDs including SCNA, SFD and Crochet in about 80% cases.

Dasgupta *et al.* (1973), in a study involving 1670 cases of microwave bursts at 4995 MHz during the flat peak of solar cycle 20 and SIDs of different types found varying correlation with the types of bursts as given in Table VIII. The burst types were simple impulsive (code types 1 and 3), (ii) simple impulsive with fluctuations (code types 2 and 4), (iii) gradual rise and fall (code type 22), and (iv) complex (code types 45 and 46). Dasgupta and Sarkar (1972) had earlier concluded that certain aspects of the disturbed Sun, namely the occurrences of microwave bursts, S-component flux values and spectral peak of a majority of microwave bursts are most pronounced around 5000 MHz, and consequently their statistical study used 4995 MHz solar observations at Sagamore Hill.

Deshpande *et al.* also examined how far the radio noise burst measurements can be utilized to infer whether the associated X-ray enhancement will be able to produce an SID effect. For this purpose, the number of occurrences and average peak power of cm radio noise bursts accompanying X-ray flares which do not produce SIDs were considered. It was found that in 80% cases of complex radio noise bursts and 62.5% of impulsive bursts, the accompanying X-ray enhancements are able to produce SID effects. For GRF and simple type of radio bursts the accompanying X-ray flares are able to produce SIDs in only 48% and 40% cases respectively. Moreover, the average peak power of radio noise bursts associated with SID-producing X-ray enhancements is 3 to 5 times higher than the average peak power for the bursts associated with X-ray events not capable of producing SIDs.

From these results it can be concluded that the X-ray flares associated with most of the complex cm radio noise bursts (80%) and with the impulsive and GRF bursts of size above about 60 and 20 flux units respectively invariably produce an SID effect. Examination of a large number of power spectra of radio bursts did not reveal any particular spectral features of the spectrum.

In another detailed work involving a correlation of 207 microwave burst events at 7 GHz and SPAs observed at Sao Paulo, Brazil, SPA coincidence was as high as 99% (Kaufmann and Mendes, 1970). 96% of the simple 2 impulsive events (79 bursts) were simultaneous with SPAs. For simple 3 bursts (48 bursts), Complex (44 bursts) and Great Bursts, the percentages were 98, 100 and 85. Even long-enduring Simple 3 bursts, which are believed to have a physical origin different from that of the others, showed an excellent correlation.

4.1.2. *F*-REGION SIDs AND EUV FLUX

The comparison of SFDs with X-rays as above is not quite appropriate, since the SFD is mainly an *F*-region effect, with contributions coming principally from the EUV. A more appropriate analysis is that relating to SFDs with EUV flux changes. Richards (1971) examined 78 EUV burst events with associated SFDs. The EUV events were compiled by Hall from telemetered OSO-3 data. Since the experiment measured the EUV flux integrated over the whole solar disc, it was not possible to identify the location of the EUV burst on the solar disc. A definite conclusion was reached regarding change in EUV flux (ΔEUV) and the occurrence of an SFD. If ΔEUV > 3% at λ304 or λ630, there is an 85% probability of an SFD occurring. If ΔEUV > 4%, the probability is $\sim$100% (Figures 57a, b). In addition there was some

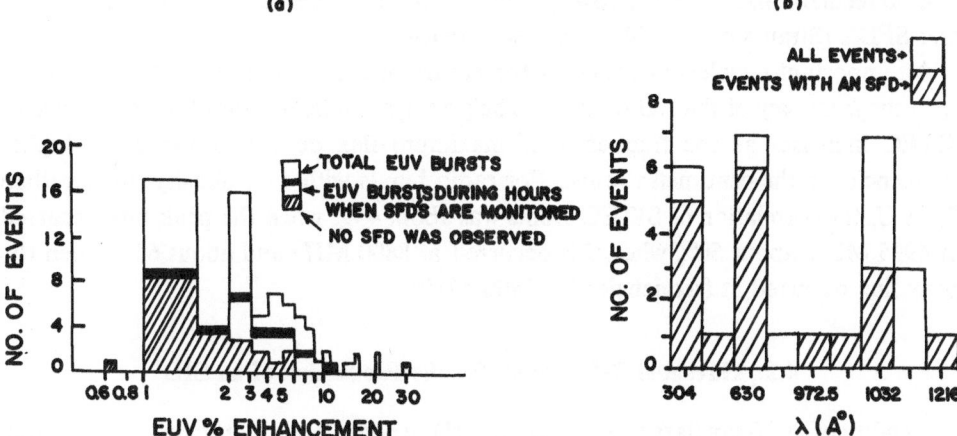

Fig. 57. Comparison of SFD events with observations of EUV enhancements (a) number of EUV bursts vs per cent enhancement, (b) EUV events with SFD per cent vs wavelength of observation (after Richards, 1971).

evidence that radio bursts of spectral type A1 were associated with SFDs with large Δf values.

Another F-region effect for which some occurrence statistics are available is SITEC, although in this case only 19 cases were examined. The results were reported by Matsoukas *et al.* (1972) who used 2 yr of TEC data recorded at AFCRL with an electronic polarimeter that measured continuously the VHF polarization angle of the signals received from the synchronous satellites ATS-3 and ATS-5 at 137.35 and 136.47 MHz respectively. Although a more appropriate analysis would have involved the use of measured EUV fluxes, the authors used solar radioburst flux intensities measured at eight discrete frequencies (245, 606, 1415, 2695, 4995, 8800, 15400 and 35000 MHz) at AFCRL.

The correlation of solar radio bursts with $S > 500$ flux units in centimeter range with SITEC events increased with increasing peak flux density as well as with increasing peak frequency. The correlation weakened for flares closer to the central meridian of the Sun and for events occurring during the local early morning hours. They concluded that a flare event which appears within 45° of the central meridian of the Sun, which occurs at noon or early afternoon, and which is accompanied by a radio burst with a peak flux density greater than 100 FU in the high frequency range of the centimeter region, will produce a readily detectable SITEC event in the ionosphere with a probability of nearly 100%. The two SITEC events earlier observed and analysed by Garriott *et al.* (1967) agree with this result, since they were accompanied by radio bursts with a peak flux density in the centimeter range of 2000 and 8100 FU respectively, and occurred near local noon. A flare event which produces only a type-G (peak flux density decreasing with increasing frequency) burst, even if this burst is of an extremely high flux density, seems to be incapable of generating X-ray and EUV emissions sufficient to produce a detectable SITEC event. This is in agreement also with the low correlation observed between type-G radio bursts and SFD's (Strauss *et al.*, 1969; Donnelly, 1970).

An important conclusion concerns the nature of correlation of the SITEC events with the *frequency* of the radio bursts. Their analysis indicates that the occurrence of SITEC increases as the frequency of maximum flux density advances to higher frequencies in the centimetre range. For radio bursts with flux density greater than 500 FU, the correlation of SITEC events was about 45% when the peak flux occurred at 4995 MHz, about 50% when this occurred at 8800 MHz and about 60% when the peak flux occurred at frequencies $\geqslant 15400$ MHz.

4.2. Threshold X-Ray Flux for the Production of an SID

Capability of an X-ray flare to produce an SID depends both on the flux value as well as on the magnitude of spectral hardening. The latter may be expressed in terms of the ratio of 0–3 Å and 0–8 Å flux levels (the so-called hardening ratio).

Earlier estimates referred only to the flux and no consideration was given to the spectral composition. Thus Kreplin *et al.* (1962) reported a threshold flux of 2×10^{-3} erg cm^{-2} s^{-1} (for the production of SPA) in the 0–3 Å band. Similarly Sengupta and Van Allen (1968) deduced a threshold flux of 1×10^{-3} erg cm^{-2} s^{-1} for 2–12 Å band from Explorer-33 observations of X-ray flares – a somewhat low value.

Both flux level and spectral hardening were considered by Deshpande *et al.* (1972a). From an examination of 122 SID-producing X-ray flares, they found that for only

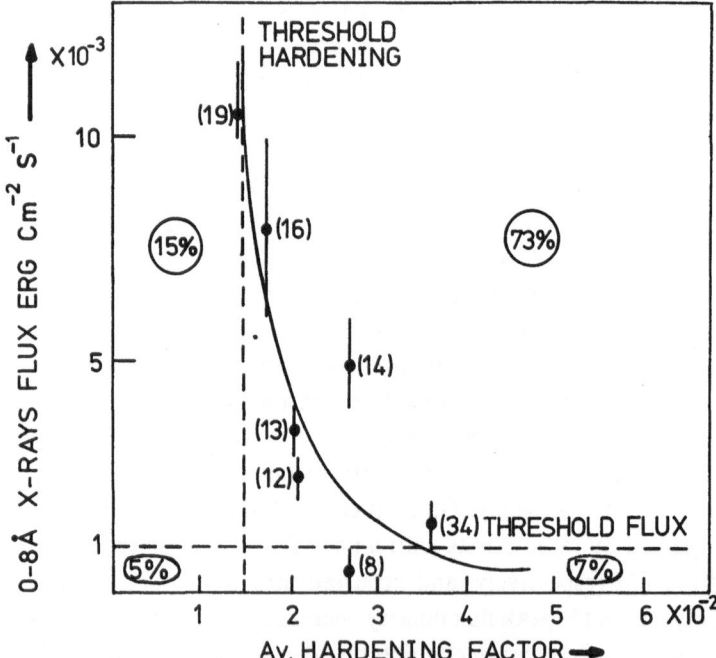

Fig. 58.   Threshold flux and hardening factor (after Deshpande *et al.*, 1972a).

6 cases the 0–8 Å flux level and the hardening ratio were below 1×10^{-3} erg cm^{-2} s^{-1} and 1.5×10^{-2} respectively. These were taken as thresholds for the 0–8 Å flux level and spectral hardening for the production of an observable SID effect. The threshold flux level for 0–8 Å band obtained from the independent examination of 0–3 Å data was found to be 2×10^{-5} erg cm^{-2} s^{-1} ($T = 1 \times 10^7$ K) consistent with the above threshold hardening ratio (Figure 58). About 73% of X-ray flares that produced SIDs had both flux level and spectral hardening above the thresholds. In about 15% cases the hardening ratio was below the threshold of 1.5×10^{-2}. In only 7% cases the 0–8 Å flux was below the threshold of 1×10^{-3} erg cm^{-2} s^{-1} but the hardening ratio was high, close to 3×10^{-2}.

4.3. Level of Solar Radio Flux in the 3000 MHz Region and Its Relation to the Occurrence of Sudden Ionospheric Disturbances

It has been suggested that the centimetric burst emission and the ionizing radiation (X-rays) responsible for D layer originate from the same region in the solar atmosphere, and are both generated by similar emission mechanisms. However, all solar flares do not cause terrestrial disturbances; Mitra *et al.* (1964) found a curious requirement of a threshold radio flux needed for an SID to occur.

In their work they considered both IGY (solar maximum) and the IQSY (solar minimum). During the IGY it has been found that out of 490 optical flares of class 2⁻ and above, 140 flares were not followed by centimeter bursts. During the period of low solar activity there were 44 flares of class 2⁻ and above during 1962 and 18 during 1963 for which no SID phenomena were observed. For all the above cases of flares for which no SID phenomena were observed, it was found that these occurred during an unusually low level of solar radio emission in 3000 MHz. The periods examined were:

(i) August 1 to September 25, 1957

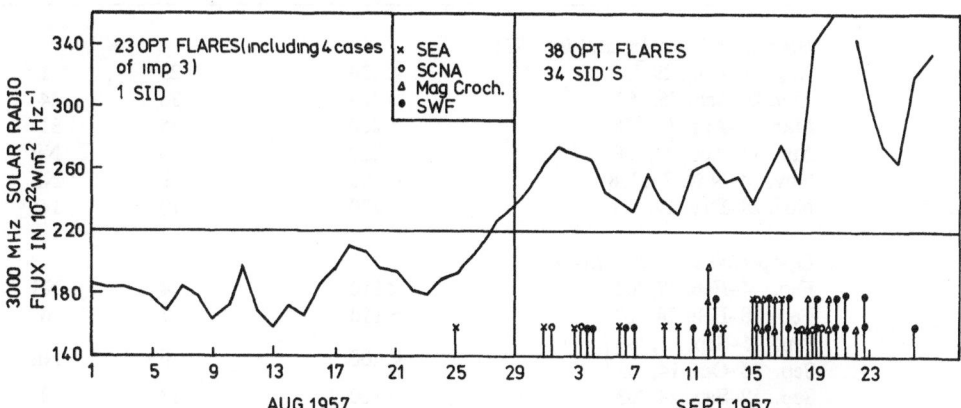

Fig. 59.   Mean daily values of solar radio flux at 3000 MHz against days during the period August 1–September 25, 1957, along with indications of SIDs occurring during this period (after Mitra *et al.*, 1964).

(ii) March 22 to April 27, 1958

(iii) November 4 to December 14, 1958

(iv) February 4 to February 28, 1962

(v) September 1 to October 14, 1963.

In Figure 59 the intensity of solar radio flux on 3000 MHz is plotted against time in days for one of the periods. On the same is also plotted the number of SEAs, SCNAs and SWFs occurring during the period. This period (August 1 to September 25, 1957), can be divided into two parts in the first, upto 29th August, the solar radio flux was below 220×10^{-22} Wm^{-2} Hz^{-1}. In the second part after 29th August, it increased to a high level and maintained upto 25th September. It was found that during the first half there was only one SID (an SEA on 25th August), even though there were 23 optical flares, out of which four were of importance 3. In the second part of the period, there were 38 optical flares, and 34 SIDs of different types. Similar results were obtained for the periods March 22–April 27, 1958, and November 4 to December 14, 1958 – in both cases the threshold flux appeared to be 220 FU.

During the low solar activity period (period of February 4–28, 1962 and September 1–October 14, 1963), similar results were obtained. During the period 12th September to 25th September, 1963, the flux was much above the average value of 100×10^{-22} Wm^{-2} Hz^{-1}; 13 optical flares of class 2$^-$ and above, and 9 SID events were observed. On the other hand, during the remaining periods of 1st September to 12th September and 25th September to 14th October, 1963, six optical flares were observed, out of

TABLE IX

Summary of the level of radio flux on 3000 MHz, during IGY and during period of low solar activity, and number of observed optical flares and SIDs
(after Mitra and Subrahmanyam, 1966)

Period	Level of radio flux in 10^{-22} Wm^{-2} Hz^{-1}	No. of optical flares	No. of SIDs
High period of solar activity (IGY)			
Aug. 1 –Aug. 29, '57	<220	23[a]	1
Aug. 29–Sep. 25, '57	>220	38	34
Mar. 22–Apr. 10, '58	>220	35	31
Apr. 11–Apr. 27, '58	<220	5	Nil
Nov. 4–Nov. 24, '58	<220	9	Nil
Nov. 29–Dec. 14, '58	>220	10	14
Low period of solar activity			
Feb. 4–Feb. 17, '62	<110	4	Nil
Feb. 18–Feb. 28, '62	>110	7	6
Sep. 1–Sep. 11, '63 ⎫ Sep. 25–Oct. 14, '63 ⎬	<100	6[b]	Nil
Sep. 12–Sep. 24, '63	>100	13	9

[a] Out of 23 optical flares, 4 are of importance 3.
[b] One of them is a class 3 optical flare.

which one was class 3, and these were not followed by any SID event. Table IX summarises the above results.

It is clear that for the production of an SID, the primary criterion is that the general level of microwave solar flux should continuously and abnormally exceed a threshold value. For 3000 MHz this threshold value appears to be 220×10^{-22} Wm^{-2} Hz^{-1} during solar maximum period. During the descending part of solar activity this threshold value was reduced to 110 flux units during 1962, and 100 flux units during 1963. When the level of solar flux is below this value, the chance of occurrence of an SID is small even though an optical flare may occur. During the period August 1 to August 29, 1957 there were even four optical flares of importance 3 which did not produce an SID, because the value of flux was lower than the threshold value. The same was the case during the period September 1 to September 11, 1963 when there were 6 optical flares, out of which 1 was of class 3, but no SIDs were produced. During 1964, the solar radio flux was almost steady around 70 units without any abnormal rise. The number of flares, class 2^- and above was very small with no corresponding SIDs.

4.4. The Size Correlation

From flare to flare there are large differences in the time development of the intensity and spectral distribution of X-ray enhancements, Hα intensification and size of centimeter radio noise bursts.

It was observed that on the whole the X-ray emission corresponds in size with Hα and radio noise rating of the flare event. High X-ray source temperatures are associated with class 2 and 3 flares as well as complex and impulsive bursts. These are

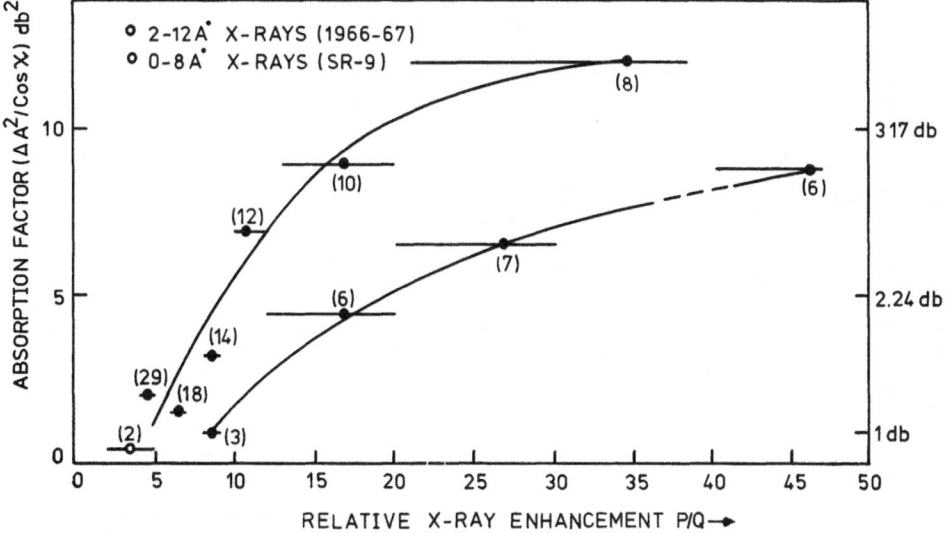

Fig. 60. SCNA size, defined by $\Delta A^2/\cos\chi$, related to X-ray flux enhancements (after Deshpande *et al.*, 1972a).

also the findings of Teske's (1969), who compared OSO-3 soft X-ray fluxes with Hα developments.

The SCNA measurement of excess absorption (ΔA) during the maximum phase of the event may be taken as a measure of size of an SID. The quantity ΔA^2 is proportional to the flux of ionizing X-rays. Deshpande *et al.* (1972a) reduced all SCNA measurements to 18 MHz by using an inverse frequency square relationship and normalized to $\cos\chi=1$; the absorption factor ($\Delta A^2/\cos\chi$) is thus computed. The correlation plots of relative enhancements in the bands 0–3 Å, 0–8 Å and 2–12 Å against ($\Delta A^2/\cos\chi$) show a large scatter. They, therefore, obtained the average of the relative enhancement and of the factor $\Delta A^2/\cos\chi$ for events divided into different ranges of the former. These are graphically shown in Figure 60 for 2–12 Å and 0–8 Å bands. It can be seen from this figure that absorption factor increases at first rapidly with the increase in X-ray enhancements ($P/Q=20$). It may also be noted that a large SCNA event of size above 3 dB for $\cos\chi=1$, is associated with relative enhancement of above 10 in 2–12 Å band, but the converse is not always true.

Kaufmann and Barros (1969) found an exponential correlation between the SPA phase advance and X-ray flux levels in the three bands namely 0–3 Å, 0–8 Å and 8–20 Å. They also found that the SPA size increases with the hardening of the spectrum. The SCNA size, however, does not show a simple relation with the hardening ratio. Deshpande *et al.* (1972b) specifically examined the SCNAs in the selected cases from Solrad-9 X-ray flares, where 0–8 Å flux level remained the same but the hardening factor changed. It was found that the role of hardening is not decisive. In some cases large values of hardening ratio was associated with high absorption but not in others. It may however be noted that all X-ray flares producing SCNAs have very high flux levels in the 0–8 Å band in the range 2×10^{-2} to 5×10^{-1} erg cm^{-2} s^{-1}.

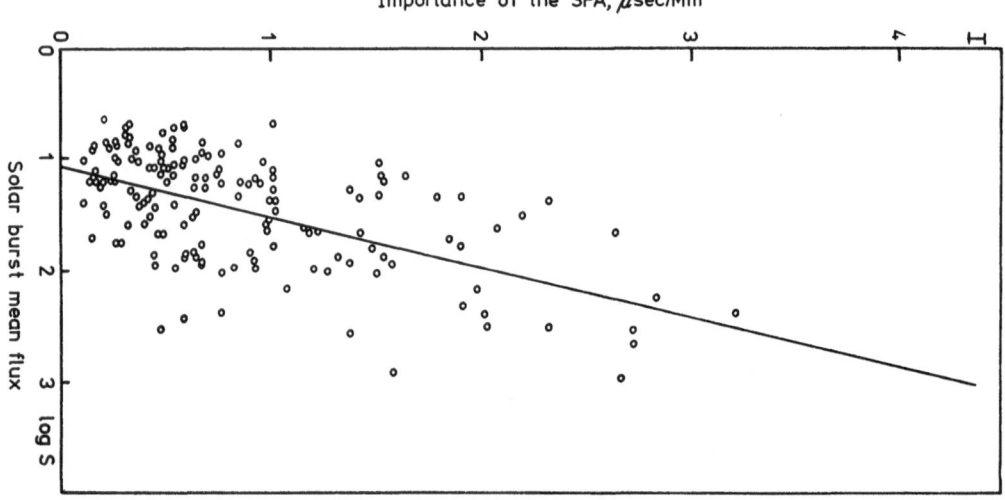

Fig. 61.   Correspondence between mean flux of solar radio burst and SPA magnitude (after Kaufmann and Mendes, 1970).

A size correlation also exists between the D-region SIDs and microwave bursts. Kaufmann and Mendes (1970) found an excellent correspondence between $\log S$ (S being the mean flux of the radio burst in 10^{-22} Wm^{-2} Hz^{-1}) and SPA magnitude defined by an apparent 'importance' parameter I (in μs Mm^{-1}) given by:

$$I = \Delta\phi'/d, \tag{25}$$

where $\Delta\phi'$ is the measured maximum deviation in microseconds and d the great circle length of the propagation path. The relationship is shown in Figure 61.

For SFDs, Richards (1971) has examined the relationship between SFD sizes with the intensity distribution of solar radio bursts. SFD being principally an F-region SID,

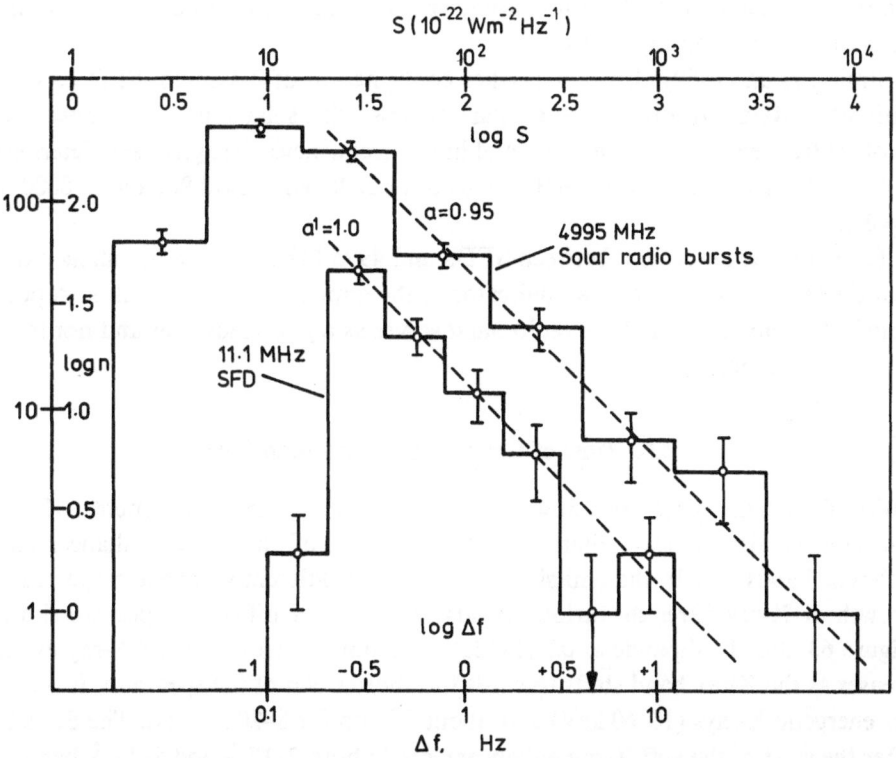

Fig. 62. Probability distributions of solar radio burst fluxes at 4995 MHz and of the SFD magnitudes. The identical natures of the probability distributions may be noted (after Strauss *et al.*, 1969).

use of X-rays is not appropriate. Since it has been seen that radio burst at 4995 MHz show a close correlation with SFDs, the size correlation has been limited to 4995 MHz. For both, the probability distributions given by $P(S)$ and $P(\Delta f)$ were found to be essentially identical. Figure 62 gives the variation of $\log n$ with $\log S$ and $\log \Delta f$, where n is the number of bursts with intensities in the range $\log S$ to $\log S + d(\log S)$ in the first case, and $\log(\Delta f)$ to $\log(\Delta f) + d(\log \Delta f)$ in the other case, with proportionality constants $a \approx 0.95$ and $a^1 \approx 1.0$ in the two cases respectively. The probability equations were:

$$P(S) = \frac{\text{Constant}}{S^{1+a}} \propto S^{-1.95} \tag{26}$$

$$P(\Delta f) = \frac{\text{Constant}}{(\Delta f)^{1+a^i}} \propto (\Delta f)^{-2}. \tag{27}$$

In the analysis of the data, Strauss *et al.* (1969) used SFD events of only one frequency and one ray path – the one hop WWI transmission at 11.1 MHz from Havana, Illinois to Boulder, Colorado, and further limited the data to 94 events occurring within the zenith angle interval $40° < \chi < 70°$. The last choice was determined by the fact that for $\chi > 70°$ the ionospheric effects of solar X-ray bursts are greatly diminished, and for $\chi < 40°$, there is a higher probability of Es occurrence or complications due to two-hop ray paths.

Strauss *et al.* (1969) also find further confirmation of maximum SFD correlation with 4995 MHz radio bursts by similar analysis with solar radio noise observations at other frequencies. At both lower and higher frequencies the agreement deteriorates: at 606 MHz, $a=0.75$; at 1415 MHz $a=0.85$; at 2695 MHz $a=0.90$; and at 8800 MHz $a=0.85$.

The complete agreement between SFDs and 4995 MHz solar radio bursts indicates that the latter can be used as an indicator of the impulsive solar EUV flux. A point to note is that only the impulsive radio burst serves as a good indicator and not the later μ-IV solar outburst.

4.5. Time of Growth and Relaxation Time

SIDs of different types are known to show different time developments. X-ray enhancements also indicate that time development of the X-ray enhancements in different bands and in the emission lines proceed differently. The average times of growth of X-ray flares in various bands and SIDs of different types are shown in Figure 63 after Deshpande *et al.* (1972a). The time of growth of an X-ray event decreases as the X-ray band shifts towards the shorter end of the spectrum. It is 3.5 min for energetic X-rays (10–60 keV) and about 15 min for 8–20 Å band. The SIDs begin after the start of the soft X-ray enhancements in both 2–12 Å and 8–12 Å bands. The delay is on the average 3 min for SFD, 3–4 min for SPA, SEA and SES and 4–5 min for SCNA, SWF and Crochet. Secondly, the X-ray flux in the 0–3 Å band peaks earlier than in the 0–8 Å band (Kreplin *et al.*, 1968). However, the VLF–LF effects believed to be mainly controlled by 0–3 Å band reach their maximum later than HF–SID effect like SCNA and SWF.

Deshpande *et al.* (1972a) have determined the relaxation time τ for different types of SIDs with references to X-ray flare maximum in five different wavelength bands. The results are given in Table X.

It was found that SFDs reach maximum earlier than X-ray flux in all bands by 1–3 min. For all the X-ray wavelength bands except 10–50 keV, the maximum of SID

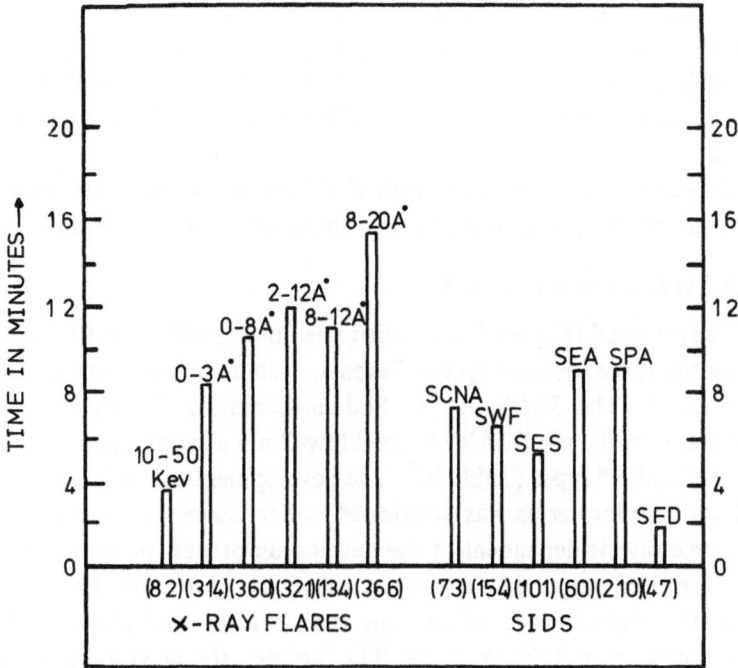

Fig. 63. Time of growth of SIDs of different types and of X-ray flares in different bands (after Deshpande *et al.*, 1972a).

TABLE X

Relaxation time of SIDs of different types with respect to X-ray flares in five bands and with EUV for SFD events (after Deshpande *et al.*, 1972a for first five columns and Richards, 1971 for last column)

SID effects	Relaxation time in minute					
	Satellites and X-ray observation bands					
	OGO-I-III 10–50 keV (1966)	Solrad-9 0–3 Å (1969)	Solrad-9 0–8 Å (1969)	Explorer-33 2–12 Å (1966–67)	OSO-3 8–12 Å (1967)	EUV events (1967)
SPA	+5 (15)	+3 (130)	+2 (133)	+2 (218)	+2 (42)	
SEA	+7.5 (8)	+5 (30)	+3 (38)	+2 (84)	+3 (15)	
SES	+5 (12)	+3 (67)	+2 (65)	+2 (110)	+2 (20)	
SWF	+3.5 (16)	+3 (78)	+2 (85)	+1 (191)	+1 (30)	
SCNA	+2.5 (10)	+2 (17)	+1 (26)	+1 (115)	+1 (12)	
SFD	−1.5 (10)	–	–	−3 (81)	−2 (28)	−1 (21)

frequency occurs before that of the integrated flux level. This is true for almost all kinds of SIDs. The relaxation time with respect to soft X-rays of 0–3 Å, 0–8 Å, 2–12 Å and 8–12 Å bands is 2–3 min, whereas with hard X-rays of 10–50 keV band it is 4–5 min. The relaxation time for LF/VLF effects is about 2–3 min and for HF effects about 2 min.

For SFDs for which comparison with EUV events is more appropriate, average of 21 events as given by Richards (1971) is included in the last column.

4.5.1. SEA TYPES AND COSMIC RAYS

Sakurai (1968) showed (Figure 7) that when one observes SEAs at 10, 21 and 27 kHz, one can identify three separate types of events, each associated with specific types of cosmic ray events. The SEA's of type Sudden-C are mainly associated with GeV, cosmic-ray flares, while the SEA's of type Slow-C are accompanied by MeV cosmic-ray flares of F and F^* types (Table XI). The development of the latter is always slow compared with the former as was mentioned earlier. Since the acceleration efficiency of solar cosmic rays is dependent of the developing process of cosmic-ray flares the above mentioned correspondence between cosmic-ray flares and the types of SEA's implies that the characteristics of ionizing radiation are strongly dependent on the developing process of cosmic ray flares. The fact that the magnitude of the intensity decrease of atmospherics at 10 kHz is much greater for the SEA's of type Sudden-C than that for the SEA's of type Slow-C (always < 2 dB), indicates that the power flux of the ionizing radiation of the former is much higher than that of the latter.

4.6. SDA Characteristics and Implications

Tříska and Laštovička (1972), in a detailed examination of SDAs at 5 kHz, find that:

TABLE XI

Solar cosmic-ray flares and associated SIDs

Cosmic-ray flare	I	II
Cosmic rays	GeV particles (UI or SI) and MeV particles of type F and F^* (PCA)	MeV particles of type S (PCA)
Spectra of type IV burst	*[plot: F vs Freq., IV m rising to IV μ]*	*[plot: F vs Freq., IV m descending to IV μ]*
SID		
SWF (type IV spectra)	Sudden drop-out	Slow drop-out
SEA 10 kHz	Decrease	Increase or no change
21 kHz	Increase	Increase
27 kHz	Type C (Sudden or slow-C)	Types A and B
SFD	Yes	No
f_0F_2	Increase	No

(i) The relative frequency of SDA occurrence is rather low – about 11.3% of all SIDs on the average.

(ii) SDA occurrence shows both solar cycle and annual variation. No SDA events were recorded during low solar activity; the relative occurrence frequency was maximum during medium solar activity (above 22%). The annual variation of SDA occurrence showed a summer and autumn maximum and a winter minimum.

(iii) No case occurred with an SDA occurring alone in the absence of any other SIDs, but there were cases when SDAs occurred but not SEAs at 27 kHz.

During a flare there is a decrease in the waveguide height (from about 70 km to typically about 65 km), as well as a change in conductivity of the upper waveguide boundary. When the first effect predominates, an SDA occurs; when the second predominates an SEA occurs. Field (1970) has made theoretical analysis of the changes in propagation effects of VLF and ELF radio waves due to changes in the X-ray spectral studies. He found that the computed attenuation rates, expressed in decibels per thousand kilometers of path length (dB Mm^{-1}), for frequencies between 100 Hz and 30 kHz, showed that the general behaviour i.e. the occurrence of enhancement in the upper VLF band and a decrease in the lower VLF band, is quite insensitive to the details of the assumed X-ray spectrum; its main requirement is a sufficient concentration of X-ray energy in the 1–8 Å wavelength band.

While the SDAs offer an important insight into the changes in the lower ionosphere, when combined with SEA observations, quantitative studies are still lacking, excepting some analysis involving the attenuation factor, following Volland (1966):

$$A = a_i \pi^2 c^2 / 4\omega^2 h^3 + b_e / h, \tag{28}$$

where $a_i = 0.34 f^{2/3}$, and $b_e = (\varepsilon_0 f / \sigma_e)^{1/2}$ (f in kHz) where σ_e is the ground conductivity. Tříska and Lăstovička (1972) find that for a source distance of 1000 km and a waveguide height decrease of 5 km from an initial value of 70 km, SDA at 5 kHz is 15%; but rises sharply to 75% for a source distance of 8000 km.

4.7. Examinations of Complete Time Profiles of X-Ray Flares, Radio Noise Bursts and SID Effects

Complete time profiles of X-ray flares in different wavelength bands and those of SIDs have been compared by various authors. The bands usually considered are: 10–50 keV (Arnoldy et al., 1968), 2–12 Å (Van Allen, 1967), 0–3 Å, 0–8 Å and 8–20 Å (Kreplin et al., 1968) and selected EUV lines (Neupert, 1967; Hall and Hinteregger, 1969; Richards, 1971). Early comparison of the time curves of cm radio noise bursts and SIDs made by Hachenberg et al. (1959) and by Hakura (1966) showed excellent agreement between the two. In Hakura's study, for example, two typical SWFs occurring around noon, but of different types – one S–SWF occurring on October 16, 1957 with growth-time of only 1.5 min, and the other slow-SWF occurring on 11 September, 1957 with growth time as long as 26 min have remarkable resemblances with 3750 MHz bursts, the one on 11 September rising and falling very gradually without the usual first phase. Mitra (1966), Chilton *et al.* (1965), and Subrahmanyam (1967) found excellent agreement between the time development of X-ray flares monitored by Vela and Ariel 1 satellites and SIDs corresponding to the X-ray flares. More detailed examinations have more recently been done by Deshpande *et al.* (1972a) with the time profiles of X-ray flares, different types of SIDs and radio noise bursts. They find that the nature of X-ray SID relationship is complex; the time and wavelength dependence of X-ray energy as well as height and time dependence of ionization and electron density are involved. Therefore, no simple correspondence is expected. Some typical examples given by Deshpande *et al.* are shown in Figures 64a–f. In these

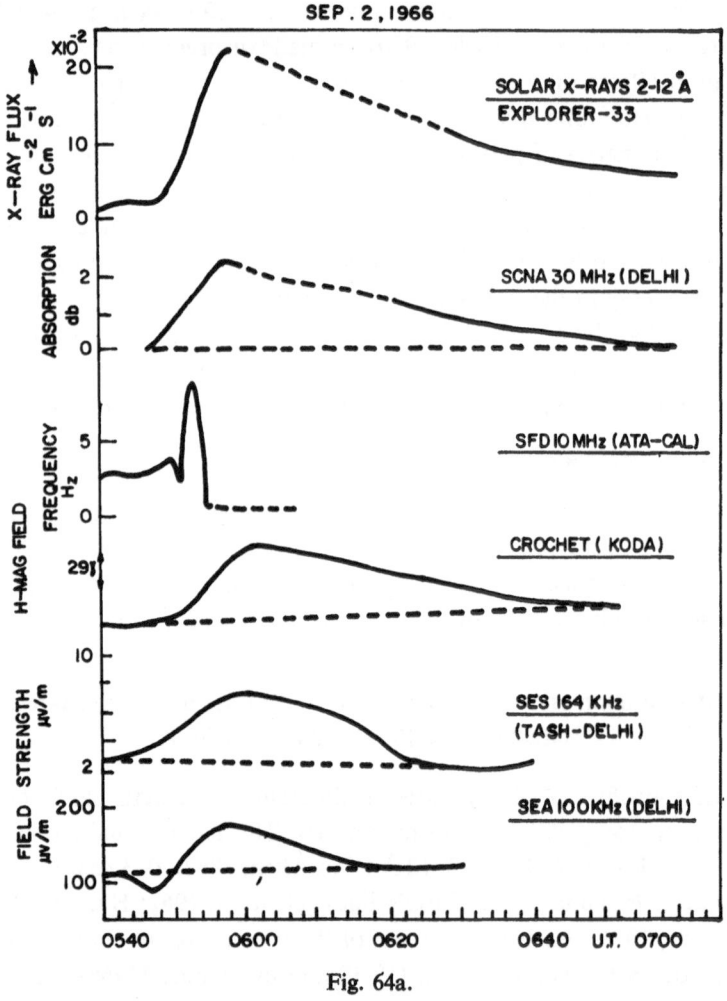

Fig. 64a.

Fig. 64a–f. Time curves of selected SIDs and associated X-ray events in different energy bands: (a) 2–12 Å; (b) 0–8 Å and 0–3 Å; (c) X-ray emission lines at 1.87 Å and 11.8 Å; (d) 10–50 keV energetic X-rays; (e) after Jones (1971), shows difference in the recovery times between long and short paths; (f) Comparison of rate of change of total electron content recorded along Boulder-ATS-1 ray path for August 7, 1972 flare event with radio burst at 35000 MHz recorded at Sagamore Hill Radio Observatory, Mass, U.S.A. (after Mendillo *et al.*, 1974).

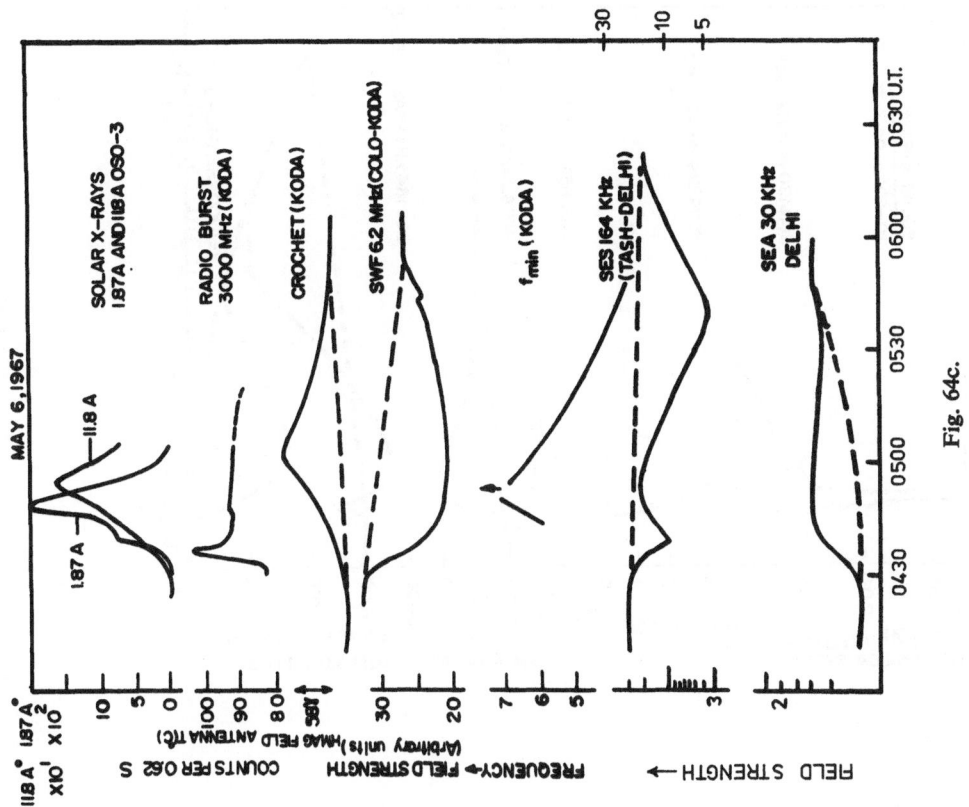

Fig. 64c.

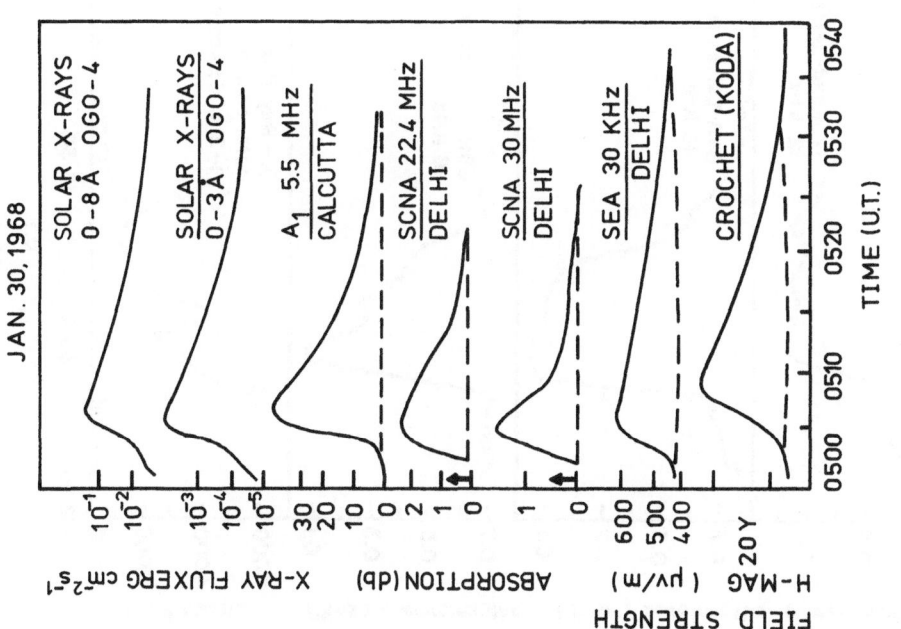

Fig. 64b.

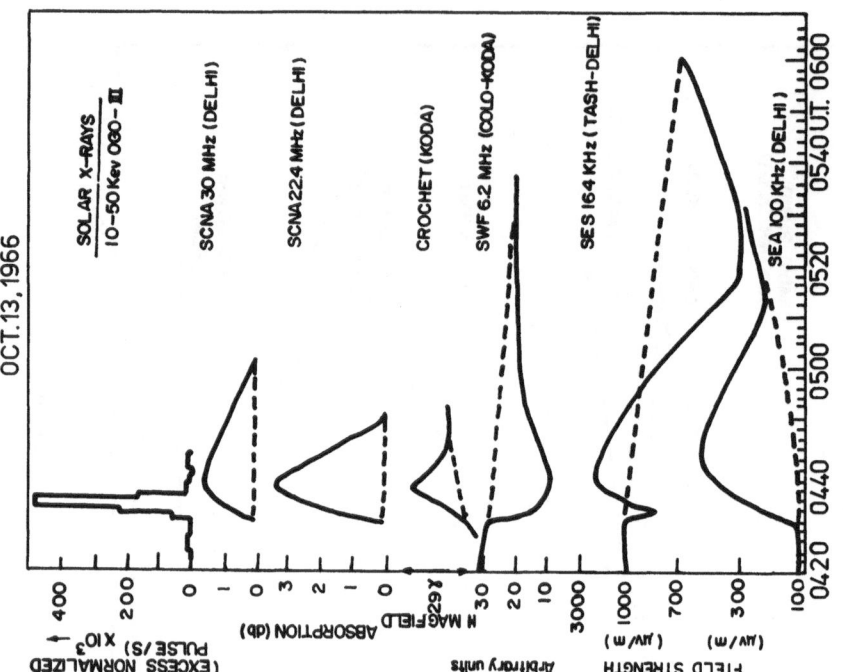

Fig. 64d.

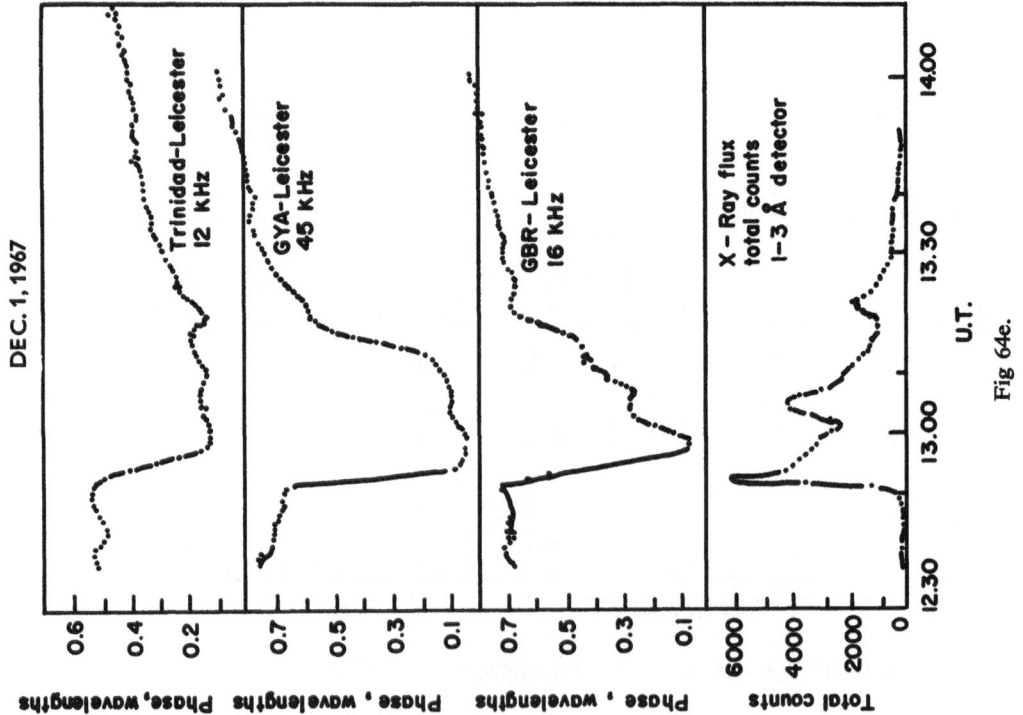

Fig 64e.

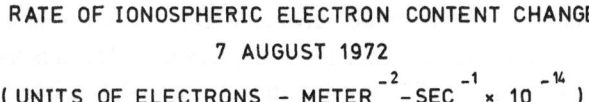

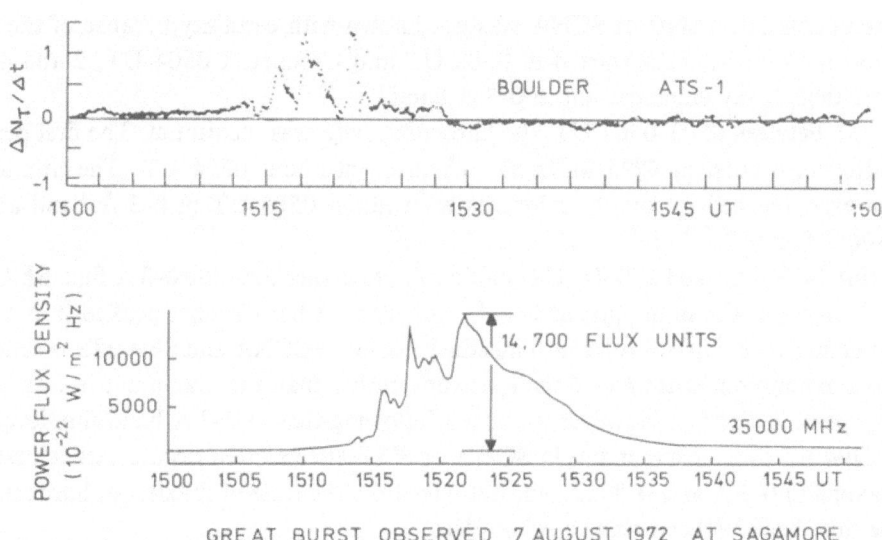

Fig. 64f.

examples, X-ray time profiles are considered in the following sequence:

(i) Time curves of 2–12 Å X-ray flux and associated SIDs for the flare of 2 September, 1966 (Figure 64a)

(ii) Time curves of X-ray flux in 0–3 Å and 0–8 Å bands and associated SIDs for the flare of 30 January 1968 (Figure 64b)

(iii) Time curves of X-ray emission lines at 1.87 Å and 11.8 Å and associated SIDs for the flare of 6 May, 1967 (Figure 64c)

(iv) Time curves of 10–50 keV energetic X-rays and associated SIDs for the flare of 13 October 1966 (Figure 64d)

(v) X-ray time curve in 1–3 Å band and phases of VLF radio signals monitored at Leicester by Jones (1971) (Figure 64e).

It can be seen from Figure 64a that the time curves of SCNA, SES, crochet and SEA are similar to those of X-ray flux in 2–12 Å band, but the SFD time curve is quite different. There are the following differences in details:

(i) SEA and SES start one minute after the start of 2–12 Å enhancement. SCNA and Crochet start three minutes after the X-ray flare begins.

(ii) SCNA and SEA reach their maximum almost at the time of peak in X-ray flux.

(iii) SEA and SES effects end early; SCNA and crochet end about half an hour later. 2–12 Å X-ray flux level continues to be above the pre-flare level even after the end of SCNA and Crochet.

In Figure 64b, comparison is made between SIDs and X-ray fluxes in two bands namely 0–3 Å and 0–8 Å. Here again the time curves of HF absorption, SEA and Crochet were similar to those of X-ray flux in 0–3 Å and 0–8 Å bands. Furthermore:

(i) 0–3 Å X-ray enhancement started at 0500 UT while that of 0–8 Å X-rays started one minute later. Start of SCNA was not known with certainty because of the solar radio noise burst. SEA started at 0502 UT and Crochet at 0504 UT, 2 and 4 min later than X-ray enhancement in 0–3 Å band.

(ii) Between 0503–0504 UT, the hardening ratio was maximum. The first peak of radio noise burst at 4995 MHz also occurred at about 0504 UT. The flux levels, however, reached maximum a little later at about 0505 UT in 0–3 Å band and at 0506 UT in 0–8 Å band.

(iii) SCNA peaked at 0505 UT, half a minute earlier than the 0–3 Å flux. SEA pea- ked almost at the same time as 0–3 Å enhancement but Crochet peaked 2 min after the enhancement in 0–8 Å band. Thus the peak in the SCNA and SEA effects followed the maximum in *hardening* of the spectrum, rather than the maximum in flux levels. The peak in Crochet, however, occurred following that in 0–8 Å band flux level.

That the time of maximum in SCNA or SWF effects often occurs earlier than the maxima in 0–3 Å and 0–8 Å band, but later than the time of maximum hardening of the flux, has also been seen in other cases.

In Figure 64c SID time curves are compared with those in two emission lines at 1.87 Å and 11.8 Å. The enhancements in 1.87 Å and 11.8 Å fluxes started at about 0430 UT, at about the same time as a Crochet, slow-SWF and SES effects; but SEA started about 5 min earlier. The X-ray time curve at 1.87 Å had two stages in its growth, but these were not evident in the SIDs. Of the different SIDs associated with this event, the SEA had the most rapid growth maximising 2 min before the peak in 1.87 Å. The maxima in SWF and SES were reached at 0552 UT, 2 min before than that in 11.8 Å. Crochet showed a slow rate of growth and reached its maximum only after the 11.8 Å line.

The time curves of energetic X-rays in 10–50 keV band shown in Figure 64d for the flare of 13 October, 1966, corresponded to those of SIDs only in the growth stage. These X-rays contributed little to the SID ionization because of low flux levels (10^{-5} erg cm^{-2} s^{-1}), but the sudden onset and rapid growth in the X-ray flux level was also reflected in the sudden start and rapid rate of growth of SCNA, Crochet, SWF and SES effects.

The case of the event of 1 December 1967 (Figure 64e) reported by Jones (1971) falls in a somewhat different category. This event was rather unusual in that marked structures were evident in the phase anomalies; these corresponded with structures in 1–3 Å X-ray flux variations monitored with higher time resolution by the detectors aboard the OSO-4 satellite. In addition to the initial phase advance associated with the initial flare burst, there was a clearly evident second phase advance on all three propagation paths immediately following the second X-ray burst at 13.05:00 UT. (GBR at 13.07:00 UT; GYA at 13.06:30 UT, Trinidad at 13.09:00 UT). More remarkable was the great difference in the recovery time of the phase on the long

(Trinidad–Leicester, 7200 km) and short (GBR–Leicester: 27 km; GYA–Leicester: 186 km) propagation paths. The phases of the latter had recovered by 1330 UT, but the phase of the Trinidad path had not recovered even 30 min after the X-ray flux had returned to its pre-flare value at 1345 UT. It appears, therefore, that at the lower heights from which the 12 kHz signals are reflected the loss rate is slower.

Figure 64f falls in a somewhat different class. Here we are concerned with the time variations of rate of change of the total electron content for which the contributions come from a wide range of heights in the F-region. It is not appropriate, therefore, to compare these variations with those in X-rays; one should instead look at the time variations in EUV fluxes. Mendillo *et al.* (1974), from whom this diagram is taken, have, as a substitute for the EUV fluxes, looked for a comparison with solar radio fluxes in the centimeter and the millimeter regions. The parameter dN_T/dt is plotted rather than N_T, because the former relates directly to the production function q. The dN_T/dt values given are those of Boulder for which alone a high time resolution was available. It will be recalled that SITEC observations were made for this event by as many as 17 stations (Section 2.10). Mendillo *et al.* compared the time changes of dN_T/dt with flux values at 35000 MHz recorded at the Sagamore Hill Radio Observatory in Hamilton, Massachusetts (Castelli *et al.*, 1973). The similarity between the two, even in the small-scale structure, is remarkable. There is a clear correspondence in the precursors near 1524 UT and the ledge near 1519 UT. There was no such correspondence with 2800 MHz fluxes; for this there was only a single broad maximum from 1527–1529 UT. Mendillo *et al.* point out that this remarkable similarity with the mm region and lack of similarity in the fine structures with 10.7 cm flux must be kept in view, since there is wide use of 10.7 cm radio flux in describing and predicting the variations of quietday E and F-region ionization parameters. The apparent conflict of this view with the statistical result reported in Section 4.4 (i.e. the SFDs are best correlated at 4995 MHz) arises perhaps because of the fact that while statistically the latter is true, there is no optimum frequency for all types of flares or one that would represent both EUV flare estimates and nonflare variations. It is possible that the best correlation is provided essentially fortuitously at frequencies near those where the maximum increase in radio intensity occurs. Many small flares have 10.7 cm bursts with detailed time agreement with SFDs, although for larger burst intensities, the detailed time agreement diminishes.

CHAPTER 5

SOME SIMPLE ANALYSIS TECHNIQUES

Some simple analysis techniques suitable for use for synoptic observations of SIDs are given in this chapter. Before discussing these simple techniques, however, some background information on photoionization rates and on the equations of continuity for electrons and ions are outlined.

5.1. Photoionization Rates and Equations of Continuity

Electron production rate $q(\chi, z)$ of the solar electromagnetic radiation incident upon the Earth's atmosphere with a zenith angle χ at altitude z is expressed by the sum of the individual electron production rate $q_i(\chi, z)$ over the respective wavelength ranges, that is,

$$q(\chi, z) = \sum_i q_i(\chi, z) \qquad\qquad 29(a)$$

$$q_i(\chi, z) = \phi_i(z) \sum_j \eta_{ij}\sigma_{ij}n_j(z) \qquad\qquad 29(b)$$

$$\phi_i(z) = \phi_i(\infty) \exp\left[-\,\mathrm{Ch}\,(\chi, x, \beta) \sum_j \sigma_{ij} \int_z^\infty n_j(z)\,\mathrm{d}z\right] \qquad\qquad 29(c)$$

where suffixes i and j indicate the wavelength range and the neutral constituent species, respectively and

$n_j(h)$:   number density of neutral atmospheric constituent j at altitude h,

σ_{ij}:   absorption cross section of a neutral atmospheric constituent j for electromagnetic radiation over a wavelength range i,

η_{ij}:   photoionization yield, i.e. the number of electron-ion pairs formed per photon with the wavelength range i absorbed by a neutral atmospheric constituent j,

$\phi_i(\infty)$:   photon number flux intensity of solar electromagnetic radiation with the wavelength range i outside the Earth's atmosphere,

$\mathrm{Ch}\,(\chi, x, \beta)$: Chapman function or optical depth factor, $x = a/H$, (a is the distance from the Earth's center, H is the scale height of the atmosphere), and $\beta = \mathrm{d}H/\mathrm{d}h$ i.e. the scale height gradient.

$q_i(\chi, z)$ varies with $\phi_i(z)$ which depends on the solar activity and the atmospheric situation. The ratio $q_i(\chi, z)/\phi_i(z)$ represents the number of electron-ion pairs produced by one solar photon per unit path length in the Earth's atmosphere. The quantity is called '*photoionization efficiency*', which is implicitly dependent on the solar activity, only through the atmospheric densities at an altitude z. Moreover, it is desirable to introduce 'local photoionization efficiency' $Y_i(\chi, z)$ as a more practical quantity *in situ* concerned with the height distribution of neutral atmospheric density above the altitude z;

$$Y_i(\chi, z) = \frac{q_i(\chi, z)}{\phi_i(\infty)} = \sum_j \eta_{ij}\sigma_{ij}n_j(z)\exp\left[-\mathrm{Ch}(\chi, x, \beta)\sum_j \sigma_{ij}\int_z^\infty n_j(z)\,\mathrm{d}z\right],$$

(30)

where the exponential factor denotes the atmospheric effect on the solar electromagnetic radiation reaching the altitude z. Values of $Y_i(\chi, z)$ represented in unit of electron-ion pairs (photon cm)$^{-1}$ have been given by Ohshio *et al.* (1966), by Chakrabarty *et al.* (1968) and by others.

The actual value of the electron production rate q depends, of course, on the atmospheric model assumed. The atmospheric distributions given in CIRA 1972, or the models of Jacchia (1970, 1971) are generally used for such computations. Since atmospheric density in the mesosphere varies widely with season and latitudes and in the thermosphere with solar activity and diurnal time, it is important to select the appropriate density model from these sets. A convenient way of obtaining an album

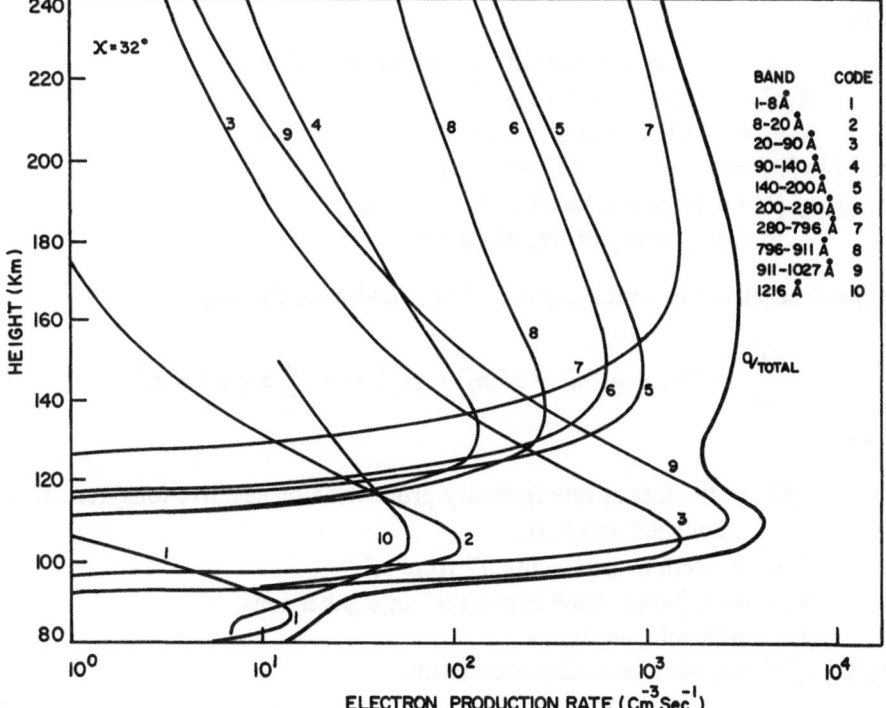

Fig. 65. Ionization contributions from different wavelength bands in the XUV radiation.

of q-profiles for different conditions is to compute for different exospheric temperatures from 600 to 2000 K, and then selecting the appropriate model on the basis of the appropriate exospheric temperatures. One must also have not merely the total production rates, but also the individual production rates $q(O)$, $q(N_2)$, $q(O_2)$ and $q(NO)$.

Figure 65 gives the relative contributions of the different wavelength bands and the heights at which they dominate. If the enhancements during a flare in the different bands are different, as we have seen they are, the total production rate assumes a different shape, and the relative emphasis in the production of atomic and molecular ions also changes.

The density of each ionic constituent may be calculated from a knowledge of the reaction rates and the equation of continuity

$$\frac{dn_j^{\pm}}{dt} = q(n_j^{\pm}) - L(n_j^{\pm}) \qquad (31)$$

for each species of ion of density $n_j^{\pm}$ at each altitude. $q(n_j^{\pm})$ is the production term and $L(n_j^{\pm})$ is the loss term. These normally depend on $n_j^{\pm}$ and $n_k^{\pm}$, $k \neq j$, so the problem becomes the solution of a system of first order non-linear differential equations.

A more explicit form of the continuity equation for a typical positive ion is

$$\frac{dn_j^{+}}{dt} = q_j + K_1 n_k^{+} - K_2 n_j^{+} - \sum_k \alpha_i(jk) n_j^{+} n_k^{-} - \alpha_D(j) n_j^{+} N_e, \qquad (32a)$$

where

$$q_j = \text{primary production rate (photoionization)},$$
$$K_1 n_k^{+} = \text{reactions producing } n_j^{+} \text{ from } n_k^{+},$$
$$K_2 n_j^{+} = \text{reactions which convert } n_j^{+} \text{ into other ions},$$
$$\alpha_i(jk) n_j^{+} n_k^{-} = \text{ion-ion recombination terms},$$
$$\alpha_D(j) n_j^{+} N_e = \text{dissociative recombination terms},$$
$$N_e = \text{the density of free electrons}.$$

A typical negative ion continuity equation would have the form

$$\frac{dn_j^{-}}{dt} = \beta N_e + K_1 n_k^{-} - K_2 n_j^{-} - \gamma(j) n_j^{-} - \sum_k \alpha_i(jk) n_j^{-} n_k^{+}, \qquad (32b)$$

where

$$\beta N_e = \text{attachment rate (primary production of negative ions) (see also Sections 9.2 and 9.3)},$$
$$K_1 n_k^{-} = \text{reactions producing } n_j^{-} \text{ from } n_k^{-},$$
$$K_2 n_j^{-} = \text{reactions which convert } n_j^{-} \text{ into other ions},$$
$$\gamma(j) n_j^{-} = \text{detachment terms},$$
$$\alpha_i(jk) n_j^{-} n_k^{+} = \text{ion-ion recombination terms}.$$

The quantities α_D, α_i, β, and γ are referred to as the dissociative recombination,

ion-ion recombination, attachment, and detachment coefficients, respectively (see Sections 9.2 and 9.3).

The relationship

$$N^+ = N^- + N_e \tag{33}$$

is assumed to hold on a macroscopic scale, where N^+ is the *total* positive ion concentration and N^-, the *total* negative ion concentration.

One can consider equations for the total ion densities in order to gain some insight into the processes operating.

$$\frac{\mathrm{d}N^+}{\mathrm{d}t} = q - \alpha_i N^+ N^- - \alpha_D N^+ N_e \tag{34a}$$

$$\frac{\mathrm{d}N^-}{\mathrm{d}t} = \beta N_e - \gamma N^- - \alpha_i N^- N^+ . \tag{34b}$$

In this case the parameters α_i, α_D and γ are weighted averages taken over the ions present. Thus, for example,

$$\alpha_D = \sum_j \frac{\alpha_D (j) \, n_j^+}{N^+} . \tag{35}$$

Some simplifications may be made in these equations as outlined below.

Defining $\lambda = N^-/N_e$, we have

$$(1 + \lambda) \frac{\mathrm{d}N_e}{\mathrm{d}t} + N_e \frac{\mathrm{d}\lambda}{\mathrm{d}t} = q - (\lambda \alpha_i + \alpha_D) (1 + \lambda) N_e^2 \tag{36}$$

which is an equation describing the electron density in terms of weighted recombination coefficients and the negative ion density. In some cases (especially at high altitudes) $\lambda \cong 0$ so that

$$\frac{\mathrm{d}N_e}{\mathrm{d}t} = q - \alpha_D N_e^2 . \tag{37}$$

Even if λ is not zero, $\mathrm{d}\lambda/\mathrm{d}t$ may be zero, so that

$$(1 + \lambda) \frac{\mathrm{d}N_e}{\mathrm{d}t} = q - (\lambda \alpha_i + \alpha_D) (1 + \lambda) N_e^2 . \tag{38}$$

If $\mathrm{d}N_e/\mathrm{d}t$ is zero also, the resulting equations are

$$q = \alpha_D N_e^2 \tag{39}$$

and

$$q = (\lambda \alpha_i + \alpha_D) (1 + \lambda) N_e^2 = \psi N_e^2 . \tag{40}$$

These are two commonly used forms of the electron continuity equation under equilibrium conditions; Equation (39) if no negative ions are present and Equation (40) if negative ions are present.

Another form of the electron continuity equation may be derived as follows:

$$\frac{dN_e}{dt} = \frac{dN^+}{dt} - \frac{dN^-}{dt}$$

giving

$$\frac{dN_e}{dt} = q - \left[\beta - \frac{\gamma N^-}{N_e} + \alpha_D N^+\right] N_e = q - B' N_e. \tag{41}$$

This gives a third form of the continuity equation in terms of weighted attachment, detachment, and dissociative recombination parameters.

Which of these equations is most useful in a practical sense depends on whether α_D, B', or ψ is more nearly independent of q and N_e. Equation (41) reduces to (39) if negative ions are not important (i.e., $\beta \cong 0$, $N^- \ll N_e$).

The approximations 39, 40 and 41 are extensively used in D region work because they apply (under steady state conditions) as long as (32a, b) are valid. Much confusion has arisen in the past by the indiscriminate use of the symbol 'α' to refer to either α_D or ψ as used in 39 and 40. A further complication is introduced by the fact that, if only steady state electron densities are measured in an experiment, only q/N_e^2 or q/N_e can be determined.

5.2. Use of Relaxation Time

The diurnal curve in N_e usually lags behind that in q by an amount τ, called relaxation time which in its simplest form, is obtained by using the first order solution of Equation (37), namely:

$$N_e = \left(\frac{q}{\alpha}\right)^{1/2} - \frac{1}{4\alpha}\frac{q'}{q} \tag{42}$$

with a quadratic form of q given by:

$$q(z, t) = Z(z)\left[q_0 - at^2\right], \tag{43}$$

where $Z(z)$ is a function which contains the height variation and q_0 is the value of q at local noon ($t=0$).

Using Equations (42) and (43), and setting dN_e/dt for the time of maximum one obtains

$$\tau = (2\alpha N_m)^{-1} \tag{44}$$

in which N_m is the value of the maximum electron density.

The same concept of 'relaxation time' can also be applied during a solar flare, if the time of maximum intensity of ionizing radiation is known. In this case τ is the difference between this time and that of the maximum of the excess ionization.

The above equation is, however, only an approximate one, and is strongly dependent on the nature of time curve of the perturbing radiation. Mitra (1958) showed that it is possible to include a wide variety of time curves by using a production function of the form:

$$q(t) = q_n + q_m \left(\frac{t}{t_g}\right) \exp\left[b\left(1 - \frac{t}{t_g}\right)\right], \tag{45}$$

where $q(t)$ is the rate of electron production at any time during the flare event, and q_n its preflare value. The exponent b determines the shape of the time curve. The expression for the corrected relaxation time is then given by (Mitra, 1958):

$$\tau_f + (b - 1) t_m = \frac{1}{2\alpha N_m} \left[\frac{t_g}{t_m} - \frac{q_n (t_g - b t_m)^2}{q_m t_m t_g}\right], \tag{46}$$

where $\tau_f = t_m - t_g$.

The above reduces to the simple relation of Equation (44) if

(i) $\qquad \dfrac{q}{q_m} \dfrac{(t_g - b t_m)^2}{t_m t_g} \ll 1$ $\qquad\qquad\qquad\qquad\qquad$ (47)

(ii) $\qquad b \approx 1.$ $\qquad\qquad\qquad\qquad\qquad\qquad\qquad\qquad\quad$ (48)

Computations show that in most flares assumption (i) is met, but assumption (ii) is sometimes not fulfilled. It is thus necessary to retain the exponent b in the equation

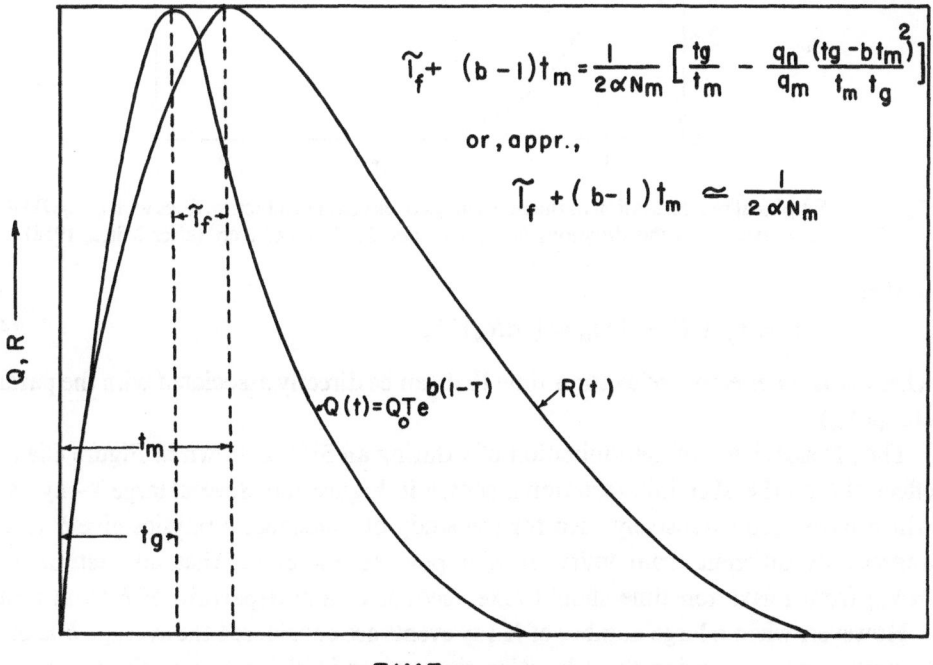

Fig. 66a. Diagram giving physical basis for the determination of the relaxation time during an SID. $R(t)$ represents a parameter proportional to N_e.

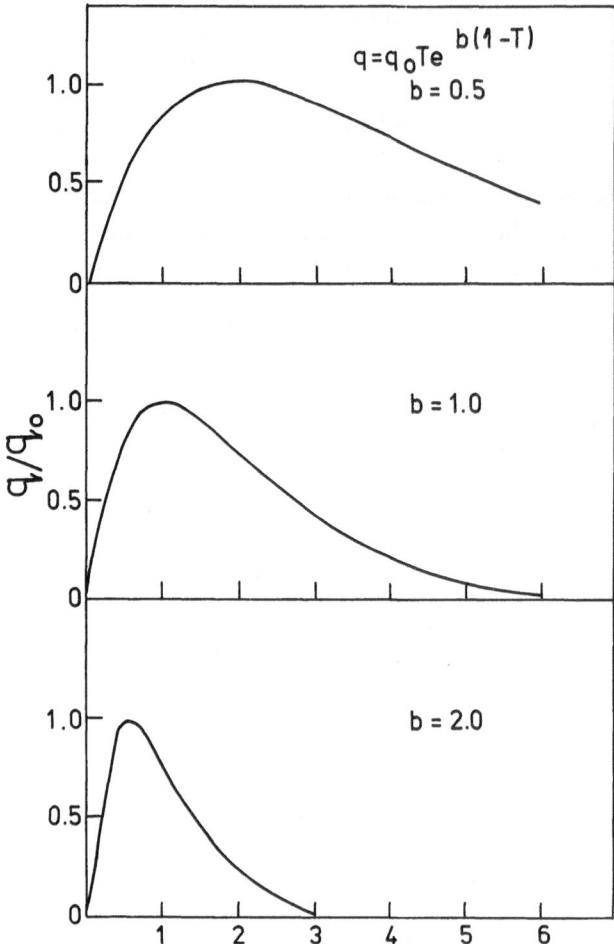

Fig. 66b.   Empirical representation of the time curves of electron production rates with $q=q_0 Te^{b(1-T)}$
The three curves show the situations $b<1$, $b=1$ and $b>1$ respectively (after Mitra, 1958).

so that:

$$\tau' = \tau_f + (b-1)\, t_m = (2\alpha N_m)^{-1}, \qquad (49)$$

where τ' is an effective relaxation time that can be directly associated with the parameter (αN_m).

The physical basis of determination of τ during an SID is shown in Figure 66a and effect of b on the SID time variation is shown in Figure 66b. Several large X-ray flares which have been extensively used for the study of ionospheric physics give b values appreciably different from unity. It is important, therefore, that any estimate of (αN_m) from relaxation time should take account of any departure of b from unity.

However, when a large number of X-ray events are considered the average b is close to unity, and then, using the relaxation times given in Table X, one obtains $(\alpha_{\text{eff}} N_m)$ value around $(25$–$40) \times 10^{-4}$ s^{-1} for HF effects. This should be contrasted with the value of $(2$–$4) \times 10^{-4}$ s^{-1} that one derives at the end of an SID.

5.3. Decay Curve Analysis

5.3.1. PRINCIPLE OF THE DECAY CURVE ANALYSIS

Methods of analysis have been given by Appleton and Piggott (1954) and by Mitra (1958).

The equation for electron density during the decay of the flare is:

$$\frac{d(\Delta N_e)}{dt} = \Delta q - \alpha(\Delta N_e)^2 - 2\alpha\Delta N_e \cdot N_0,\tag{50}$$

where ΔN_e is the excess electron density at any time during the flare. If one leaves out the initial portion of the decay curve so that for the period of analysis Δq is zero or small, then

$$\frac{1}{\Delta N_e}\frac{d(\Delta N_e)}{dt} = -\alpha(N_f + N_0) = -Q.\tag{51}$$

If moreover $N_f \gg N_0$, then the reciprocal of the excess electron density varies linearly with time according to:

$$\frac{1}{\Delta N_e} = \frac{1}{\Delta N_0} + \alpha t,\tag{52}$$

where ΔN_0 is the excess electron density at an arbitrary time t_0.

The same relationship will hold for any parameter directly proportional to ΔN_e. Consequently if the level of the absorbing layer is not greatly changed, one can write:

$$A_f/A_0 = N_f/N_0 = K\tag{53}$$

and

$$\frac{1}{\Delta A} = \frac{1}{\Delta A_0} + \frac{\alpha t}{G},\tag{54}$$

where G is a constant relating absorption with electron density.

If, however, ΔN_e is small in comparison with N_0 so that it is not possible to ignore $(\Delta N_e)^2$ compared with $2N_0\Delta N_e$

$$\Delta q = 0,$$
$$\Delta N_e = \Delta N_0 \exp[-2\alpha N_0 t].\tag{55}$$

In the more general case where ΔN_e and N_0 are comparable

$$\Delta N_e = \Delta N_0 \left[\frac{2N_0}{\Delta N_0(e^{2\alpha N_0 t} - 1) + 2N_0 e^{2\alpha N_0 t}}\right].\tag{56}$$

Equation (56) may be rewritten as:

$$(2\alpha N_0)^{-1} = \tau_0 = t/P,$$

where

$$P = \log_e\left[\frac{\Delta N_0}{\Delta N_e} + \frac{\Delta N_0}{2N_0}\right] - \log_e\left[1 + \frac{\Delta N_0}{2N_0}\right].\tag{57}$$

Equation (57) gives a method for determining τ_0 as a function of time during the decay of the flare:

5.3.2. DECAY CURVES ANALYSIS FOR ABSORPTION EVENTS

Consider the decay curves of excess absorption for medium and short waves and for SCNAs. It is found that in accordance with Equation (54), the reciprocal of the excess absorption, in the majority of cases, varies linearly with time during most of the decay period. Figure 67 shows the time variations of ΔA^{-1} for a medium wave'fadeout

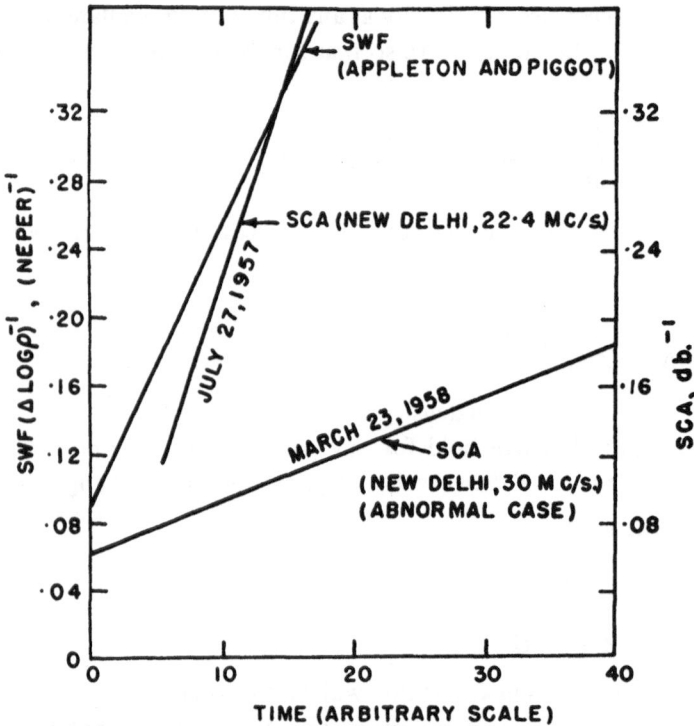

Fig. 67. Variation of the reciprocal of absorption with time during the decay period of a flare.

reported by Appleton and Piggott (1954) and two SCNAs at 22.4 and 30 MHz recorded by Sarada and Mitra (1962). Such linear relationships are compatible only with major absorption occurring over a short height range for which χ is nearly constant. It was also found that in most cases (both SCNAs and fadeout), $\alpha(N_f+N_0)$ decreased asymptotically to a value of about $5 \times 10^{-4}\,\mathrm{s}^{-1}$ near the end of the flare giving a relaxation time of 3.3 min for the peak absorption level (Figure 68). Ionization enhancement factor at this level, using the very approximate relationship of Equation (53) comes out to be about 2–3 for a weak flare, 3–5 for a moderate flare and 5–10 for a large flare (Shain and Mitra, 1954; Sarada and Mitra, 1962; Appleton and Piggott, 1954). The major assumption in these estimates is that the peak absorption level remains unchanged during flares. This assumption is certainly not valid for flares with appreciable hardening of the soft X-ray flux.

We now consider the observations at low frequencies. In the Pennsylvania State University, measurements of absorption were made several years ago on 150 kHz, 75 kHz and 60 kHz. While normally the fadeout was complete at the time of maximum of the flare, there were cases, especially during morning and late afternoon hours, when

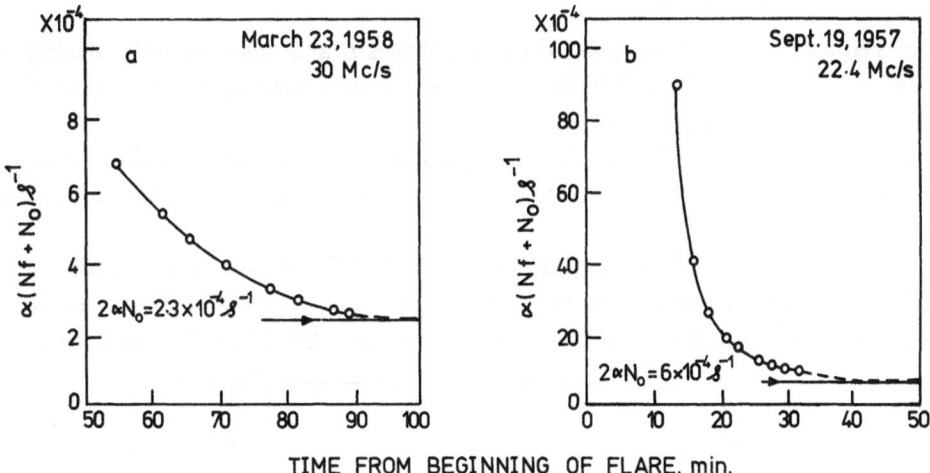

Fig. 68. Time variation of $\alpha(N_f+N_0)$ during SCNAs (after Sarada and Mitra, 1962).

it was possible to follow the course of excess absorption for the entire duration of the flare. As one would expect, in these examples of low frequency fadeout absorption, the assumption that A varies linearly with N_e during the entire course of the flare, becomes doubtful. Unlike the medium and shortwave cases, ΔA^{-1} vs time was rarely a straight line.

5.3.3. DECAY CURVES OF SPAs

An analysis of the decay curves of SPA can only be made if the relation between changes of phase height and of electron density are known.

Shain and Mitra (1954) gave an experimental relationship between the sizes of 18.3 MHz SCNAs and 16 kHz SPAs. In this they compared the observations of SCNAs (made in Australia during 1950–51) with Bracewell and Straker's observations of SPAs (made in England during 1947–48). The comparison is effected by drawing the cumulative frequency distributions of SCNAs and SPAs and then shifting the logarithmic scale of SPA sizes so that the distributions of the sizes of SCNAs and SPAs fit as closely as possible. It was found that the distributions fit well over a wide range of sizes with a linear relationship between the sizes of SPAs and SCNAs, SCNA of size 1 dB corresponding to an SPA of size 140 deg. They commented, however, that this relationship may not hold for larger disturbances.

In view of this empirical relationship Mitra (1958) wrote:

$$\phi = KN_e \tag{58}$$

and obtained the effective coefficient of recombination at any time during the decay of the flare by using

$$\alpha = \frac{1 - \phi_2/\phi_1}{N_e(t_2 - t_1)}, \tag{59}$$

where ϕ_1 and ϕ_2 are the changes of phase at times t_1 and t_2 respectively, during the decay phase of the SPA and N_e is the electron density required for reflection of the radio wave under consideration.

It has been mentioned that the linear relationship may not be valid for severe disturbances. For such cases, one may use the empirical relationship connecting $\Delta A/A$ with Δh.

Now, examination of different SPA curves at 16 kHz show that these are remarkably similar, and, when plotted in the same time curve, appear to fit one universal curve of $\phi - t$. This curve is shown in Figure 69. We may consider Figure 69 as the

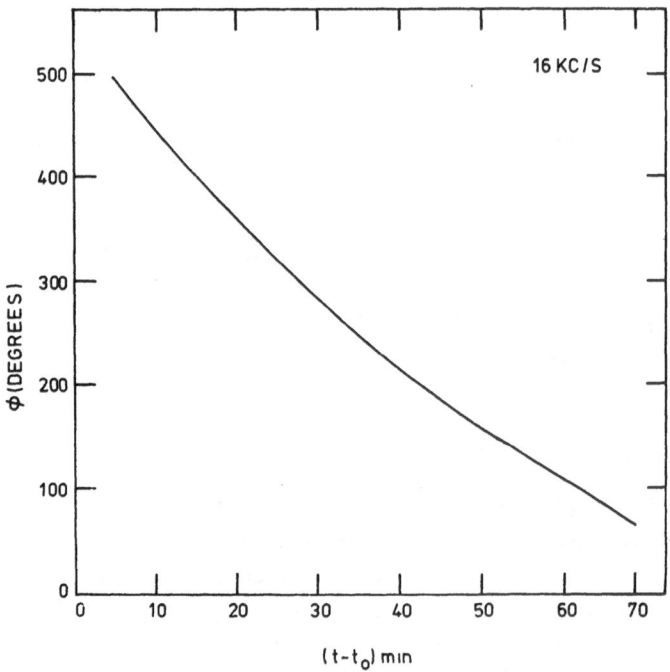

Fig. 69.   Synthesis of observed SPAs at 16 kHz.

synthesis of all observed SPAs at 16 kHz. Determination of α from this diagram using Equation (59) yields the following values: 3×10^{-6} cm^3 s^{-1} at 60 km, 1.4×10^{-6} cm^3 s^{-1} at 65 km and 4×10^{-7} cm^3 s^{-1} at 70 km, values which, as we will see later, are considerably lower than those believed to exist under normal conditions. These are, however, near the minimum distribution that will occur under the situation where water cluster ions have totally disappeared (Section 9.3) and the ions at these heights consist only of NO^+ and O_2^+.

5.3.4. DECAY CURVES OF MAGNETIC CROCHETS

Analysis of decay curves of geomagnetic flare effects was used by Mitra (1958) to test whether the flare current system is merely an augmentation of the Sq system, or is located at a different level. Although the original Crochet studies of McNish (1937) supported the augmentation hypothesis, McIntosh (1951) had found, in a survey of the geomagnetic flare effects at Lerwick and Eskdalemuir over the period 1936–49, that the V-components of the flare and the normal variations were considerably different indicating spatial difference between the two current systems. Decay curve analysis provides simple, although crude, clue to these two alternatives. If $1/\Delta H$ vs time curves are plotted, any deviation from straightline denotes a change in location of the current system; so does any change in the value of τ_0 derived through Equation (57). Mitra found that: (i) $1/\Delta H$ vs time often deviated from a straight line, (ii) the values of (αN_m) for the crochet level was twice as high as that for the SPA level but was only $\frac{1}{4}$ of the value of the fadeout level, and (iii) τ_0 varied from a value of about 30 min at the peak to a value of about 5 min at the end of the effect. Mitra concluded that the flare current system forms at a level somewhat lower than the Sq level of about 110 km.

More detailed analysis of the current systems have been made using the geomagnetic flare effects recorded simultaneously in a number of stations in the sunlit hemisphere, and in particular, the progressive delay that is observed in the time of the peak crochet effect with increase in solar zenith angle. This latter feature is particularly noticeable for flares with small flux intensity and/or of long duration, in which case the lag amounts to as much as 10 min from the subsolar point to the region of the day-night boundary. Ohshio (1964) examined in detail the progressive changes in the equivalent overhead current systems for three crochets, and came to the following conclusions:

(i) In the northern hemisphere, the main current vortex near the subsolar point is on a counter-clockwise Sq-type current-system throughout the developing and decaying stage of the geomagnetic solar flare effect.

(ii) There are some sub-vortices in the polar region and in the dark hemisphere. They have much smaller current intensity and shorter duration in comparison with the main vortex. These sub-vortices seem to change in position and size, during the course of an event.

(iii) Electric current predominant in the equatorial region forms an eastward zonal current through the sunlit and the dark hemispheres. Even from this point of view, a geomagnetic solar flare effect does not seem to be an augmentation of Sq current-system. The zonal current would seem to have longer life than sub-vortices in the polar and/or dark regions.

(iv) Time variation in the relative position of the foci of both the crochet and the

Sq vortices shows that, in general, the focus of the flare current-system approaches that of the Sq current vortex in the developing stage, and the flare current pattern becomes similar to that of Sq in the maximum and decaying stage.

(v) The flare current system is situated in about 104–105 km at $0° \leqslant \chi \leqslant 70°$, and the height increases to about 110 km at $\chi = 85°$.

(vi) The electron production ratio (ratio of maximum excess electron density N_m to the steady constant electron density N_α that would be reached if the ionizing flux at its peak could be kept constant) approaches unity at the subsolar point in a large solar flare, but is about 0.94 and is as low as 0.84 at the subsolar point in moderate and small solar flares, respectively. When $70° \leqslant \chi \leqslant 85°$, the ratio is approximately $0.92 \leqslant r \leqslant 0.98$, $0.78 \leqslant r \leqslant 0.91$ and $0.59 \leqslant r \leqslant 0.75$ in large, moderate, and small solar flares, respectively. Thus the production ratio decreases as the flare magnitude decreases.

(vii) The maximum electron density, N_m, in the effective current-flowing layer amounts to $N_m = 2.2 \times 10^6$ cm^{-3}, $N_m = 4.7 \times 10^5$ cm^{-3}, and $N_m = 9.8 \times 10^4$ cm^{-3} at the subsolar point in the non-equatorial region in large, moderate, and small solar flares, respectively. For $80° \leqslant \chi \leqslant 86°$, N_m is about 10^6 cm^{-3}, 10^5 cm^{-3} and 10^4 cm^{-3} in large, moderate and small solar flares, respectively.

(viii) The observed time lag of the crochet is a superposition of two effects (a) electron production due to the enhanced solar XUV radiation, and (b) electromagnetic induction accompanying the sudden change in electric conductivity of the ionosphere. The time lag due to latter is about a half to one minute all over the world, including the dark hemisphere.

5.4. Simultaneous Correlation of $\Delta\phi$ and ΔA

Simultaneous correlation between change of phase height and absorption at the same frequency, or at different frequencies, may occasionally provide quite fruitful results, since in this we have a method of determining the distribution of electron density, over a limited range of heights.

Three possible cases will, in general, be encountered (see Figure 70).

Case 1 (refer to point A in Figure 70) in which the change in phase height observed is essentially a change in reflection height, i.e.

$$\Delta h = \Delta h_r . \tag{60}$$

Case 2 (refer to point C in Figure 70) in which $\Delta h_r = 0$ and the observed change is almost wholly due to a change in the refractive index in the path.

Here

$$\Delta h = \int_0^{h_0} \Delta\mu \, ds . \tag{61}$$

Case 3 (refers to point B in Figure 70) in which, the observed change is caused by a

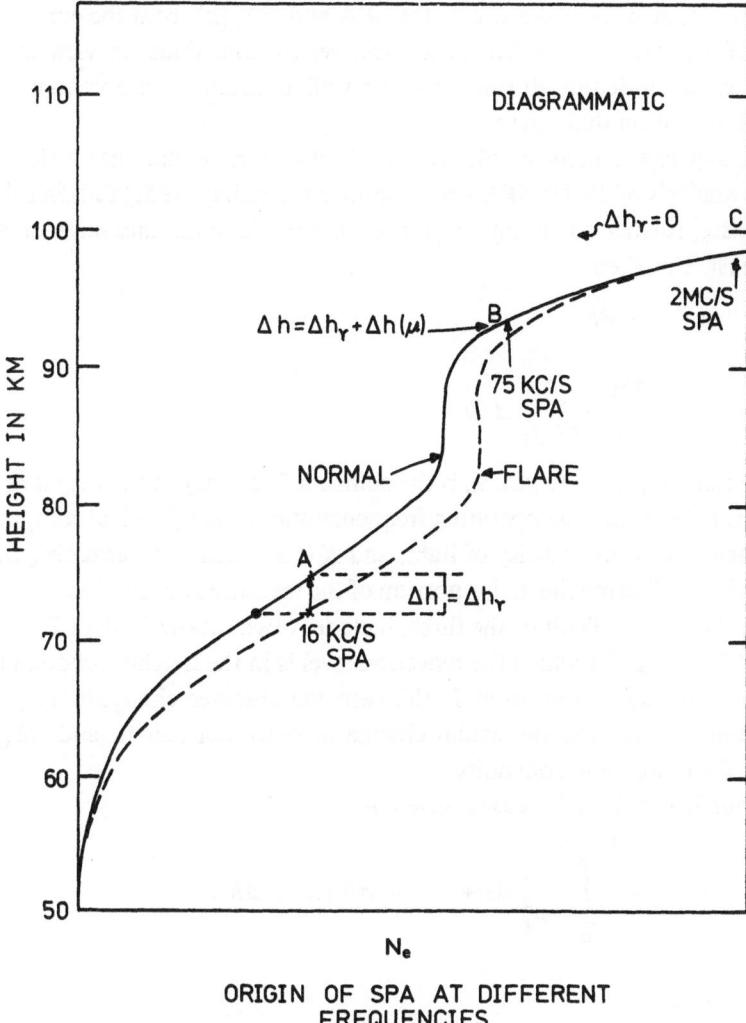

ORIGIN OF SPA AT DIFFERENT
FREQUENCIES

Fig. 70. Phase and Group Height changes for different Frequencies.

change in actual reflection height *as well as* by a change in μ. This case is, of course, the most difficult to deal with. Δh, in this case, is given by

$$\Delta h = \Delta h_r + \int_0^{h + \Delta h_r} \Delta \mu \, ds$$

$$= \Delta h_r + \Delta h(\mu). \tag{62}$$

SPA's at very low frequencies will fall under case 1. An analysis of this type was carried out by Mitra (1954) who related SPA at 16 kHz with SCNA at 18 MHz, and obtained:

$$N_n(h) = 280 - \Delta N_e(h),$$

where h is obtained from the size of the SPA and $\Delta N_e(h)$ from the curve relating the sizes of SPA's and SCNAs. The latter now seems unjustified, in view of the appreciable spectral hardening during flares as well as detailed observations of flaretime profiles of ionization that now exist.

Cases under type 2 involve reflection levels at which the flare has little or no effect. Findlay's analysis of 2 MHz SPA and absorption (Findlay, 1951) falls in this category. In such cases, further advantage is gained, if simultaneous measurements of group heights exist. For then,

$$\Delta h' = - \Delta h$$

and

$$\Delta A = \frac{K v_0}{2c} \frac{f}{f+f_L} \Delta (h' - h), \tag{63}$$

where h' refers to group height, v_0 is the collision frequency at the height of maximum ionization, f and f_L are the operating frequency and the longitudinal component of the gyrofrequency, c is the velocity of light, and K is a constant determining the location of the level contributing the major portion of the measured absorption.

Case 3, the most difficult of the three, includes observations in the LF, such as those on 75 and 150 kHz, for which the reflection level is in the neighbourhood of or within the region of enhanced ionization. In this case, the observed change in the phase height is a combination of Δh_r, the actual change in reflection height, and $\Delta h(\mu)$, that is caused by deviation of μ from unity.

The group height h' in this case is given by

$$\Delta h' = - \int_0^{h_0 + \Delta h_r} \frac{\Delta\mu}{\mu_f^2}\, ds + \Delta h_r = \Delta h'(\mu) + \Delta h_r.$$

If

$$\mu_f \approx 1$$

we again have

$$\Delta h = - \Delta h'$$

and

$$\Delta h - \Delta h' = - 2\Delta h_r.$$

However, if

$$\mu_f < 1$$
$$\Delta h'(\mu) > \Delta h(\mu)$$

and we revert to Equation (63)

(a) *16 kHz Results*

Now consider the average 16 kHz SPA sizes which are 2 km, 5.5 km and 9 km for class 1, class 2 and class 3 flares respectively. The excess ionization $\Delta N_e(\phi)$ produced at these heights for these flares are approximately 90 at 69.5 km (class 1), 210 at 66 km (class 2) and 250 electrons cm^{-3} at 62.5 km (class 3), if one uses the average active Sun profile given recently by Mechtly *et al.* (1972). The ionization enhancements needed are, 1.5, 4.0 and 9.3, respectively, for the three classes of flares, with the quiet day reflection level at 71.5 km. The corresponding ionization enhancements derived from absorption measurements are 2.5, 4.5 and 7.0 respectively-values very similar.

(b) *2 MHz Results of Findlay (1951)*

The theoretical relationship connecting changes in phase height, group height and absorption for 2 MHz radio waves is given in Equation (63). Using this equation for some 100 fadeouts observed at a frequency near 2 MHz at Cambridge, Findlay (1951) determined the quantity $K v_0$.

$$K v_0 = (4.43 \pm 0.20) \times 10^5 \, s^{-1}.$$

Findlay showed that $(K v_0)$ can be considered as the collision frequency v_{eff} of a level which, for the several ionization models he considered, fell near the level of maximum fadeout absorption. With Brigg's models of v (Briggs, 1951), Findlay deduced a height of 101 km for the fadeout level. With currently accepted models of collision frequency and remembering that for $v \ll \omega$, $v_{eff} \approx 2.5 \, v_M$, where v_M is the collision frequency for monoenergetic electrons, it corresponds to a height of 87 km.

(c) *Observations at 75 kHz*

From an examination of 64 SIDs in the phase and group height records of the Pennsylvania State University at the frequency of 75 kHz, Houston (1957) found that the average phase height drop was 2 to 4 km corresponding to a rise in group height of 5 to 8 km. As one would expect for frequencies for which the flare time refractive index is less than unity,

$$\Delta h' > \Delta h.$$

We now consider the SID occurring on January 16, 1956. In this case

$$\Delta h' = 10 \text{ km}$$
$$\Delta h = -2 \text{ km}$$
$$\Delta \log \varrho = 0.75 \text{ neper}.$$

Then using Equation (63), we obtain

$$K v_0 = 1.1 \times 10^5 \, s^{-1}.$$

This refers to a level of about 92 km, which is somewhat below but near the height of reflection. It would then appear that, for this case, most of the excess absorption was deviative.

5.5. Zenith Angle Dependence of SIDs

Chilton *et al.* (1963) examined the solar zenith angle dependences of the mean change

in reflection height for a number of VLF propagation paths by assuming that the
ionosphere is sharply bounded and that only the first-order wave-guide mode of
propagation is present. An average value of $\sec\chi$ at the time of each flare was calculated by taking the average of $\sec\chi$ values computed for 100 km intervals along the
transmission path. The results for 10 solar flares are presented in Figure 71. Each line
in this figure represents a separate solar flare and is characterized by a horizontal axis

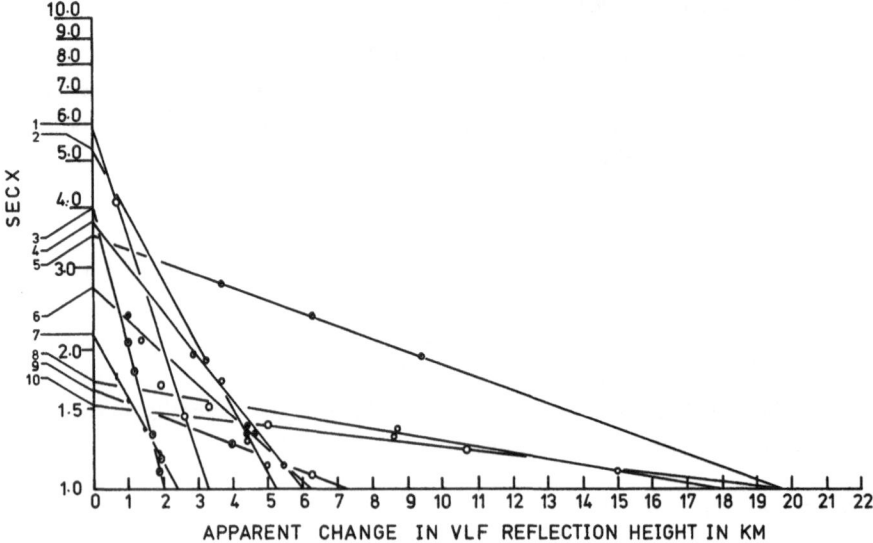

Fig. 71. Solar Zenith angle dependence for SPAs for a number of VLF propagation paths for 10
solar flares (after Chilton *et al.*, 1963).

intercept and a unique slope. The intercept represents the apparent change in VLF
reflection height that would be observed by means of a short VLF transmission path at
the subsolar point ($\sec\chi=1$). The large variation in slope from line to line indicated as
we would expect that the character of the ionizing radiation varied from flare to flare.
One may also note that flares 5, 8 and 10 indicated height changes of about 20 km at
the subsolar point, whereas results obtained from the other seven flares had markedly
steeper slopes and indicated changes in height of 7 km or less at the subsolar point.
Although flares 8 and 10 were both of optical importance 3 and flare 5 was of optical
importance 1, the *D*-region ionospheric effects of flare 5 were at least as great as the
effects of the two flares of importance 3. The other seven flares were all of importance 2
or less.

In another work involving measurements of SPAs over the paths Annapolis–Sao
Paulo (21.4 kHz), Rugby–Sao Paulo (16 kHz) and Trinidad–Sao Paulo (12 kHz),
Kaufmann and Mendes (1970) found no significant zenithal dependence (about
$1\ \mu s\ Mm^{-1}$) for SPAs of 'importance' I (see Equation (25)). They found, however,
significant zenithal dependence of the 'impulsivity' of the SPA defined by $I/\Delta t$ where
Δt is the time difference between the start and the maximum of the event.

The essential physical significance of these zenith angle relationships are that they can be used to yield the value of h_0, the height where the optical depth is unity (and consequently related to the spectral structure of the incoming radiation), as well as the value N_r/N_0, N_r being the value of electron density necessary for reflection and $N_0 (= \sqrt{q_0/\alpha})$ is the maximum electron density for $\chi = 0$. One must, however, assume certain forms of distribution for the electron production rate and of α. When the X-ray layer is produced well below the pre-existing D layer, one can take advantage of the simplicity of the Chapman photo-ionization equation given by:

$$q(z, \chi) = q_0 \exp\left(1 - z - e^{-z} \sec\chi\right)$$
$$= q(z - \ln\sec\chi, 0°)/\sec\chi \tag{64}$$

which shows that the photo-ionization profile for $q(z, \chi)$ has the same shape as $q(z, 0°)$ but is moved upward by $\ln\sec\chi$ and reduced by $1/\sec\chi$. The implicit assumption that the X-ray ionizing flux can be represented by radiation of an effective wavelength λ_{eff} is a very approximate one, but as we shall see later, is an acceptable one and is, in fact, quite simply related to the spectral distribution of the X-rays. It is possible to refine such analysis by introducing a catalogue of a variety of X-ray spectral distributions and keeping an album of curves for $\Delta h + h_0$ vs $\sec\chi$, and of the optical depth τ_d as a function of height. One can then use these curves to decide which of these sets agrees with the observed $\Delta h - \ln\sec\chi$ curve, and can therefore, obtain information on the ionizing flux and/or of the loss rate.

CHAPTER 6

SOME OUTSTANDING SOLAR FLARE EVENTS

In this chapter we describe the principal ionospheric features of a few selected out-standing solar flare events. Some of these events will be examined in greater detail in subsequent chapters.

What constitutes an outstanding solar flare event can be a matter of some contro-versy. The choice can be different depending on the particular aspects of the wave and particle emissions that one considers – hard and soft X-ray bursts, EUV emissions, radio bursts of spectral types II, III, IV and V, microwave impulsive bursts, solar cosmic ray protons, heavier nuclei and electrons, corpuscular clouds and shock waves; or depending upon the geophysical effects they produce – the SIDs, the PCAs and the magnetic storms. Although every flare is not accompanied by all of these emissions, there are some, usually a major flare of importance 4B (or 3^+) which will produce all of these emissions and the entire sequence of geophysical events.

Dodson and Hedeman (1971) has compiled a list of 'major' flares occurring during 1955–1969 from all flares in solar cycles 19 and 20 (upto 1969) based on *five compo-nents*. When taken sequentially, it constitutes a crude profile of the flare. The sum of the five components was taken to give the *Comprehensive Flare Index*. The five quan-tities are:

(1) Importance of ionizing radiation as indicated by accompanying SIDs (scale 1–3).

(2) Importance of Hα flare (scale 1–3).

(3) Magnitude of ~ 10 cm flux (characteristic of log of flux in units of 10^{-22} W m^{-2} Hz^{-1}).

(4) Dynamic spectra (type II = 1, continuum = 2, type IV, with duration > 10 min = 3)

(5) Magnitude of ~ 200 MHz flux (characteristic of log of flux in units of 10^{-22} W m^{-2} Hz^{-1}).

A *major* flare was considered to be one that satisfied any *one* of the following criteria:

SWF (or SID) importance	$\geqslant 3$
Hα flare importance	$\geqslant 3$
10 cm flux	$\geqslant 500$ units
type II burst	
type IV radio emission, duration	> 10 min

Even so the total number of 'major' flares so determined during the period 1955–1969 was very large (1116 flares in all).

However, amongst these, some flares were great in all aspects. Such truly great flares, in which the *total index* was > 10, were much fewer – a total of 158 out of these 1116 'major' flares. These occurred with a frequency of about 20 yr^{-1} near maximum in cycle 19 (1957–59) and 10 yr^{-1} near maximum in cycle 20 (1967–69).

We are concerned primarily with the ionospheric effects, and consequently with those that have an ionospheric importance of 3 at the least. Such effects are many; the total number in this list amounts to approximately 300. The distribution of these events is shown in Figure 72. Of these, those events which have a total index of 13 or more, are marked with heavy lines.

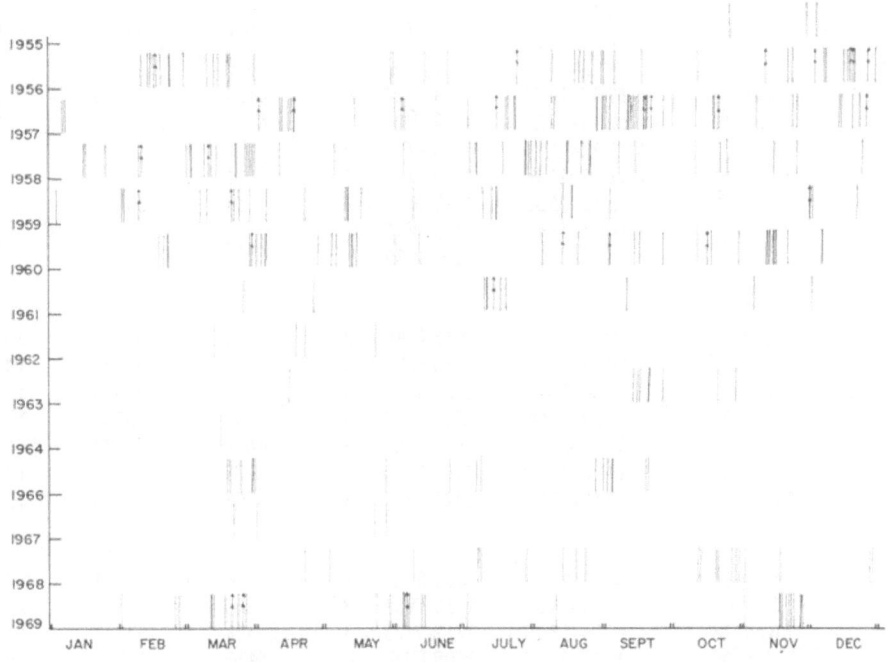

Fig. 72.   Distribution of 'major' SID events during 1955–1969.

Events with a total index of 13 or more, and with ionospheric index of 3, are listed in Table XII.

Unfortunately much of the ionospheric measurements concerning earlier flares in the list are too sketchy or are, in many ways, incomplete. To be of value, observations should exist on the ionizing radiation (at least X-rays) and on SIDs covering *both* the lower and the upper ionospheres. These considerations preclude in-depth portrayal of outstanding SID events before early 1960s. The ones selected for detailed description (and all of these are also proton events) are:

(1) July 7, 1966:   Dodson-Hedeman Index = 13
(2) May 21-25, 1967:   Dodson-Hedeman Index for May 23 = 16
(3) July 8, 1968:   Dodson-Hedeman Index = 17
(4) August 2–9, 1972:   Dodson-Hedeman Index ⩾ 16.

TABLE XII

Outstanding solar flares (index ⩾ 13 with ionospheric importance ⩾3), 1955–1969
(compiled from Dodson-Hedeman table)

Date	Time (UT) of flare or event	Position	Hα imp.	McMath plage	Profile abcde	Index
1956						
Feb. 23	0331–0510	N23 W80	3	3400	333–4	⩾13
1957						
Apr 17	2000–2300	N20 E69	3+	3941	33332	14
Jul 24	{ 1712–1801 1801–2025	S24 W27	3	4070	33332	14
Aug 28	0913–1404	S31 E33	3	4125	33332	14
Aug 31	1257–1455	N25 W02	3	4124	33333	15
Sep 11	0236–0722	N13 W02	3	4134	33333	15
Sep 18	{ 1722– 1815–2110	N23 E08 N20 E03	3+ 3+	4151	33232	13
Sep 19	0350–0555	N23 E02	3	4151	33333	15
Oct 20	{ 1637–1644 1644–1804	S26 W45 S26 W35	3+	4189	33332	14
1958						
Mar 3	1005–1411	S16 E60	3	4445	33334	16
Mar 23	0947–1445	S14 E78	3+	4476	33332	14
Jul 7	0020–0414	N25 W08	3+	4634	33333	15
Jul 29	0259–0408	S14 W44	3	4659	33332	14
Aug 16	0433–0831	S14 W50	3+	4686	33334	16
Aug 22	1428–1717	N18 W10	3	4708	33333	15
Aug 26	0005–0124	N20 W54	3	4708	33333	15
1959						
May 11	2006–2150	N10 E41	3	5148	33233	14
Jul 16	2114–2430	N16 W31	3+	5265	33333	15
Aug 18	1014–1350	N12 W23	3	5323	33232	13
Nov 30	0247–0356	N08 E16	2+	5476	32333	14
1960						
Feb 22	1352–1520	N08 E41	3	5581	33232	13
Apr 5	0215–0530	N12 W63	2	5615	32332	13
May 13	0519–0733	N30 W67	3	5654	33333	15
Sept 3	0037–0154	N18 E88	2+	5837	32333	14
Nov 11	0305–0428	N28 E12	2	5925	32334	15
Nov 12	1315–1922	N27 W04	3+	5925	33333	15
Nov 14	0246–0520	N27 W20	2+	5925	32332	13
Nov 15	0207–0427	N25 W35	3	5925	33433	16
Dec 5	1825–2350	N26 E74	3+	5959	33233	14
1961						
Jul 11	1615–2040	S07 E32	3	6171	33332	14
Jul 12	1000–1300	S07 E23	3	6171	33334	16
1963						
Sep 20	{ 2314– 2351–2601	N10 W09	2	6964	32334	15

Table XII (Continued)

Date	Time (UT) of flare or event	Position	Hα imp.	McMath plage	Profile abcde	Index
1966						
Mar 30	1241–1423	N28 E50	2n	8223	32234	14
Jul 7	0025–0135	N35 W48	2b	8362	32332	13
Sep 2	0542–0800	N24 W56	3b	8461	33313	13
Sep 4	0407–0579	N21 W87	3n	8461	33233	14
1967						
May 23	⎰ 1802– ⎱ –2200	N29 E25	3b	8818	33334	16
1968						
July 8	1708–1815	N13 E58	3b	9503	33335	17
Oct 29	1222–1300	S16 W12	2b	9740	32234	14
1969						
Mar 12	1739–1809	N12 W80	2b	9966	32333	14
Mar 21	1312–1430	N19 E09	2b	9994	32333	14
Jun 5	0952–1050	N12 E64	2b	10134	32333	14
Nov 18	1633–1756	N14 E40	2b	10432	32333	14
Nov 27	1928–2010	N16 W82	2b	10432	32224	13

6.1. Ionospheric Effects of the Flare of July 7, 1966

Principal details of this flare and the associated SIDs and the PCA events are summarised in Table XIII (A).

This flare had a Dodson-Hedeman Index of 13, and is, therefore, not as outstanding

TABLE XIII

Major features of four outstanding flare events

(A) July 7, 1966 event

Optical: 0025–0135/2b; position N35 W48; McMath Plage 8362

X-rays: NRL 0144–0150

0–3 Å $> 9.0 \times 10^{-4}$ erg cm^{-2} s^{-1}
0–8 Å $> 2.0 \times 10^{-2}$
8–20 Å $> 1.8 \times 10^{-1}$
Preflare values: 1.5×10^{-4} (0–8 Å)
$\qquad\qquad\qquad 1.5 \times 10^{-2}$ (8–20 Å)

Explorer 33   0023–0042–0353
(composite in three phases:
0023–0025, 0026–0035, 0035–0042)
2–12 Å: 0.191 erg cm^{-2} s^{-1} peak at 0042
6.4×10^{-3} preflare at 0000
Subsidiary peaks:
0035 UT   0.115 erg cm^{-2} s^{-1}
0219 UT 3.17×10^{-2} erg cm^{-2} s^{-1}
0257 UT 2.6×10^{-2} erg cm^{-2} s^{-1}.
(Figure 73a)

SCNA: Vertical: 0035 UT   dB 6.5 ± 1.3 (10 MHz), 2.0 ± 0.2 (18 MHz), 1.4 ± 0.2 (22.25 MHz)
absorption
at Boulder 0042 UT   dB 8.9 ± 1.7 $3.6 \pm^{7}_{0.5}$ $2.0 \pm^{7}_{0.3}$
(estimated by Donnelly, 1968c)

Table XIII (Continued)

SPA: 0025–0055/99
SFD: 0026–0027–0042/3.0   (Figure 18)
            Boulder: WWV 10–3.0

$N_e - h$ profiles: $\dfrac{\Delta N_e}{N_e} \times 100 \approx 50$ at 200 km   (Figure 73b)

Remarks: 1. 2–12 Å flux discussed by J. F. Drake, J. Gibson and J. A. Van Allen (1968)
          2. SFD, SPA, SCNA data analysis by Donnelly (1968c)
          3. *F*-region ionosonde study by Bhattacharyya and Balakrishnan (1967)
          4. Extensive observations reported in *Annals of the IQSY* 3, *The Proton Flare Project (1969)*

(B) May 23, 1967 event

A sequence of three consecutive flares

Optical: Flare 1: Start 1803 Max ~1817 UT   Imp 3b
         Flare 2: Start 1835 Max   1843–1845 Imp 3b
         Flare 3: Start 1932 Max ~1950   Imp 3b

Solar radio noise: 606 MHz: Simple 3   Start 1755.3 Max 1813.8 2.9 units at peak
(AFCRL, Sagamore Hill)

	Great burst Start 1835.8	Max 1839.5	534.0 units
		1948.5	19200 units
	Postburst increase	Start 2115 Max 2115	125 units

8800 MHz: Precursor   Start 1802.5 Max 1809.6 130.8 units
          Great burst Start 1835   Max 1839.7 8100 units
                                      1947   23000 units
          Postburst   Start 2027   Max 2027   109 units
          increase

X-rays: 2–12 Å: Three X-ray flares clearly detectable, as follows:
                  First: Start 1759 UT Max 1817 ~0.04 erg cm^{-2} s^{-1} peak
                  Second: Start 1834 UT Max 1846 ~0.65 erg cm^{-2} s^{-1} peak
                  Third: Start 1936 UT Max 1953 ~0.3 erg cm^{-2} s^{-1} peak
                  Preflare flux $= 7 \times 10^{-3}$ erg cm^{-2} s^{-1}
                  (Figure 75a)

Ionospheric effects: Incoherent scatter: Peak enhancement (94–304 km)
                       $= 2.34 \times 10^{16}$ el m^{-2} (total content)
                       200% at 94 km
                       25% at 190 km
                       Beacon satellite: N_t peak enhancement: 4.4×10^{16} el m^{-2} (Stanford).
                       (Figure 114)

          SPA: 1840–1852–1943/99 deg: Sao Paulo, Brazil–GBR   16 kHz.
          SCNA: 1809–1819/34% absorption
          SFD: 1808–1809–1820/1.9 Hz Boulder–WWI 8 MHz
                 1835 –   1840/16.4 Hz Boulder–WWI 13 MHz
                 (Figure 75c)

Remarks: Comprehensive data published in WDCA Report UAG-5, Feb. 1969
          2–12 Å X-ray observations discussed by Van Allen (1968)
          Sagamore Hill radioburst data discussed by Castelli *et al.* (1967)
          Incoherent scatter $N_e - h$ profiles discussed by Thome and Wagner (1971) and
          Garriott *et al.* (1969)
          Beacon satellite N_t variations by Garriott *et al.* (1967)

Table XIII (Continued)

(C) July 8, 1968 event

Optical:	Two consecutive phases 1708–1815/3b: Position N13 E58, McMath Plage 9503
X-rays: 2–12 Å data:	Start 1707: Max 1723. Peak flux 0.4 erg cm^{-2} s^{-1} e-folding time 27 min. Three precursor flares with onset times 1520, 1559, 1630 UT With peak fluxes of 4×10^{-3} erg cm^{-2} s^{-1} for first two, and 9×10^{-3} erg cm^{-2} s^{-1} for third Preflare background flux at 1520 UT $= 2 \times 10^{-3}$ erg cm^{-2} s^{-1}.
1–8 Å data:	Start 1703 $\pm$ 1 min. Max 1712. Peak flux 0.49 erg cm^{-2} s^{-1}, remained practically constant till 1744. After 1744 decay e-folding time 17 min.
8–20 Å data:	Start 1707 $\pm$ 1 min. Estimated spectral distributions: Figure 76b.
Radio burst:	Sagamore Hill: 8800 MHz Start 1704.9 Max 1711.3 5100 units peak

Ionospheric effects:	(a) *Accompanying Principal Flare at 1707*
	SPAs: All start within 1 min after 2–12 Å flux
	        $\Delta h(\chi = 0°) = 16$ km.
	SCNA: Start (Boulder) at 1709 Max: 1715 UT (18 MHz)
	                                       1718 UT (30 MHz)
	        Corrected ΔA (vertical path): 15.5 dB (30 MHz),
	                                                  5.5 dB (18 MHz)
	SFD: Start 1707 Max. 1710/9.5 Hz Boulder–WWI, 11 MHz
	$N_e - h$ profiles: Figure 33a.
	
	(b) *Accompanying Precursor at 1630 UT*
	SPAs: All started 1630. Max in some cases preceded 2–12 Å peak
	        $\Delta h(\chi = 0°) = 3.8$ km.
	SFD: Start 1632 Max 1634/0.4 Hz Boulder–WWI 9 MHz

Remarks:	Ionospheric effects studied by Sengupta (1971) $N_e - h$ profiles for D-region with partial reflection equipment by Montbriand and Belrose (1972).

(D) August 1972 events

A whole series of severe interplanetary and terrestrial disturbances.
Four largest are marked: $F1$, $F2$, $F3$, $F4$ (Figures 77a, b). McMath Plage: 11976.
Optical/Radio/X-ray effects:

		10 cm burst		X-ray flare (1–8 Å) estimated (erg cm^{-2} s^{-1})	Optical		Proton burst (Earth)			Peak flux 10 MeV (protons cm^{-2} s^{-1} ster^{-1})
		Rise time (min)	Peak flux units		Imp.	Long.	Onset	Delay	Max. rigidity	
F-1	Aug. 2 0310UT	55 (slow)	2600	0.15 (0.3–0.1)*	3n	E34	2/10	7h slow	0.07 GeV soft	> 76
F-2	Aug. 2 1959UT	109 (slow)	9700	0.18 (0.2)*	2b	E28	3/03	7h slow	0.07 soft	> 1070
F-3	Aug. 4 0619UT	17 (fast)	7600	> 0.5 (1–2)*	3b	E08	4/08 [b]	< 1.7h fast	1.5 GLE	1.1×10^6
F-4	Aug. 7 1505UT	23 (fast)	4500	> 0.5 (0.8–4)*	3b	W37	7/1540	0.6h fast	3 GLE	3530

Table XIII (Continued)

Ionospheric effects: SITEC (17 stations, global) $F4$ ΔN_T from 1.8 to 8.6×10^{16} el m^{-2}
              (Figure 37)   (15–30% range)
              Incoherent scatter $F4$ $\Delta N_e - h$ from 125–1200 km
              (Millstone Hill) $\Delta N_e/N_e = 100\%$ at 125 km
              (Figure 39)   = 60% at 200 km
                                                = 20% at 300 km
                                             = 12.5% at 525 km
                            ΔN_T (upto 2000 km)$= 3.8 \times 10^{16}$ el m^{-2} (30%)
                            $\Delta T_e \approx 400$K (ΔT_i negligible)

Dramatic upward surge preflare drift negative (downwards) with little height dependence; during flare drift increased nearly linearly with height.

Remarks: (i) Event selected for special study by URSI and IAGA. Observational data compiled by WDCA: UAG 28 (Parts I, II, III), 1973.
         (ii) Starred values of X-ray flux are estimates made by Hakura *et al.* (1973) from SPAs.
      (iii) SITEC effect on Aug. 7 discussed by Mendillo *et al.* (1974)
              Incoherent scatter effect on Aug. 7 discussed by Mendillo and Evans (1974)

---

as those of July 8, 1968, May 23, 1967 or August 7, 1972, for which total index was 16 or more.

The X-ray observations by Van Allen and his group (Drake *et al.*, 1968), reproduced in Figure 73a show the onset of the X-ray flare (in 2–12 Å and observed on satellite Explorer 33) at 0023, reaching its maximum intensity at 0042. There is a small but significant pause in the increase of intensity at 0025 and a more prominent one at 0035. Thus the increase of intensity to its maximum value was a composite of three phases:

        0023–0025
        0026–0035
        0035–0042.

The increase during the second phase, from 0026 to 0035, may be represented by $[1 - \exp(-t/3.8 \text{ min})]$ with $t = 0$ at 0026. The decline of the excess flux from 0042 to 0200 was accurately exponential in form with an e-folding time at 36 min. Further decay after 0217 was complex, but had a similar e-folding time. Some specific absolute values are:

Preflare flux at 0000 UT of July $7 = 6.4 \times 10^{-3}$ erg cm^{-2} s^{-1}.

Maximum flux at 0042 UT $= 0.17$ erg cm^{-2} s^{-1}.

Time integral of flux from 0023–0345 UT $= 560$ erg cm^{-2}.

Integral X-ray (2–12 Å) energy $= 8.1 \times 10^{29}$ erg.

Solar radio emissions observed during this event, as observed in Mitaka, Toyokowa and Hiraiso in Japan on frequencies 17000 MHz, 9400 MHz, 3750 MHz, 2700 MHz, 2000 MHz, 1000 MHz, 500 MHz and 200 MHz are also shown in the same diagram as the X-rays. The peak radio burst occurred a few minutes earlier than the X-ray burst; also the major peaks occurring in the solar radio emission at frequencies less than 2700 MHz around 0120 UT were not accompanied by any X-ray event (or any ionospheric event).

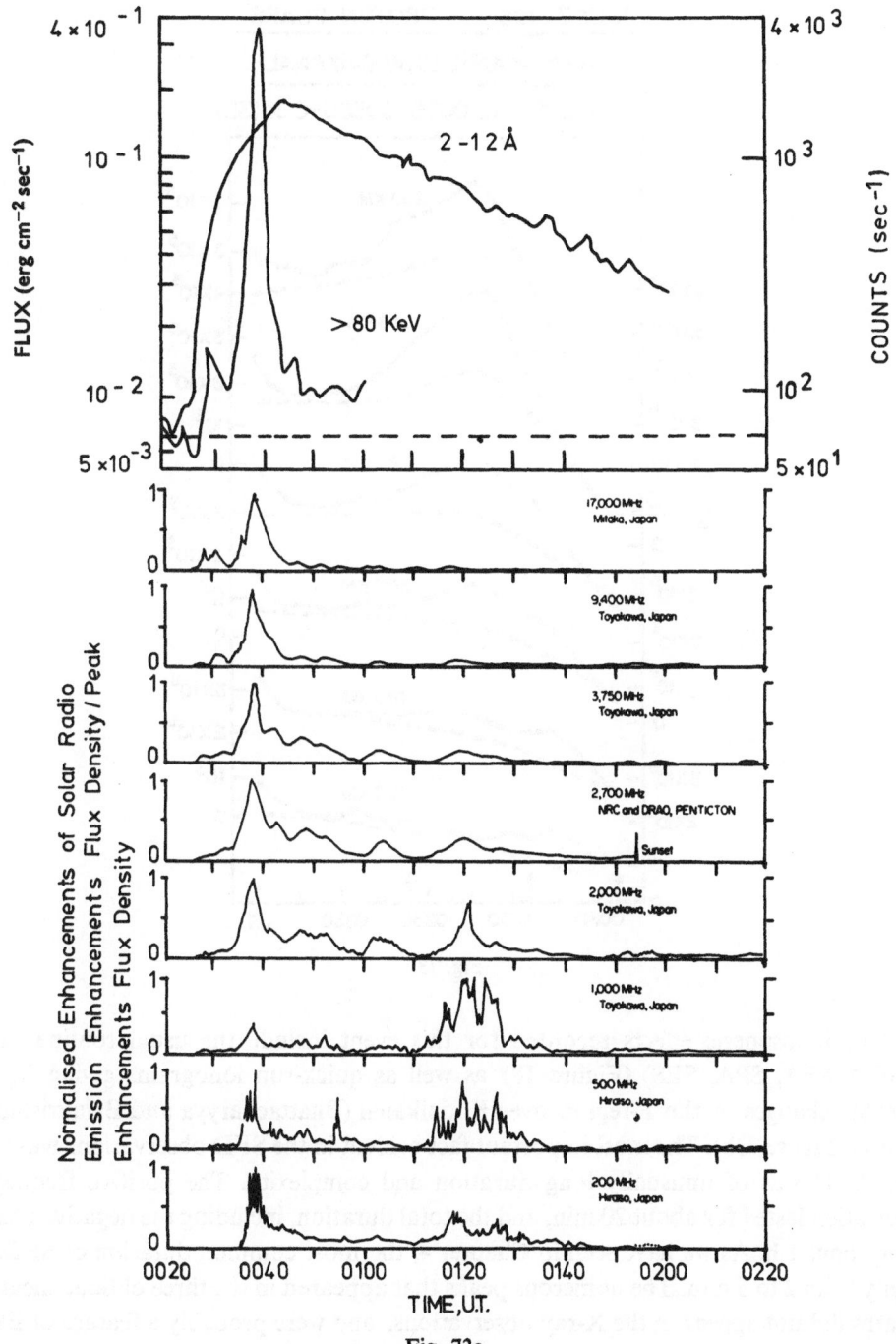

Fig. 73a.

Fig. 73a–b. Major features of the SID events, along with radio and X-ray enhancements for the outstanding solar flare event of July 7, 1966. (a) Shows the 2–12 Å X-ray enhancements recorded by the University of Iowa (after Van Allen, 1967), the influx of high energy particles (> 80 keV) and the enhancements in solar radio flux density recorded in Japan and Canada (reproduced from Donnelly, 1968c). (b) Depicts the $N_e - h$ profiles obtained from quick-run ionograms by Bhattacharyya and Balakrishnan (1967). SPAs recorded at Boulder for this event are shown in Figure 18.

JULY 7 , 1966   PROTON FLARE

Ne-h PROFILES, KODAIKANAL

Hα FLARE 0025-0038U-0135(3B)

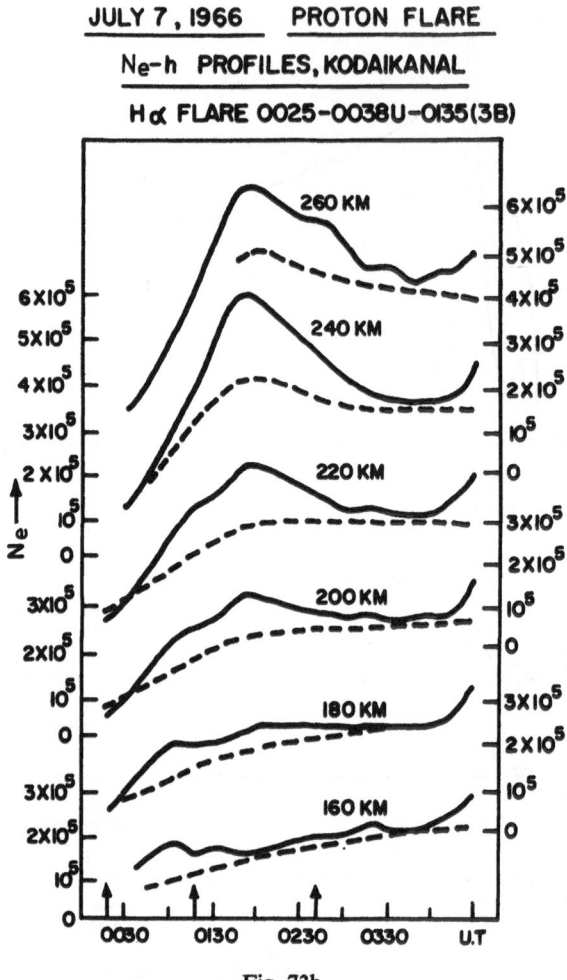

Fig. 73b.

The ionospheric effects recorded for this event include the usual routine SIDs (SFD, SEA, SPA, SES) (Figure 18) as well as quick-run ionograms giving $N_e - h$ profile changes in the F-region over Kodaikanal (Bhattacharyya and Balakrishnan, 1967) (Figure 73b). The most important fact evident in the SFD observations was that the SFD was of unusually long duration and complexity. The positive frequency deviation lasted for about 20 min, and the total duration, including the negative phase, for about 1 h. As we have seen in Chapter 4, the most common duration of SFDs is only from 2 to 5 min. The numerous peaks that appeared in the three oblique incident paths did not appear in the X-ray observations, and were probably a feature of EUV fluxes for which measurements did not exist.

Shortwave fadeouts, sudden cosmic noise absorption and sudden phase anomalies all corresponded to the first X-ray event. No pronounced ionospheric effects were evident at or near 0120 UT at the time of large peaks in solar radio emission below 2700 MHz; nor can one see any feature resembling the second peak in X-ray at 0257, (excepting in the 13.6 kHz Haiku to Boulder path, which was, however, not repeated on 10.2 kHz). Presumably, the ionospheric relaxation time in the lower ionosphere does not permit reproduction of such fine variations in the ionizing flux. For the observed SCNAs and SWFs, Donnelly (1968c) estimated the absorption for a vertical path over Boulder after making appropriate corrections for the antenna aperture (Chapter 2, Section 5); these values are 8.9 ± 1.7 dB for 10 MHz and 2.0 ± 0.3 dB for 22.25 MHz (see Table XIII (A)).

6.2. Ionospheric Effects of the Solar Flares of May, 1967

Solar activity charts for the entire year of 1967 and specifically for May 1967 as prepared by Japanese scientists are presented in Figures 74a and 74b respectively. The large activity during May and the very large burst of solar protons during this period ($E \geqslant 10$ MeV on board Explorer 34 satellite) are noticeable.

The year 1967 began with a series of large events. As early as January 28 and 29 neutron monitors at ground level indicated that a proton event had occurred, but optical correlation was questionable. A 4B flare accompanying type IV (meter wavelength was observed on February 13, but no PCA followed. On February 27 the microwave radio emission accompanying the flare of importance 2B–3B was very intense, but no proton event was recorded. On March 11 large optical and microwave bursts were observed and an increase in proton flux was detected in satellite. However, the first decisive correlation between PCA (proton events) and radio/optical data in 1967 occurred during the last ten days of May.

A large and concentrated region of sunspots of at least three separate groups located approximately at N26 produced flares of importance 1 or larger from May 18 through 28, 1969. On May 23 a series of three solar flares was observed and recorded in Hα light starting approximately at 1803, 1834 and 1932 UT. The second flare was observed in white light, a rare event in the whole history dating back to 1858. There have been only about two dozen flares observed in white light till the May event.

The radio, X-ray, ionospheric and particle aspects of the flare were so great that special studies were conducted on the effects of this flare, and the World Data Centre compiled a report on the data of the event of May 23, 1967, the largest during this period (WDC-A Report UAG-5).

Of the major events beginning May 21, the following exceeded Dodson-Hedeman total index of 10:

May 21	1919–2025 UT/2b	N24 E39	12
May 23	1808–2200 UT/3b	N29 E25	16
May 25	1039–1225 UT/1b	B25 W06	12
May 28	0529–0700 UT/2b	N28 W32	12

Ionospheric observations of a wide variety, including some with sophisticated systems are available for the first three flares.

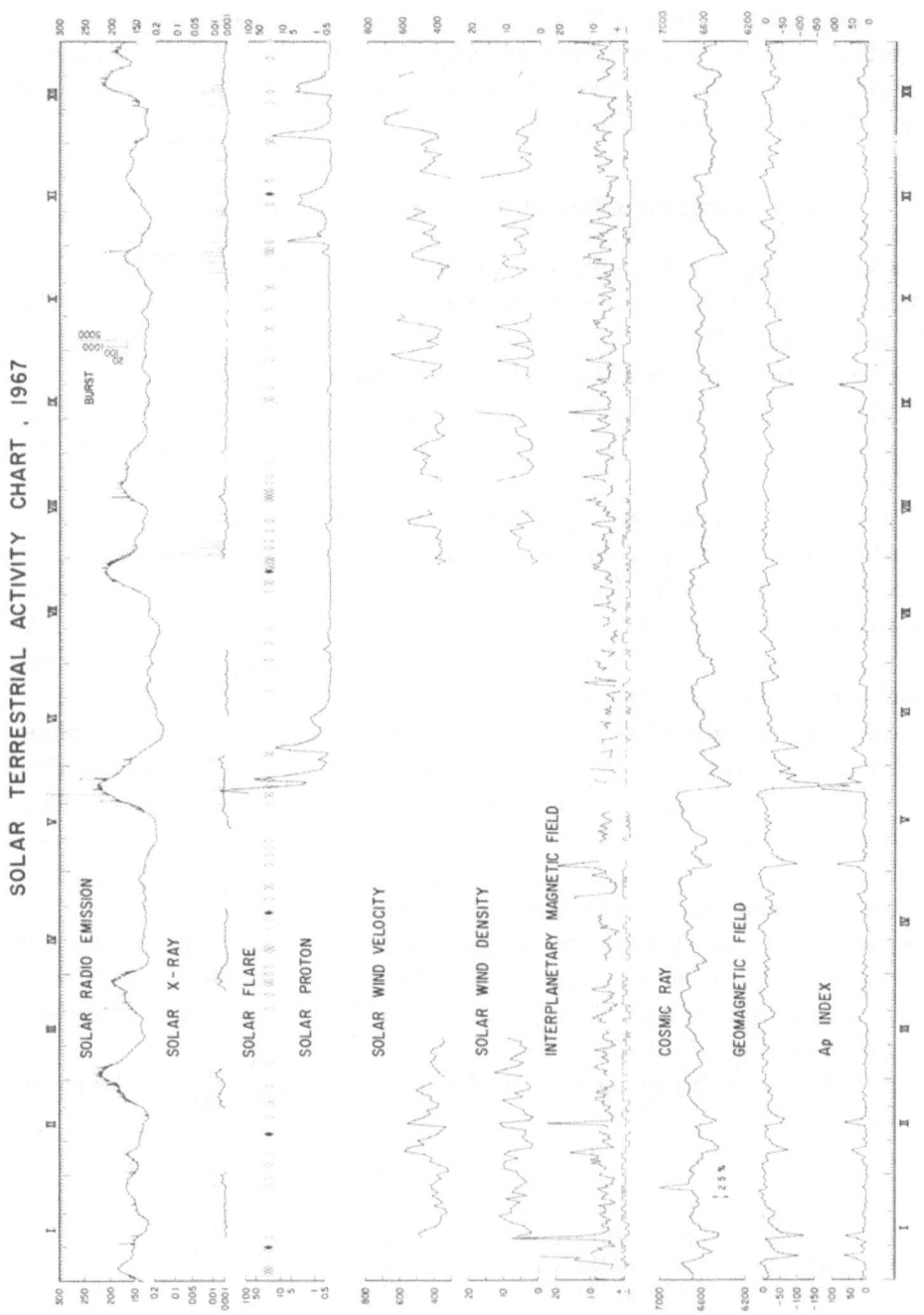

Fig. 74a–b. Solar activity charts for the entire year 1967 (a) and for May 1967 (b) (from Solar Terrestrial Activity Chart of Japan).

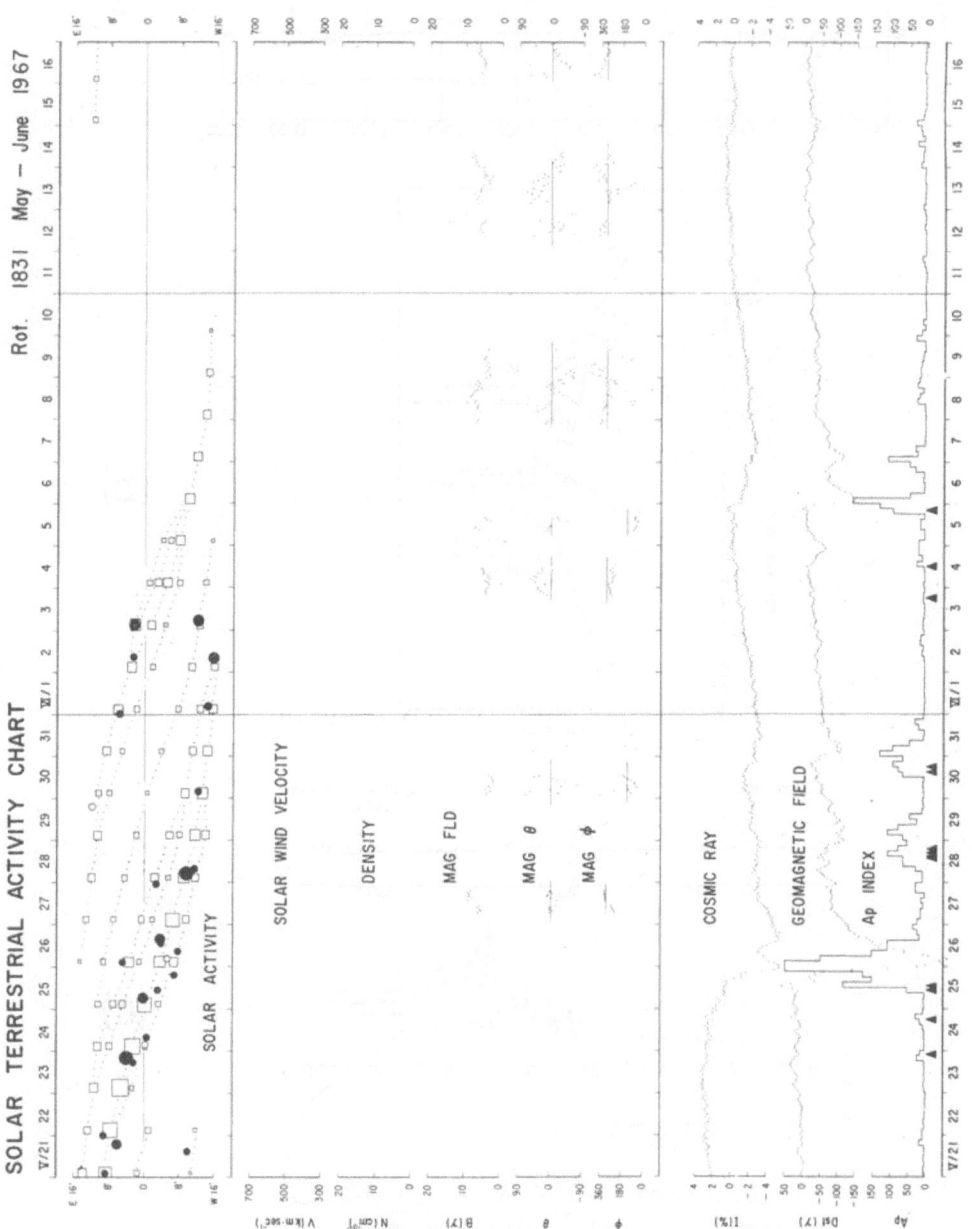

Fig. 74 b.

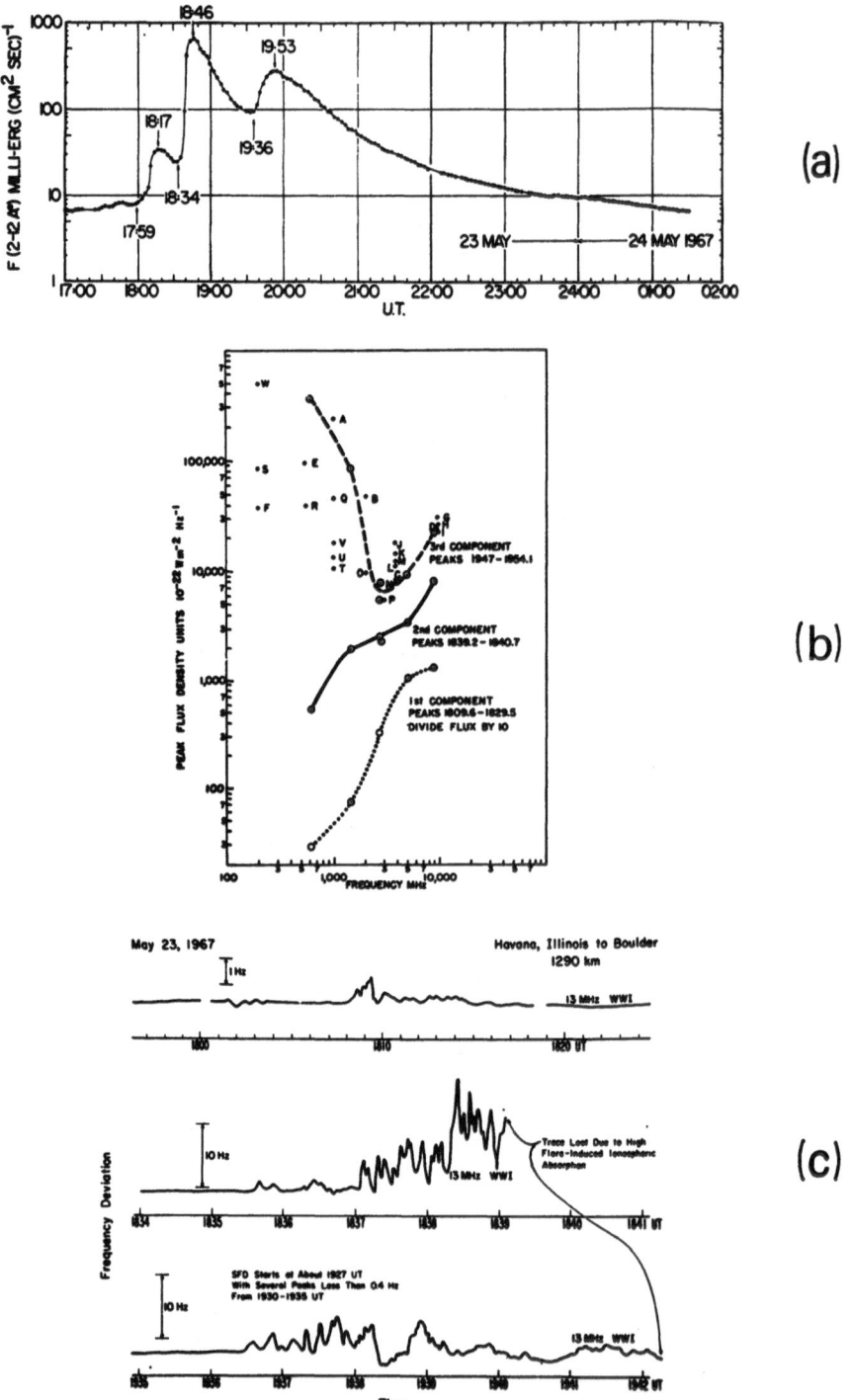

Fig. 75.   (a) 2–12 Å X-ray enhancement for the outstanding solar flare event of May 23, 1967 from the University of Iowa (after Van Allen, 1969), (b) the *U*-type radiospectrum for this event recorded at Sagamore Hill, U.S.A., (c) SFD observations at different frequencies at Boulder (after Donnelly, 1969c). A comparison of the bottomside (from incoherent scatter) and topside electron content is given in Figure 114.

The one beginning at 1808 on May 23 was an unusually large flare, with Dodson-Hedeman total index of 16. Apart from the fact that this flare was accompanied by some of largest radio, optical, X-ray and ionospheric effects, the time variation of the flare itself was somewhat unusual. It consisted of a sequence of three closely spaced flares, all three classified 3b optically, and with onset times at 1803, 1835 and 1932 respectively. The second flare was accompanied by unusually large X-ray flux enhancement (0.65 erg cm^{-2} s^{-1} in 2–12 Å band) (Figure 75a). This was larger than the X-ray enhancements during July 7, 1966 (0.19 erg cm^{-2} s^{-1}) and July 8, 1968 (0.4 erg cm^{-2} s^{-1}) and has only been exceeded during August 1972 events.

The radio events were also unusual. Prior to the events, the microwave solar emission had been lower than at any time in 1967, with a minimum on May 11–13. Nine consecutive days passed before May 18 without a single burst. On May 18, 19, 20, flare – associated burst activity began to increase, reaching a temporary 'high' on May 21: a complex burst was observed around 1922 UT in McMath Plage 8818. A type IV decameter event, lasting for one hour from 1922 to 2034 UT, was observed over the frequencies 19–41 MHz. This was a 'great' burst (≈ 1000 flux units). The following day was unusually quiet. On May 23 things broke loose. There were three bursts, approximately coinciding with the three optical flares, with maxima at 1809.6 UT, 1839 UT and 1947–1954 UT (depending on the frequency). *The greatest radio burst* was third peaking at 1947 UT (with flux of 23 000 units on 8 800 MHz); while the X-ray flares were the largest for the second flare, and so were the ionospheric events. A special characteristic of the frequency distribution obtained with 5 frequencies ranging between 600 and 8800 MHz in Sagamore Hill was that the spectrum was the *U*-type (Figure 75b) i.e. the high flux density in the meter and centimeter range and lower flux density in the decimeter range. This is a characteristic that is found to accompany most white-light flares.

As mentioned before, the ionospheric effects associated with the *second* of the May 23 flare sequence were unusually large. More important is, however, the fact that during this event for the first time coordinated ionospheric observations were made for the *F*-region with the incoherent scatter radar in Arecibo (Figure 38) and with beacon transmissions from Geostationary Satellite ATS (Figure 35b). The incoherent scatter results showed that ionization enhancement was as large as 200% at 54 km and at 190 km was still as large as 25%. The integrated electron content during the peak of the event was 2.34×10^{16} el m^{-2}. The total electron content as measured with the beacon satellite ATS, was considerably larger (4.4×10^{16} el m^{-2}), showing that the contribution above 300 km was quite considerable (Figure 114). The SFD, shown in Figure 75c was the largest observed since Boulder started measuring SFDs (= 16.4 Hz), and had no distinct negative stage, implying that the 10–1030 Å radiation remained at a relatively high level after its initial fast increase. Both the second and the third events showed a chain of smaller fast bursts; the third event also occurred before the second event ended. Some of the other SIDs are given in Figure 35b.

There were some curious features in the VLF/LF observations made by the Boulder scientists. While the *phase* disturbances for the three flares were in the ratios 1, 3 and 1, the amplitude enhancements were in the ratios of 1, 0.2 and 0.1 or smaller.

6.3. Ionospheric Effects of the Solar Flare of July 8, 1968

This flare consisted of two phases and was caused by McMath Plage 9503. The Dodson-Hedeman total index for this flare was 17. This flare had two consecutive phases as can be seen from the 2–12 Å flux measurements from the University of Iowa (Figure 76a). The main flare occurred at 1707 and the precursor at 1630 UT. The peak flux of this flare was 0.4 erg cm^{-2} s^{-1} which was somewhat lower than that for May 23, 1967. Three precursor flares with the onset times of 1520, 1559 and 1630 UT with peak

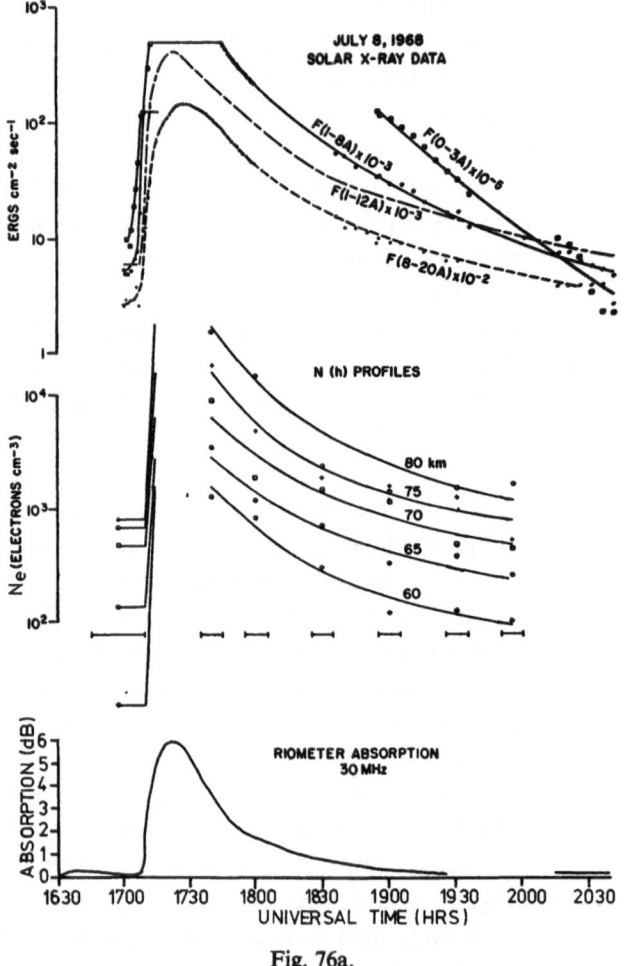

Fig. 76a.

Fig. 76a–c.   The outstanding flare event of July 8, 1968. (a) 2–12 Å flux enhancements from University of Iowa; ionization enhancements in the *D*-region (after Montbriand and Belrose, 1972). (b) Spectral distributions in X-ray flux below 12 Å radio emissions. (c) SIDs recorded for this event in U.S.A. by several observatories (after Sengupta, 1971).

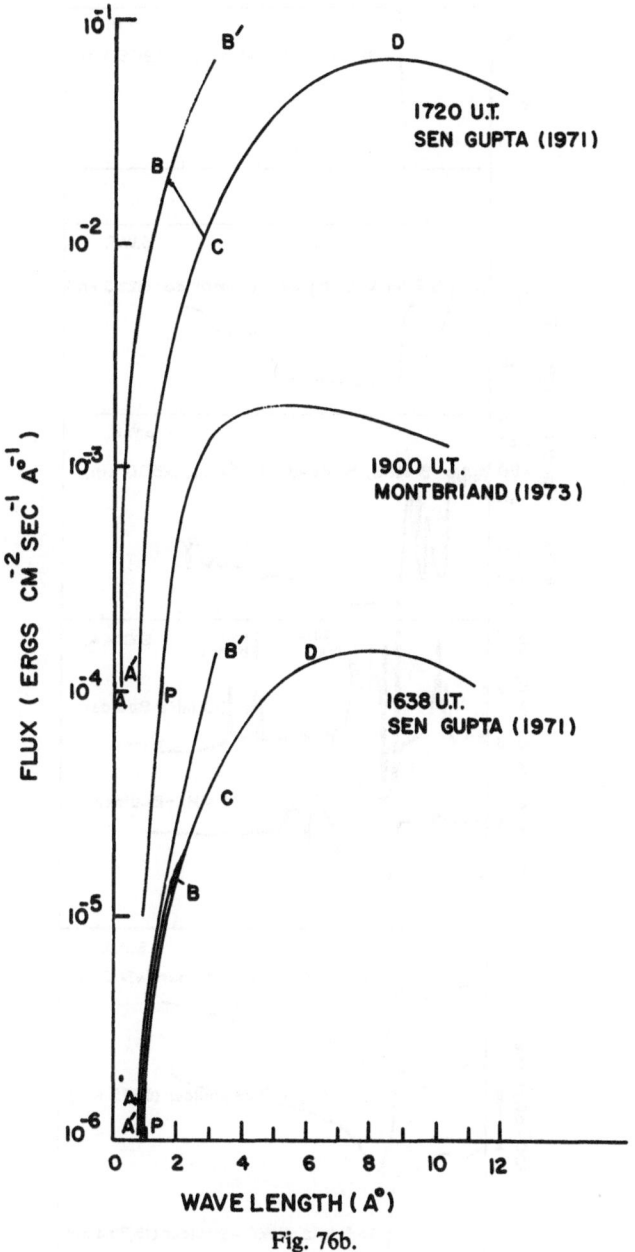

Fig. 76b.

fluxes of about 4×10^{-3} erg cm^{-2} s^{-1} for the first two and 9×10^{-3} erg cm^{-2} s^{-1} for the third were observed. Ionospheric effects accompanied only the last one. Measurements are also available of the NRL experiment on Explorer 37 on 1–8 and 8–20 Å. 1–8 Å flare started at 17.03 ± 1, 4 min earlier than 2–12 Å flare. This reached a value of 0.49 erg cm^{-2} s^{-1} at 1712 and remained apparently constant upto 1744. The flux values between 1712 and 1744 are, therefore, uncertain. The somewhat apparently larger value in 1–8 Å wavelength is, according to Sengupta (1971), due to the assumption

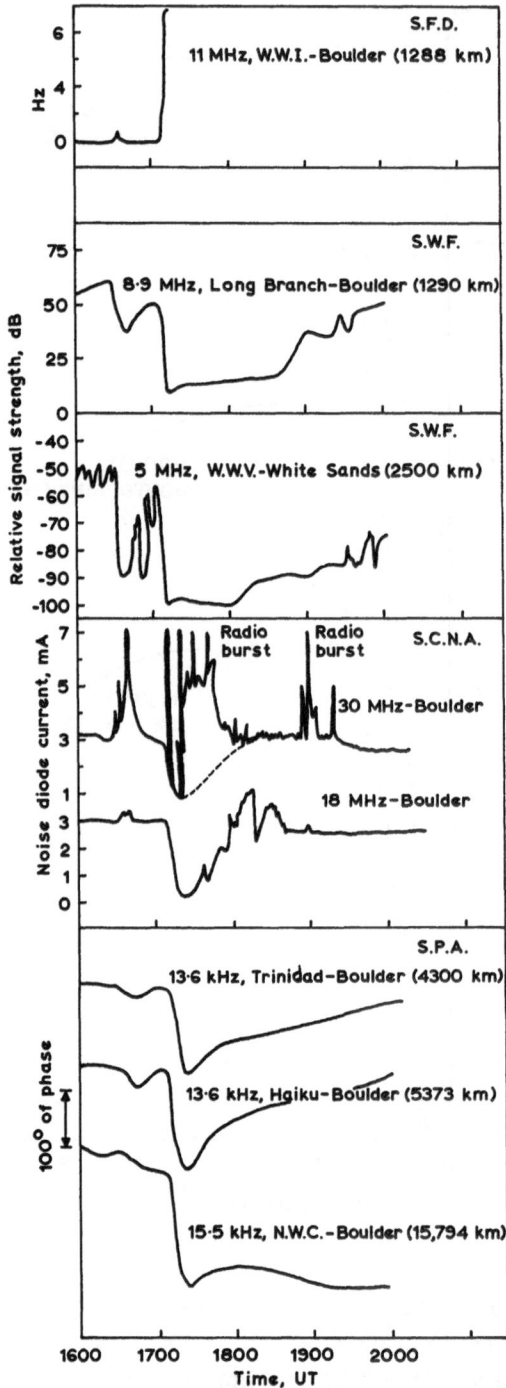

Some of the SID's associated with the flares.

Fig. 76c.

of a lower effective temperature for the Sun than obtaining during the flare. The spectral distribution of the solar flux has also been computed for this flare. Computations made by Sengupta for 1–12 Å and by Montbriand for 1–8 Å are given in Figure 76b.

The accompanying SIDs were large for the principal flare beginning at 1707 h. Most of the SPA events recorded started within 1 min after the 2–12 Å flare but reached peak values before the X-ray flare. On the other hand, the SPAs accompanying the third of the precursor flares beginning at 1630 h started simultaneously with the 2–12 Å flare. In 3 cases the maximum phase advance occurred at the peak of 2–12 Å flare, while in 6 other cases the SPAs preceded X-ray flare by 2 min. Sengupta (1971) showed by plotting the SPAs in terms of the lowering of the reflection height against the corresponding average $\sec \chi$ values over the propagation path during the flare, that there is a linear relationship between $\log \sec \chi$ and Δh. The intercepts give the reflection height Δh_0 that would be observed at the sub-solar point; these were found to be 16 km for the principal flare and 3.8 km for the precursor flare.

While no observations are available for the F-region electron density profiles, for the D-region Montbriand and Belrose (1972) have reported observations of changes of the $N_e - h$ profiles during the flare. These are shown along with the X-ray observations in Figure 76a.

6.4. Ionospheric Effects of August 1972 Flare Events

The first ten days of August 1972 witnessed a series of catastrophic solar flare events that have aroused great interest for many reasons. The events occurred during the declining phase of the solar cycle and were the largest recorded since November 1960. The principal region McMath 11976 which caused these flares was outstanding in solar cycle 20 because of the complexity of its magnetic field as well as because of its high flare productivity. Observations were available from spacecrafts, balloons, lunar based instruments and rockets as well as from ground based observatories and included X-ray, γ-ray and energetic charged particle emission measurements, radio emission from the sun and measurements of the ionospheric, cosmic ray, geomagnetic and other solar-terrestrial effects.

The Solar Activity Chart for these events is given in Figure 77a.

Four principal events marked $F1$, $F2$, $F3$ and $F4$ can be identified. These are discussed and outlined in Table XIII (D). The first two occurred on August 2 beginning at 03 10 and 19 59 h, the third occurred on 4 August at 06 19 and the fourth on August 7 at 15 09 UT (see also Figure 77b).

The flares were accompanied by major radio outbursts with complex features spread over a wide frequency range. An important feature was the absence of type II bursts in the microwave range excepting for the event of August 7 (Figure 77c). All the four events are classified as major type IV outbursts which have been pointed out before, are closely connected with the proton ejection from the Sun. There were, however, major differences in spectral features between the first two flares and the last two. In the first two, the radio flux in the outbursts peaked in the cm wavelength region, and their development was slow and gradual. The risetime to peak at 10 cm was 55 min and 109 min for F1 and F2 flares respectively. On the other had, the last two outbursts showed one peak in the millimeter and a second peak in the decimeter wavelengths, forming the so called *U*-shaped spectrum (see May event). The peak values at 10 cm were reached within 20 min. During the initial phase of these flares, the emission of γ-ray lines were observed by satellite OGO-7. This probably implies the production of solar protons starting at the very beginning of the flares.

The X-ray flux enhancements for all the flares were large. The flux values for the second group of flares were especially large, of the order of 1 erg cm^{-2} s^{-1}.

The ionospheric effects associated with these flares were very large and have been reported from all over the world. The most important observations for the purpose of this volume are, however, the observations made with satellite radio beacon trans-

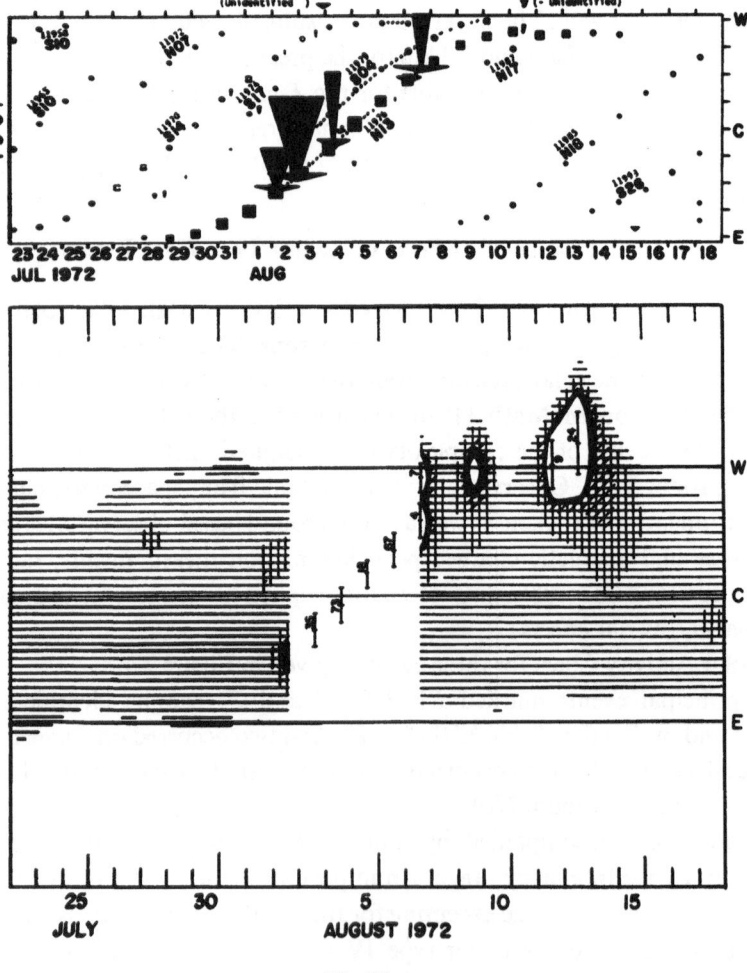

Fig. 77a.

Fig. 77a–c. The outstanding solar flare event of August 7, 1972. The solar activity for this event is shown in (a), the sequence of August flares in (b), frequency distribution of radio bursts in (c), changes in total electron content for this event are shown in Figure 37 and Figure 39. Figures a, and c are from World Data Centre A (Boulder) Report UAG-28.

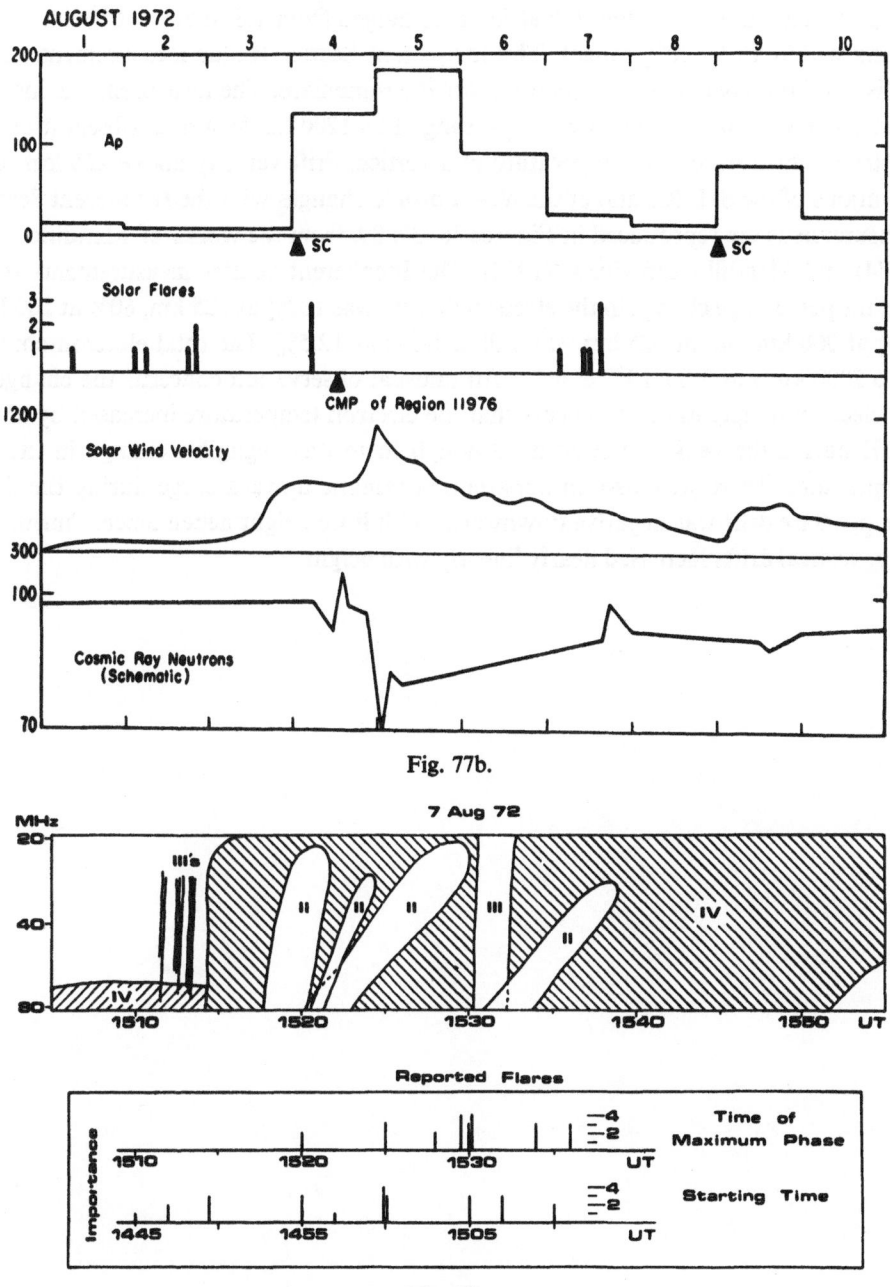

Fig. 77b.

Fig. 77c.

missions with the ATS satellites during the August 7 flare event and observations made with Incoherent Scatter Radar at Millstone Hill and in Chatanika, Alaska. Faraday rotation observations have been reported from 17 stations from North America, Europe and Africa spanning over 10 h in local time and over 70° in latitude (Mendillo *et al.*, 1974). These constitute the first truly global morphology of a flare induced

F-region. The sizes of the individual SITECs ranged from 1.8 to 8.6×10^{16} el m^{-2} (all falling within 15 to 30% range). The Incoherent Scatter Radar measurements from Millstone Hill have also been reported for the same flare. The measurements include electron density profiles over the height range 125–1200 km as well as observations of electron temperature, ion temperature and vertical drift velocity above 225 km. Observations of the SITECs and of the $N_e - h$ profile changes with the Incoherent Scatter measurements are reproduced in Figures 36, 37, 39, from the works of Mendillo *et al.* (1974) and Mendillo and Evans (1974). The Incoherent Scatter measurements show that the percentage change in the electron density was 100% at 125 km, 60% at 200 km, 20% at 300 km and at 525 km was still as large as 12.5%. The total electron content upto 2000 km was 3.8×10^{16} el m^{-2}. An unusual observation concerns the change in the electron temperature. It appears that the electron temperature increased by about 400 K during the peak of the event although there was negligible change in the ion temperature. There was also an apparently dramatic upward surge during the flare. The pre-flare drift was negative downwards with little height dependence. During the flare, vertical drifts increased nearly linearly with height.

CHAPTER 7

DEVELOPMENT OF ELECTRON DENSITY PROFILES
FROM SYNOPTIC SIDs

While detailed profiles of ionization are, and have in fact been obtained, for the *D*-region during solar flares with what one may call '*direct methods*', such as the use of wave interaction technique as in Penn State, partial reflection as in Ottawa, incoherent scatter and rocket soundings, (discussed in Chapter 2 and in some of the later chapters), these are necessarily limited. Consequently, there have been several attempts to use some of the more conventional SID techniques for profile studies. These are discussed in this section.

Two approaches are used. In one (and this has been used with VLF observations on 16 and 70 kHz over distances of 90 and 100 km for a flare of 7 October 1948) a search is made for electron density distribution whose calculated reflection properties are in agreement with observations; in May's calculations (May, 1966) these profiles were obtained with the aid of full wave integration technique. In another approach, assumptions are made regarding the shape of the profile and the constants are derived from the observed data. This procedure was adopted by Ondoh and Kishida (1966) who assumed a layer shape of the form $N_e(h) = a(h-h_0)^2$, where 'a' is a constant and h_0 the base-level of the enhanced layer. Ondoh and Kishida used SCNAs at different frequencies. This approach has been used more extensively by Deshpande and Mitra (1972c), and is discussed below.

7.1. Profiles from SCNA Measurements

When electron density distribution is represented by

$$N_e(h) = a(h - h_0)^2 \tag{65}$$

the absorption $A(f_e)$, suffered by the cosmic radio noise signal at an effective frequency f_e ($f_e = f \pm f_L$) in passing through the lower ionosphere is given by

$$A(f_e) = a\left[\int_{h_0}^{90} k_0 (h - h_0)^2 \, dh + \frac{2\pi e^2 v_{90}}{\omega_e^2} \times\right.$$

$$\left.\times \int_{90}^{\infty} \exp\left[-0.18(h - 90)\right] (h - h_0)^2 \, dh\right] = aK(f_e, h_0). \tag{66}$$

The integral K is a function of the effective frequency f_e on which absorption is measured and also of the base level h_0. k_0 is the absorption per unit path length for unit electron density at the altitude h and should be computed with the generalized magneto-ionic theory of Sen and Wyller (1960). Above 90 km, k_0 is replaced by the collision frequency model $v_{90} \exp[-0.18(h-90)]$, where v_{90} is the collision frequency at 90 km. ω_e is the effective angular frequency $(=2\pi f_e)$. Under flare conditions, when the absorbing layer shifts to the lower levels, the heights of maximum absorption for radio waves at different frequencies are often well resolved. In this situation it is possible to determine the base level h_0 as well as the constant 'a' for the profile from the ratio of absorption at a pair of frequencies. The ratio is a function of the base level h_0 and is given by:

$$\frac{A(f_{e1})}{A(f_{e2})} = \frac{K_1}{K_2} = f(h_0). \tag{67}$$

Thus the level where theoretical ratio K_1/K_2 equals the ratio of measured absorption at these frequencies is taken as the base level h_0. The constant 'a' is then evaluated from Equation (66) with this value of h_0 and from the observed absorption at any one frequency.

In Figure 78 (a) values of K calculated from Equation (67) are shown as a function of h_0 for the effective frequencies of 20, 22.4, 30 and 60 MHz. The base levels are taken from 30 to 70 km. The collision frequency distribution of Thrane and Piggott (1966) is used. The ratio K_1/K_2 for the pair of frequencies 20/30 and 22.4/30 MHz are shown against h_0 in Figure 78b.

When SCNA measurements are available at more than two frequencies, all the pairs are considered. Resulting profiles are not entirely identical and a compromise profile is obtained. K ceases to be a function of the base level h_0 for frequency exponent $\to 2$; and consequently, for such conditions absorption values cannot be used for the construction of ionization profiles.

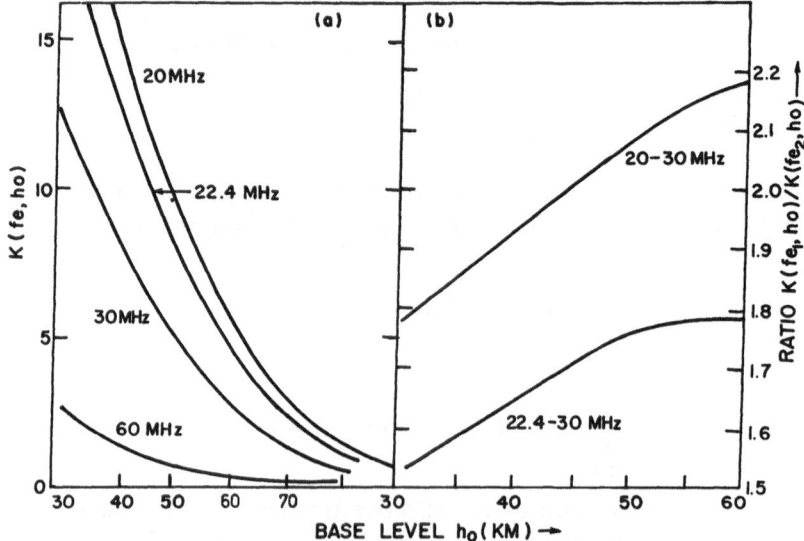

Fig. 78.   (a) Absorption integrals K; and (b) Ratio of absorption integrals K_1/K_2 for a pair of frequencies vs baselevel h_0 (after Deshpande and Mitra, 1972c).

Profiles for several events have been derived by Deshpande and Mitra (1972c) with the above method. An example is the solar flare that occurred on 30 January 1968 at 0503 UT. This was one of the largest events registered at the National Physical Laboratory, New Delhi. The profiles were based on SCNAs at two frequencies (22.4 and 30 MHz) and a simultaneous absorption measurement with the A1 pulse technique operating on 5.5 MHz at Calcutta.

Plots of the ratio of flare time absorptions at three pairs of frequencies, 22.4–30 MHz, 5.5–22.4 MHz and 5.5–30 MHz showed a rapid reduction of the frequency exponent. This, as it has often been pointed out, is a result of a rapid drop in the absorbing layer as the event progresses.

Profiles were obtained separately for the pairs 5.5 and 22.4 MHz, 5.5 and 30 MHz and 22.4 and 30 MHz. The three profiles thus deduced were seldom identical. These were then suitably averaged and then the average profile was modified by an iterative procedure till the computed absorption for the modified profiles agreed with the observed absorption at all the frequencies. $N_e(h)$ profiles finally deduced are shown in Figure 79a. It is interesting to note that during the peak of the absorption event, i.e.

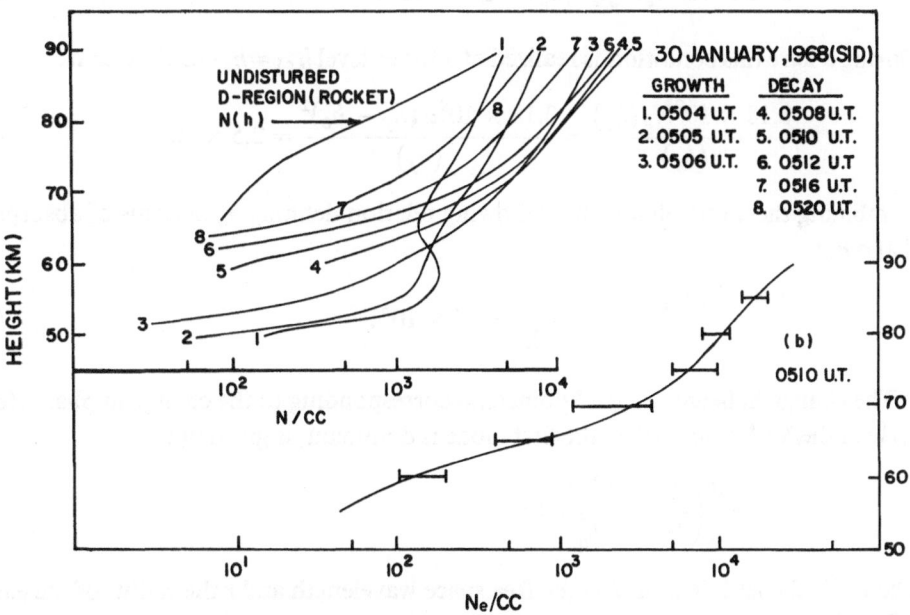

Fig. 79. (a) Electron density profiles derived from multifrequency absorption measurements during a flare effect of 30 January 1968 (after Deshpande *et al.*, 1972b). (b) Ranges of profiles obtained from three pairs of frequencies and the final choice (after Deshpande *et al.*, 1972b).

at 0504 UT, 0505 UT and 0506 UT, the $N_e(h)$ profiles derived in this manner from absorption measured on different frequencies agreed reasonably well. However, during the later part of the event the profiles were not identical. These are shown in Figure 79b in which ranges of values obtained from three pairs of frequencies are shown during the decay of the event. The spread in the three profiles is an indication of the range of errors in the determination of the profile. As the flare progressed the base level h_0 did fall as expected, with the gradual hardening of the incident radiation.

It is important to realise that this method of profile evaluation is only a first order approximation. Its limitation arises mainly through the assumption regarding the layer shape. The electron density values are also subject to an uncertainty in the base height determination. An uncertainty of ± 2 km in h_0 leads to an uncertainty of about 45% in electron density at 10 km above the base level.

7.2. Profile Determination from Simultaneous Use of SCNA and SPA Measurements

Absorption measurements can be combined with a concurrent SPA effect registered on a VLF signal to obtain the values of both h_0 and 'a' in a manner somewhat more adequate than given in Section 7.1. During normal times a VLF signal is reflected from the ionospheric level h_n approximately between 70–75 km, such that (Wait, 1962):

$$3.18 \times 10^9 N_e(h_n)/\nu(h_n) = 2.5 \times 10^5 . \tag{68}$$

During a flare this condition is realized at a lower level $h_f (= h_n - \Delta h)$, so that:

$$\frac{3.18 \times 10^9 N_e(h_f)}{\nu(h_f)} = \frac{3.18 \times 10^9 a (h_f - h_0)^2}{\nu(h_f)} = 2.5 \times 10^5 . \tag{69}$$

Combining these one obtains the following relation defining h_0 in terms of absorption A and h_f:

$$\frac{3.18 \times 10^9 (h_f - h_0)^2}{\nu(h_f)} \frac{A}{K} = 2.5 \times 10^5 . \tag{70}$$

The change in height Δh (in kilometers) corresponding to the change in phase $\Delta\phi$ in cycle of the VLF signal when the first mode is dominant, is given by:

$$\Delta\phi = \frac{d}{\lambda} \left(\frac{1}{2r} + \frac{\lambda^2}{16h_n^3} \right) \Delta h, \tag{71}$$

where d is the path length, λ is the free space wavelength and r the radius of the earth.

If h_n is taken to be 72 km (Burgess and Jones, 1967), and Equation (70) is recast in terms of the excess absorption ΔA and not A, then:

$$\frac{3.18 \times 10^9 (h_f - h_0)^2}{\nu(h_f)} \left[\frac{\Delta A}{K} \right] + \omega_n(h_f) = 2.5 \times 10^5 , \tag{72}$$

where $\omega_n(h_f)$ is the quiet day conductivity at level h_f, and is represented, after Wait and Spies (1964), by:

$$\omega_n(h_f) = 2.5 \times 10^5 \exp[0.3(h_f - 72)] . \tag{73}$$

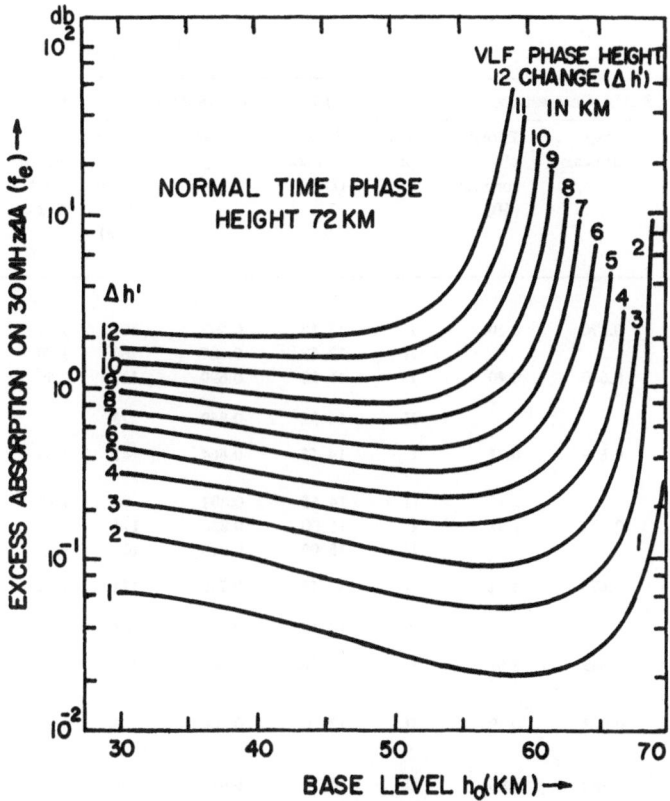

Fig. 80.   Excess absorption on 30 MHz vs h_0, for various values of VLF phase height changes Δh
(after Deshpande and Mitra, 1972c).

Equation (73) gives for 30 MHz the family of curves shown in Figure 80. Any observation of the combination $\Delta A - \Delta h$ yields directly the value of the base level h_0. The value of the constant 'a' is obtained from Equation (66) with this h_0 and the measured absorption; and the entire profile is then given by these values of a and Δh.

7.3. Profiles from Phase and Amplitude Measurements at VLF

Simultaneous observations of phase and amplitude in the VLF during flares may also be employed for constructing the ionization profiles. Usually a number of model profiles are set up and the relationships between the propagation factors, e.g. phase and amplitude, and constants of the profiles are derived adopting numerical methods for computations that are based on either the wave-guide mode theory or on the full wave theory of VLF propagation. In the computations made by May (1966) both penetrating and non-penetrating modes were combined to give the normal and abnormal components. A 'stationary phase' technique (described in detail by Piggott *et al.*, 1965) was used in order to simulate the change in amplitude and phase of the downcoming wave at fixed distance. In May's calculations agreement of the

TA

Selected SCNA-VLF SID effects and pr

Date	30 MHz SCNA Observations				VLF SID Observations			Flare time conductivity profile		
	Time of max (UT)	$\cos\chi$ at peak	Excess absorption (dB)	VLF path	Time of max (UT)	Average $\cos\chi$	Phase advance $\Delta\phi$ (deg)	Field enhancement E_f/E_n	Reference height h_f (km)	Gradient β_f (km^{-1})
1966										
8 July	17 23	0.570	0.50	I	17 18	0.597	106	1.41	68	0.348
				II	17 18	0.827	60	1.59	67.5	0.404
10 July	16 40	0.640	0.40	I	16 20	0.690	127	1.59	67.25	0.362
				II	16 22	0.849	73	1.40	66.75	0.350
26 July	14 48	0.699	0.71	I	14 55	0.864	42	1.23	70.25	0.325
				II	14 55	0.803	26	1.18	69.5	0.321
18 Sept.	15 02	0.895	1.67	I	15 00	0.828	176	1.54	65.75	0.371
				II	15 00	0.915	107	1.35	65	0.311
19 Sept.	12 15	0.670	1.12	I	12 15	0.789	174	1.82	65.75	0.407
				II	12 15	0.488	108	1.97	65	0.447
19 Sept.	15 32	0.389	1.01	I	15 32	0.804	134	1.73	67	0.394
23 Dec.	13 25	0.923	0.49	II	13 30	0.704	60	1.61	67.5	0.405
23 Dec.	15 12	0.990	0.96	II	15 13	0.879	69	1.50	67.25	0.374
23 Dec.	17 00	0.890	0.42	II	16 54	0.912	60	1.50	67.5	0.377

derived models was sought with continuous observations of $_{\parallel}R_{\perp}$ and Δh at 16 kHz and 70 kHz recorded at Cambridge at steep incidence.

Such calculations are, however, quite involved, and simple procedures, although less accurate, but suitable for routine reductions have been given by Deshpande and Mitra (1972c) and Sengupta (1971). In both, the mode theory calculations of Wait and Spies of the attenuation rate and relative phase velocities for assumed models of the conductivity parameter are used.

The flare time conductivity profile is obtained in the following manner:

(i) The quietday profile is taken as:

$$\omega_r(h) = 2.5 \times 10^5 \exp[0.30(h - 72)]. \tag{74}$$

(ii) The attenuation rate and relative phase velocity of a particular VLF signal, transmitted over the known path length, is obtained from the computations of Wait and Spies (1964) for this normal time profile.

(iii) Change in amplitude (ΔE) and in phase ($\Delta\phi$), registered as VLF SID effects, are used together with the normal time attenuation rate and relative phase velocity to obtain the flare time values of the latter factors.

XIV

of electron density based on these measurements

N_e—h profile from SCNA-SPA (type B)			Electron density at flare time ref. height h_f		Computed 30 MHz absorption for type C profiles (dB)	X-ray events (P/Q = ratio of peak flux to quiet flux)
Base level h_0 (km)	Factor 'a'	The gradient 'β' from this N_e—h profile (km^{-1})	From type B profile (cm^{-3})	From type C (VLF) profile (cm^{-3})		
65	41.5	0.51	3.78×10^2	3.97×10^2	0.23	
64	34.3	0.59	4.27×10^2	4.41×10^2	0.48	
62.5	20.0	0.495	$4.50 \times \times 10^2$	4.47×10^2	0.33	OGO I (10–50 keV) 230×10^3 NEPPS 1628–1631-X UT
62	14.7	0.50	4.98×10^2	5.30×10^2	0.39	
69.25	156.9		1.57×10^2	1.53×10^2	0.05	Explorer-33 (2–12 Å) (P/Q) = 4 1441–1450-X UT
68	99.0	0.85	2.22×10^2	2.18×10^2	0.09	
63.25	100.0	0.83	6.25×10^2	5.78×10^2	0.63	
62	78.2	0.75	7.06×10^2	7.06×10^2	0.49	
62.5	56.2	0.58	6.00×10^2	6.22×10^2	1.06	Explorer-33 (2–12 Å) (P/Q) = 7 1158–1213-X UT
61	43.9	0.58	7.02×10^2	7.20×10^2	2.56	
64.25	60.8	0.74	4.62×10^2	4.73×10^2	0.50	Explorer-33 (2–12 Å) (P/Q) = 7 1521–1528-X UT
63.5	24.0	0.55	3.82×10^2	3.79×10^2	0.34	Explorer-33 (2–12 Å) (P/Q) = 6 1309–1320-X UT
64.25	49.0	0.69	4.41×10^2	4.35×10^2	0.42	Explorer-33 (2–12 Å) (P/Q) = 8 1503–1512-X UT
63.3	23.8	0.54	4.22×10^2	4.22×10^2	0.42	

(iv) From the flare time attenuation rate and relative phase velocity, the flare time conductivity parameter β_f and h_f are deduced from the Wait and Spies results mentioned in (ii) above (see, for instance, Crombie, 1966). This gives flare time profile

$$\omega_{rf}(h) = 2.5 \times 10^5 \exp[\beta_f(h - h_f)].$$

The ionization profile is obtained from this flare time conductivity profile employing an appropriate collision frequency model.

The assumption of an ionization distribution form of the type given by Equation (65) is, in practice, found to be justified. In Figure 81 some observed flare time profiles and the corresponding best-fit $[a(h-h_0)^2]$ functions are shown. Sometimes, observed profile can be reproduced only with the superposition of two square functions; in such cases a combination of techniques is necessary. Deshpande and M (1972c) have examined about a dozen multifrequency events for profile determination (including the large flare event of January 30, 1968, previously referred to), as well nine events in 1966 in which simultaneous observations of SCNA and VLF phase amplitude changes are used. The former utilised SCNA observations of Delhi on 22.4 and 30 MHz. The latter used observations at Sao Paulo, Brazil – the SCNA

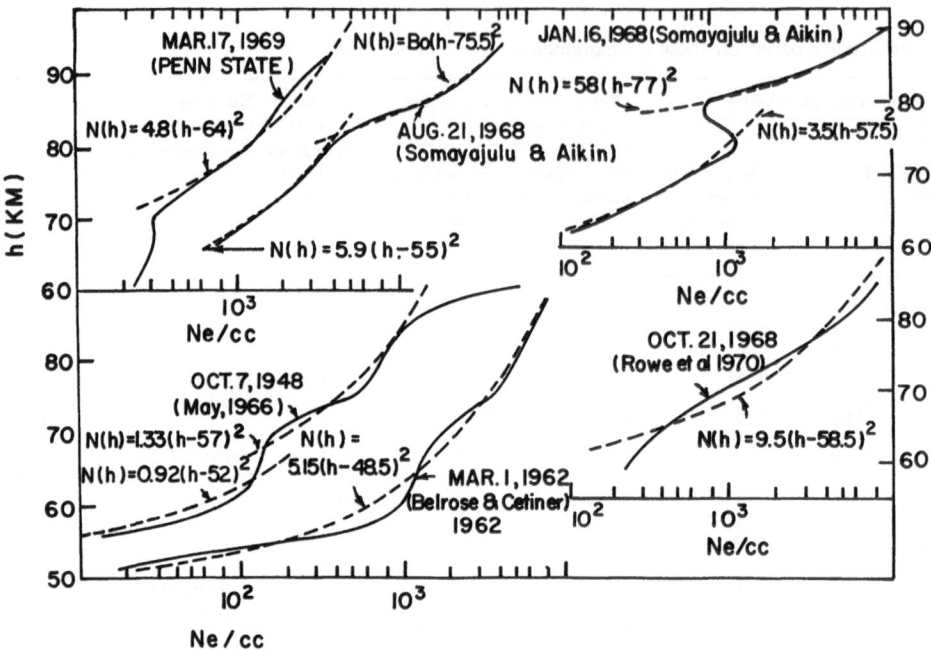

Fig. 81.   Some observed N_e-h flare profiles and corresponding best fit $a(h-h_0)^2$ functions (after Deshpande and Mitra, 1972c).

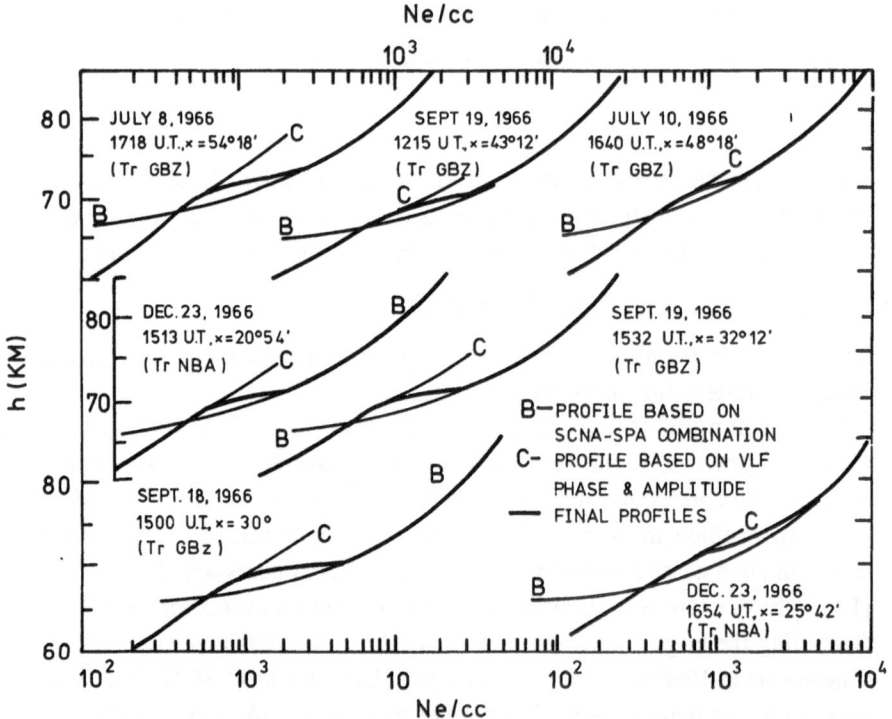

Fig. 82. N_e-h profiles based on SCNA-SPA combinations (curves B) and on VLF phase and amplitude change measurements (curves C). Final electron density profiles deduced from these are given by thick lines (after Deshpande and Mitra, 1972c).

servations were on 30 MHz, and the VLF observations, reported by Kaufmann *et al.* (1967), were for two transmission paths I and II, in which path I refers to transmission from GBZ (Great Britain) to Sao Paulo on 19 kHz over a distance of 9500 km, and path II was a transmission on 24 kHz over a distance of 5100 km from NBA (Panama) to Sao Paulo.

The selected SCNA – VLF SID effects are listed in Table XIV. In all the VLF effects, the entire path was illuminated, and the average $\cos \chi$ over GBZ – Sao Paulo (path I) transmission path and at SCNA station were quite close. Two profiles were deduced for each event. The first was deduced from a combination of SCNA and SPA (type B), the second from measurements of phase and amplitude changes at VLF (type C). The profile constants 'a' and h_0, as well as the height of signal reflection and electron concentration at this level are given in Table XIV. In this table are also given the calculated excess absorption at 30 MHz from the type C $N_e(h)$ profile, based on VLF measurements alone. Similarly the conductivity gradient β calculated from type B profiles based on SCNA–SPA combination are included. The profiles deduced in this manner are shown in Figure 82 for the events for which the average $\cos \chi$ over the VLF path was close to that of SCNA recording station.

In case of the SCNA events, the frequency exponent was usually between 1.6 and 2. For these type of profiles, the h_0 range was from 36 to 59 km, but usually ranged between 40 and 55 km. N_e at the height of maximum absorption ranged from 3×10^2 to 5×10^3 cm^{-3}. The level of maximum absorption was about 10–11 km above h_0. These SCNA events were typical cases where absorbing layer considerably shifted to the lower levels. The situation in more common events where absorbing layer did not appreciably shift downwards was represented by type B profiles. These were obtained from the SCNA–SPA combination. The base level for these profiles ranged between 61–69 km and was, on the average, 63 km. The effective height of VLF reflection was found to be about 3 km above the base level. The conductivity gradient β corresponding to these $N_e(h)$ profiles was much higher than the VLF phase and field strength measurements indicated. Thus the electron density did not decrease as sharply as shown by SCNA–SPA based (type B) profiles.

The VLF phase and amplitude measurements yielded the type C profiles which were exponential in shape. In contrast to the sharp decrease in electron density around the phase height shown by type B profiles, these showed a slow rate of decrease in electron concentrations. The excess absorption computed from $N_e - h$ steep rise to about 10^4 cm^{-3} around 80–85 km is indicated by these profiles.

Comparison of these profiles with those obtained from direct techniques (e.g. rocket, wave interaction, partial reflection) showed that profiles based on SCNA measurements gave somewhat higher electron concentrations, particularly below 75 km. When VLF data were included, the profiles come close to those obtained by direct techniques. Above about 80 km, the profiles deduced with these techniques give larger electron concentrations than those given by direct techniques.

CHAPTER 8

D-REGION IONIZATION PROFILES AND
LOSS RATES DURING FLARES

8.1. Flare Ionization Profiles

It is interesting to compare the different profiles obtained by different techniques and for different flare events. We see that the flare profiles are generally similar in shape, excepting that some have larger enhancements than the others, and that in the case of Somayajulu-Aikin rocket profiles there is an unexpected trough at 80 km that is not seen in the other cases. When one arranges the flare events according to increasing flux values (Figure 83), one finds that the electron density profiles then line up in increasing order. These confirm the reliability and consistency of these measurements as well as the predominant role played by flare X-rays in the production of the additional ionization.

If we now consider the time curves of enhanced ionization at different heights, we find that there are appreciable differences.

In Figure 84a are plotted the variation of electron density N_e, at heights of 65, 70, 75 and 80 km during the SID of 7 October, 1948, obtained by May (1966), and described in Section 7.3. The dashed lines show the estimated variation of electron density which would have occurred in the absence of the SID. It is seen that at heights in the 65–80 km range an increase in electron density occurred by factors of 1.2 at 65 km, 2.4 at 70 km, 5.8 at 75 km and 4.0 at 80 km. There is also a difference in the way the electron density varied during the SID at different heights. At 65 km, the electron density increased abruptly at 1555 UT, reached its peak within a few minutes, stayed constant until about 16 20 UT and then recovered slowly. At 75 and 80 km, the electron density started to increase sooner than at 65 km, showed a well-defined peak and a slow recovery.

The differences in the time histories for the flare event of October 21, 1968, obtained by wave interaction technique and for the event of July 8, 1968 obtained by Montbriand and Belrose (1972) with partial reflection experiment are shown in Figure 84b and Figure 76a (Section 6.3).

For the *D*-region, the increase in q comes from the increase in X-ray flare alone and, when the spectrum is carefully developed (Section 3.2), it is possible to obtain the time history of the electron production rate due to X-rays alone q_X for any height. Comparison of $q_X(t)$ values thus obtained with observed N_e-changes are given in Figures 85a, b for the flares of October 21, 1968 (80 km) and for January 30, 1968 (60, 70, 75, 80 km). In the case of October 21, 1968 event we show two consecutive flares occurring between 1600–1800 h UT; N_e changes were observable even for the

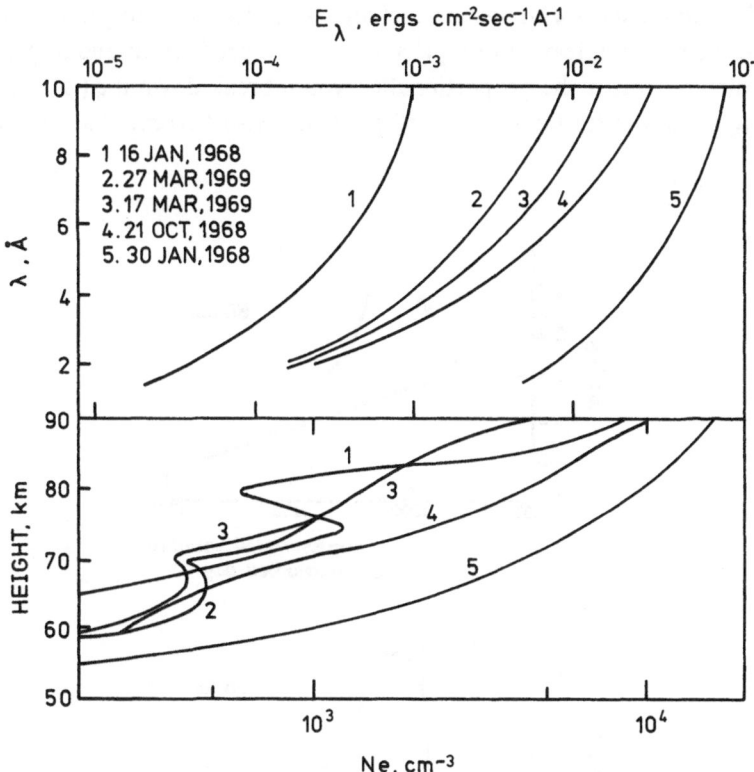

Fig. 83. X-ray spectral distributions in 2–8 Å for selected flare events along with the corresponding electron density profiles. The X-ray events are arranged in increasing order of flux values (after Mitra, 1970).

smaller flare. It is clear from these results that:

(a) $N_e(t)$ curves follow the $q_X(t)$ curves quite closely, excepting for a time lag. The time lag is surprisingly large. With an increase in N_e one would expect $\tau(=1/2\alpha N_e)$ to be considerably below the quiet time values of about 2 min in the heights 60–80 km (Figure 86). One would conclude that the increase in N_e is compensated by a decrease in α – a conclusion that has found other more quantitative support.

(b) The smaller flare of October 21, 1968, yielded at its maximum, an X-ray contribution of only 5 electrons cm^{-3}. Even this was sufficient to produce an observable change in N_e at 80 km (from about 9×10^2 to 2×10^3 cm^{-3}), and consequently the preflare production rate has to be small, around or below 5 cm^{-3} s^{-1}, even when account is taken of a decrease in recombination coefficient.

The conclusion in (b) can be further checked by plotting peak N_e^2 against Δq ($\approx q_X$) for different flares from small to very large ones so that q_X has a wide range. When this is done (Figure 87) one sees that N_e^2 is directly proportional to Δq, the proportionality extending down to the non-X-ray production rate $q(A) = 10$ cm^{-3} s^{-1}. In the two cases (October 31, 1968 and January 30, 1968) for which Δq was large, there is no doubt that such was indeed the case. If the total production rate is taken to consist of q_X and a term $q(A)$ which includes ionization due to photoionization of NO by Lα, the photoionization of $O_2(^1\Delta_g)$ as well as production due to precipitating particles, then obviously the proportionality would break down if $q(A)$ is large compared to q_X excepting for large values of q_X. Thus $q(A) \leqslant 10 \text{ cm}^{-3} \text{ s}^{-1}$ at 80 km.

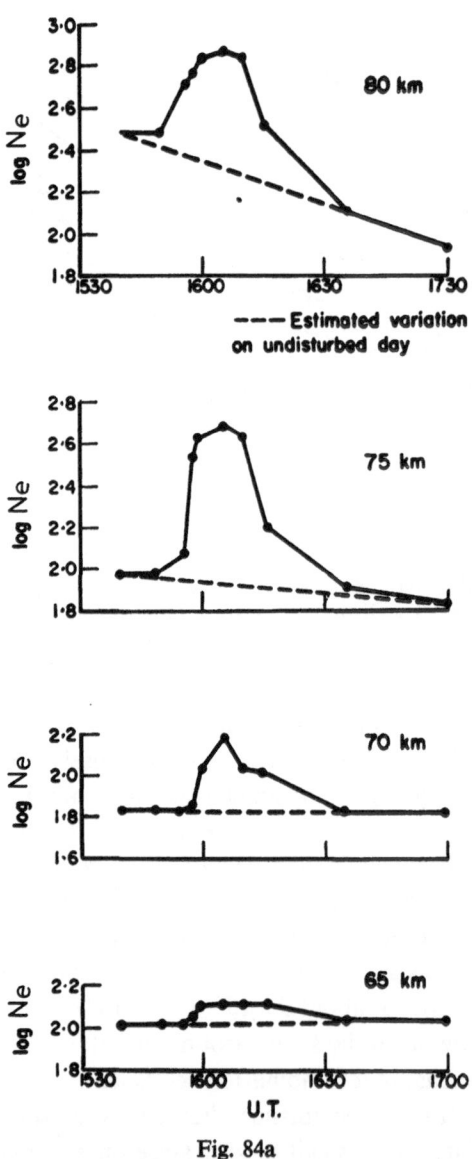

Fig. 84a

Fig. 84a–b. Time variations in flaretime electron densities at selected heights for two flare events and obtained with two different techniques: (a) From LF–VLF propagation data deduced by May (1966) for the flare of 7 October 1948. (b) From wave interaction technique deduced by Rowe *et al.*, 1970) for the flare event of October 21, 1968. Time variations with a third technique (partial reflection) is given in Figure 76a.

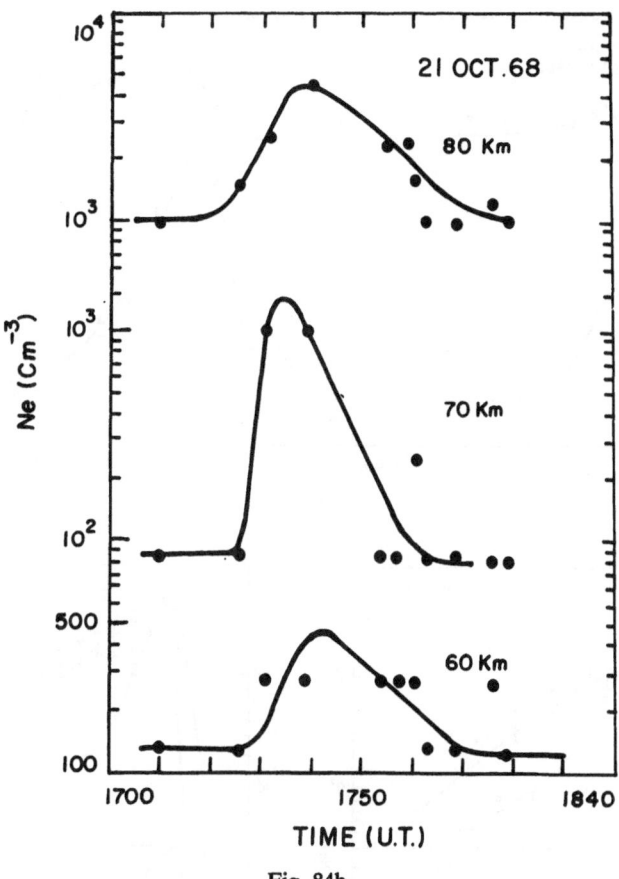

Fig. 84b.

8.2. Loss Rate During Flares

There are several convincing evidences showing that the effective loss rate decreases appreciably and sometimes drastically during flares. In this section these different evidences are presented and model recombination coefficient profiles are derived for different flare conditions.

8.2.1. FROM QUANTITATIVE RELATIONSHIP OF FLARE X-RAYS TO SIDs

The search for what can be considered a truly representative profile for the effective recombination coefficient for quiet conditions has gone on for many years. There are three approaches:

   (i) Ionospheric Method,
  (ii) Production Rate Method,
 (iii) Laboratory Method.

The ionospheric method is the oldest and most widely used. The effective recombination coefficient is determined either by putting $q=0$ in Equation (36) (and this is useful for night time and between second and third contact during an eclipse), and then

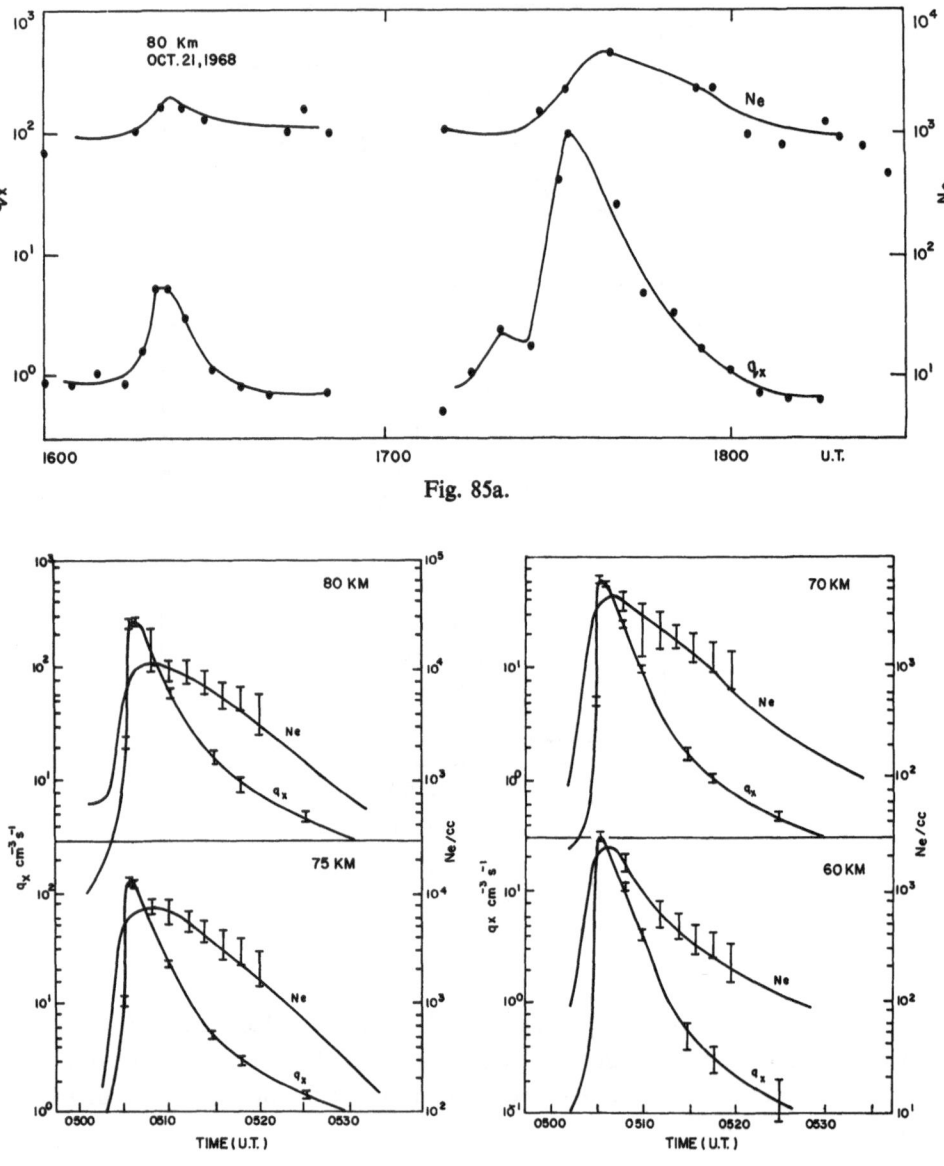

Fig. 85a.

Fig. 85b.

Fig. 85a–b.   Comparison of electron production rates due to enhanced solar X-rays and observed changes in electron density. (a) Two consecutive flares of October 21, 1968 (for 80 km). (b) Flare of January 30, 1968 (for 60, 70, 75 and 80 km).

using the resultant equation of decay curve:

$$\alpha = \left[\frac{1}{N_e} - \frac{1}{N_{e0}}\right]\Big/ t$$

or with the 'comparison method' in which two points on both sides of the maximum (or minimum) in the diurnal curve of an ionospheric parameter is used.

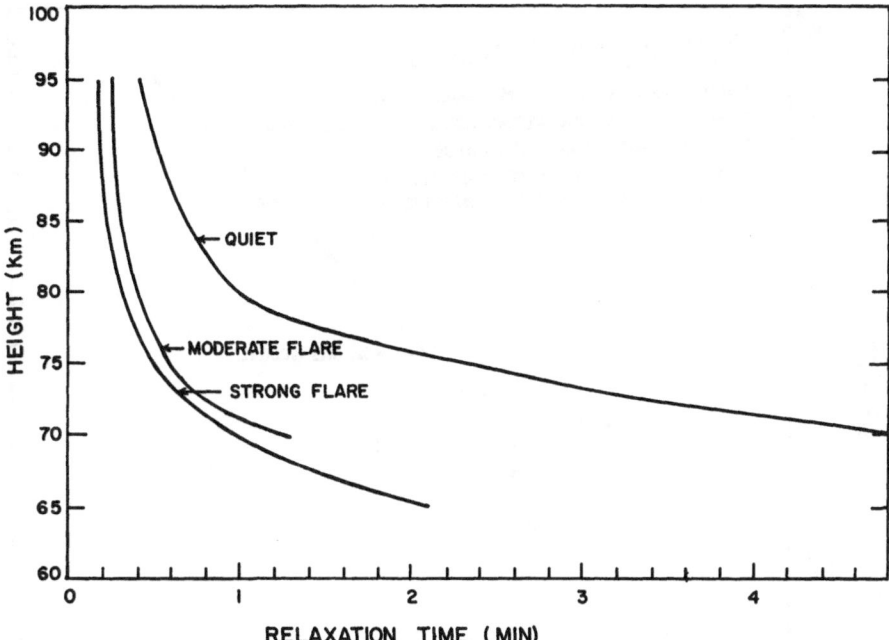

Fig. 86. Relaxation time as a function of height for quiet and flaretime conditions. Two types of flare events are considered: moderate and strong.

In the production rate method, the effective loss coefficient is determined from a comparison of measured electron density profiles with computation of electron production rate q from space measurements of ionizing radiation. Under quasi-equilibrium condition which may be assumed to exist during daytime in most of the ionospheric regions, one gets:

$$\psi = q/N_e^2 .$$

The coefficient ψ obtained by this method is not entirely identical with that of α_{eff} determined by the ionospheric methods, the former being $(1 + \lambda)$ times the other. The two are identical only if λ is small.

In the laboratory method, the many reactions that go into the production and loss of an electron, including those involving negative ions in the lower ionosphere, and charge transfer from atomic to molecular ions in the F-region, must be first identified. Such reactions are many and their reaction rates are also in many cases, known. Further discussions of these reactions are given in Chapter 9.

Some of the $\psi - h$ profiles derived in different ways are given in Figure 88. Most of these, however, refer to disturbed conditions prevailing during PCA events in high latitudes. Two of these, however, refer to *quiet* conditions: that of Reid (1970a) and of Mitra and Banerjee (1971). Reid's profile was obtained with the production rate method. The one given here is, however, a modified one in which the production rate has been calculated with Nitric Oxide concentrations as given by Meira (1971). The profile of Mitra and Banerjee (1971) is based on laboratory data on reaction rates. In a

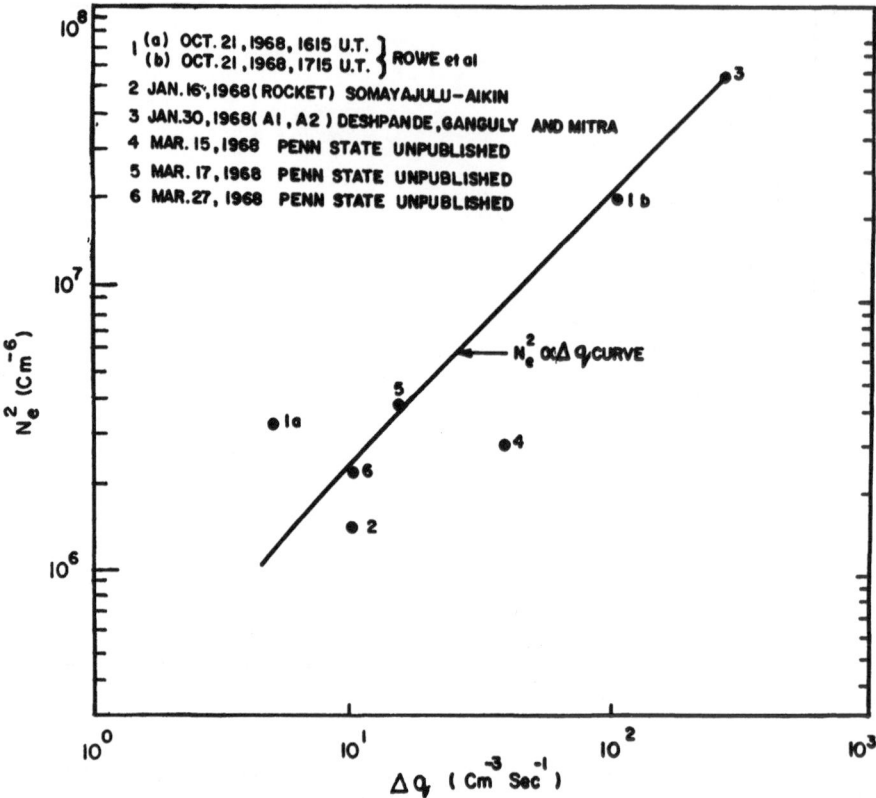

Fig. 87. N_e^2 during the peak of the flare plotted against electron production rate due to X-rays for a number of flare events (after Mitra, 1970).

more recent work Chakrabarty (1973) and Chakrabarty and Mitra (1974) have produced, on a more extensive set of data another $\psi - h$ profile calculated with the production rate method from an extensive set of $N_e - h$ profiles. This distribution, along with the standard deviation, is given in Figure 89.

The true quietday distribution is probably something between these profiles. In the subsequent discussions the distribution of Chakrabarty will be used as a reference distribution for undisturbed conditions.

(a) *Work of Deshpande and Mitra (1972b)*

Deshpande and Mitra obtained ψ from a quantitative comparison of SCNAs, and accompanying X-ray flares for a total of 50 large flare events, for which solar flux monitored aboard Solrad-9 satellite or OGO-4 satellite was available. After developing the appropriate X-ray spectra in the way outlined in Section 3.3, they computed excess absorption for 18 MHz for the zenith angle corresponding to the time and place where the SCNA was observed and for five ψ models. The five ψ models were chosen in the following way: Models 1, 2, 4 and 5 are the ψ models (see, for example, Figure 88) of Mitra (1968), Adams and Masley (1965), Mitra and Banerjee (1971), and

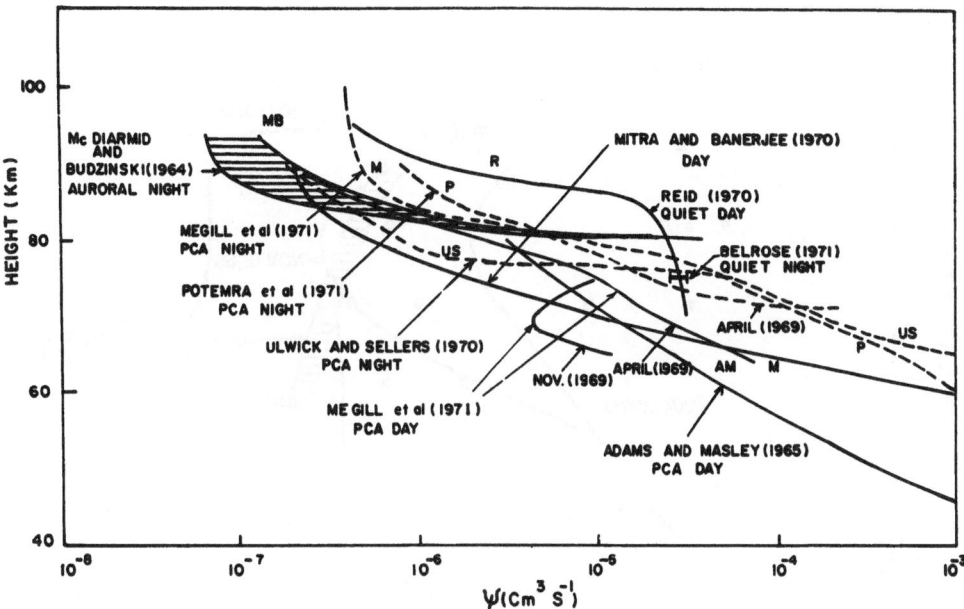

Fig. 88. Altitude profiles of effective loss rates for different conditions. Typical quiet time values are represented by the curves given by Reid (1970a) and by Mitra and Banerjee (1971).

Reid (1970a). These represent successively increasing values of ψ at the levels between 65 and 85 km. Model 3 was constructed from estimates of negative ion to electron density ratio λ based on blunt probe rocket measurements (Baker, 1969) with the dissociative recombination coefficient α_D having a value of $2 \times 10^{-7} (300/T)$ cm^3 s^{-1} and mutual neutralisation $\alpha_i = 10^7$ cm^3 s^{-1}. The model that gave a value of absorption closest to the observed SCNA effect (A_{obs}) and the percentage departure $(A_{cal}/A_{obs}) \times 100$ for this model were obtained (Figure 89). For 13 flare events (out of the total 25 cases) none of the five ψ models was satisfactory as the departure was more than 10%. In 12 cases, X-ray flares and SCNA effects showed close agreement but the best fitting ψ model differed from flare to flare. *Neither Reid's model nor that given by Chakrabarty and Mitra (Figure 89) (thick line) that represent quietday conditions was consistent with any of the events.* Thirteen flare events, for which none of the model was satisfactory, were further examined. Models for effective loss coefficients that reproduced the observed SCNA effect within $\pm 10\%$ were constructed, essentially by trial and error, and beginning with the Models 1–5 that came closest. Two assumptions were made:

(1) Between 60 and 65 km values of ψ are taken to be high, corresponding to those in Model 3. With this assumption it was found that the contribution to the absorption from 60 km and below is negligible for all events.

(2) Above 65 km ψ was assumed to decrease monotonically with height; the rate of decrease could, however, be slow or rapid, ψ distributions constructed in this manner are broadly labelled as A, B or C.

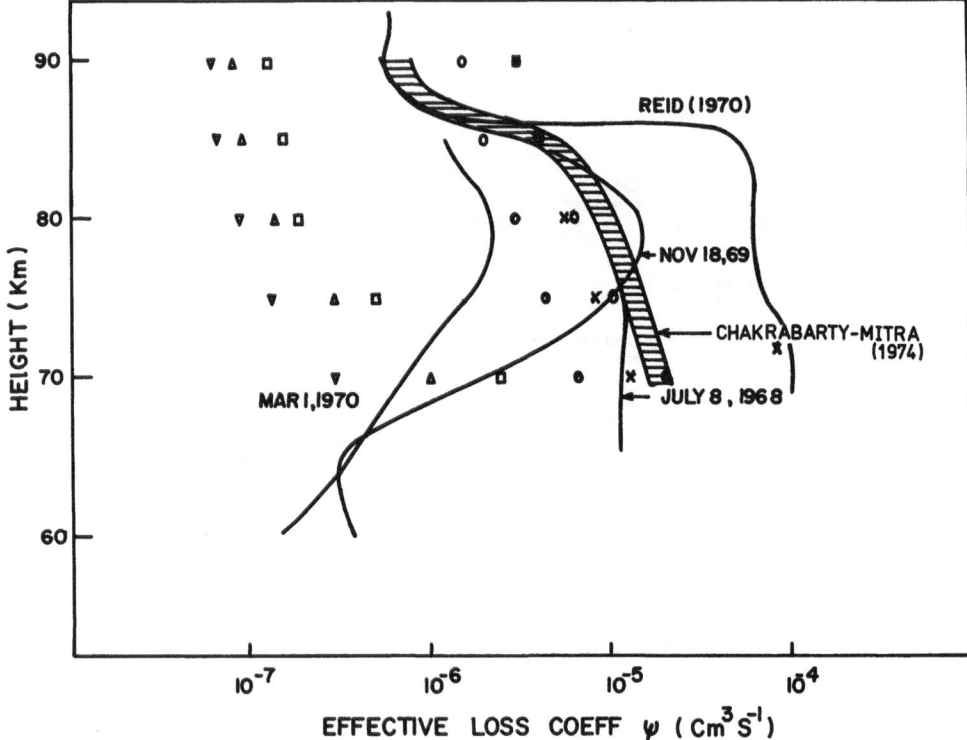

Fig. 89. Effective recombination coefficient in the lower ionosphere. Values derived from X-ray flares-SCNA analysis are shown as dots etc. ($\times$ – A1, $\bigcirc$ – A2, $\blacksquare$ – B, $\square$ – C1, $\triangle$ – C2, $\triangledown$ – C3) (after Deshpande and Mitra, 1972b). For comparison the quiet day ψ profile (with its range) is given. Also given several altitude profiles of ψ obtained by Montbriand and Belrose (1972) during several flare events with partial reflection experiment.

These distributions are shown in Figure 89, along with Chakrabarty-Mitra quiet-day model as reference. ψ-distributions obtained in this way fall in three categories. In the first category ψ has high values at all heights below 85 km (type A marked $\times$ and $\bigcirc$ in Figure 89). In the second category (type B, marked $\bullet$ in Figure 89) ψ takes medium values while in the third category (type C, marked $\blacksquare$, $\triangle$ and $\triangledown$ in Figure 89) the ψ values are very low. It can be seen that in no case was the high ψ model of Reid found satisfactory, and the highest values encountered were not much higher than Model 4.

If we take Model 5 (Reid's model) and Chakrabarty-Mitra model to represent the quiet time situation, then all the flare time models indicate a decrease in effective recombination coefficient.

It was further observed that the decrease in ψ was in proportion to the size of the X-ray flare. This is shown in Figure 90 where the values of ψ are plotted as a function of the true X-ray flux in the 0–8 Å band separately for 70 and 80 km. At both heights ψ dropped rapidly with increasing X-ray flux values for $F(1\text{–}8\,\text{Å}) > 1 \times 10^{-2}$ erg cm^{-2} s^{-1}, becoming nearly constant for $F(1\text{–}8\,\text{Å}) > 1 \times 10^{-1}$ erg cm^{-2} s^{-1}. The minimum

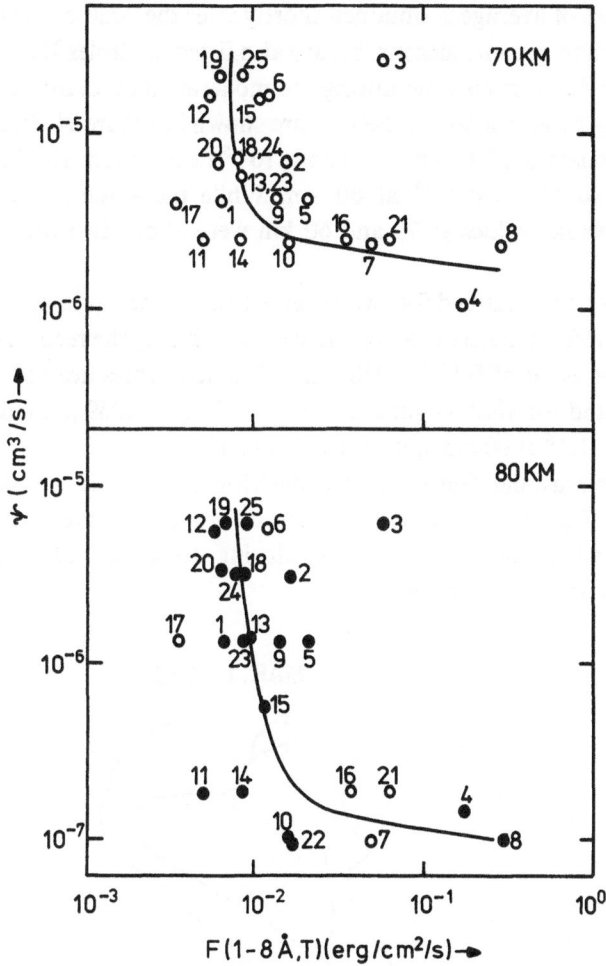

Fig. 90. Effective recombination coefficient vs 0–8 Å X-ray flux obtained from an analysis of 25 X-ray flare – SCNA events for 80 and 70 km heights (after Deshpande and Mitra, 1972b).

values reached for strong X-ray flares was 1×10^{-7} cm^3 s^{-1} at 80 km and 1.7×10^{-6} cm^3 s^{-1} at 70 km.

(b) *Work of Mitra and Rowe (1972)*

Mitra and Rowe (1972) examined 4 flare events for which $N_e - h$ profiles were obtained by the Penn State wave interaction facility and obtained ψ by using simultaneously observed X-ray fluxes obtained by NRL. These values are listed in Table XV. Here again the loss coefficient was always lower than the pre-flare values.

(c) *Work of Montbriand and Belrose (1972)*

In another work Montbriand and Belrose (1972) obtained profiles for *seven* flare events-small, moderate and large and distributed evenly over summer, equinox and winter. The values of average ψ obtained shortly after the peak of the event or shortly after $F(0.5\text{--}3 \text{ Å})$ had become measurable are also listed in Tables XV and XVI.

ψ distributions for various time during the moderate flare event of March 1, 1970, as determined by Montbriand and Belrose are shown in Figure 91. In this and another winter event (January 31, 1970) ψ was about 10^{-5} near 80 km, 10^{-6} near 70 km, and between 10^{-6} and $10^{-7} \text{ cm}^3 \text{ s}^{-1}$ at 60 km. While the value at 80 km is close to Reid's quietday value, values at 70 and 60 km were about two orders of magnitude smaller.

Similar results were obtained for two other events. In the equinox event on March 1, 1970, they observed a pattern of slowly increasing ψ during the recovery phase. During the large summer event of July 8, 1968, they obtained values near 10^{-6} at 60 km and 10^{-5} at 65 km; and for another summer event on May 2, 1969 (a moderate case) they obtained a value 10^{-6} at 60 km and 4×10^{-6} at 65 km.

An interesting seasonal factor noticed by Montbriand and Belrose was that the summer values of ψ at heights below 75 km during the recovery phase of the flare events had a height profile similar to those during the winter, but with values higher by as much as an order of magnitude.

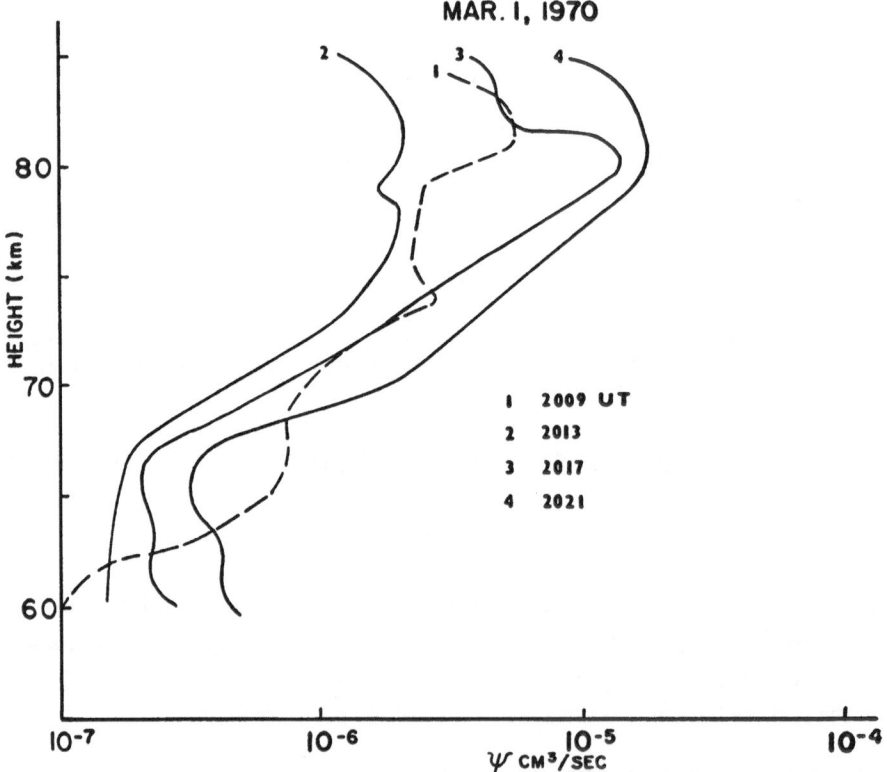

Fig. 91.   Recombination coefficient profiles derived by Montbriand and Belrose (1972) for the moderate flare event of March 1, 1970 using electron density profiles obtained with Partial Reflection Experiments (after Montbriand and Belrose, 1972).

TABLE XV

Experimentally observed values of $q/N_e^2 (=\psi)$ at 80 km (in 10^{-6} cm^3 s^{-1})

Condition	1	2	3	4	5	6	7	8	9	10	Weak	Moderate	Large	Minimum value possible
Normal	3.1	8.0	8.0	15	41	20	30*	30*	30*	30*	30*	30*	30*	30*
Flare	1.5	4.5	3.5	3.5	4.4	3.0	17	16	2.1	7.5	30	8	3	0.3
$\dfrac{\text{Normal}}{\text{Flare}}$	0.5	0.6	0.4	0.2	0.1	0.2	0.6	0.53	0.07	0.3	1	0.3	0.1	0.01

Notes:
1: Somayajulu and Aikin (1969), August 1968 event.
2-5: Penn State Observations – 21 October 1968 (1700 UT); 27 March, 1969, 17 March 1969.
6: Deshpande *et al.* (1972b) – January 30, 1968.
7-10: Montbriand and Belrose (1972), – Nov. 18, 1969 (Large, Winter), January 31, 1970 (moderate, winter) March 1, 1970 (moderate, equinox), July 8, 1968 (Large, summer).
Weak, Moderate, Large Events: Average values given by Mitra and Deshpande (1972).
Minimum Value: Value for the condition when all water cluster ions have disappeared and only simple molecular ions NO$^+$ and O$_2^+$ exist.
* Reid's quietday value (revised with current laboratory reaction rate data).

TABLE XVI

Experimentally observed values of q/N_e^2 at levels where negative ions are comparable to electrons or dominant (in 10^{-6} cm^3 s^{-1})

S.No.	Flare event	70 km	60 km	Remarks
1.	Weak	55	–	Nos. 1–3 from Mitra and Deshpande (1972)
2.	Moderate	15	–	
3.	Strong	3	–	Nos. 4–10 from Montbriand and Belrose (1972)
4.	Nov. 18, 1969 (large, winter)	2.5	0.4	
5.	Jan. 31, 1970 (moderate, winter)	3.0	0.3	
6.	March 1, 1970 (moderate, equinox)	0.7	0.1	
7.	March 1, 1970 (small, equinox)	7.2		
8.	July 8, 1968 (large, summer)	12 (a)		
		0.5 (b)	(b) 4	No. 8 (b) from Sengupta (1971)
9.	July 3, 1969 (moderate, summer)		1.1	
10.	May 3, 1969 (moderate, summer)		1.3	
	Quietday values	$\sim$ 50	50–100	

$\psi - h$ profiles for two large events are shown in Figure 89.

If these variations are attributed to λ at 70 km and below, then assuming $\alpha_i = 4 \times 10^{-7}$ cm^3 s^{-1} and $\alpha_D = 5 \times 10^{-7}$ cm^3 s^{-1}, Montbriand and Belrose obtained for λ median values of 3.1 at 75 km, 1.2 at 70 km, and to less than 0.6 at 65 km. The median values of λ for summer had a height variation similar to that for the winter, having a value near 4.5 at 70 km, below which it decreases to near 2.0 at 65 km. The height variation of λ was thus opposite to that obtained for normal conditions, and the values were considerably lower than in the normal time.

(d) *Work of Sengupta*

Similar results were also obtained by Sengupta (1971) from a detailed examination of the VLF observation during the large X-ray flare of July 8, 1968 for which X-ray fiux measurement were available from the University of Iowa experiments of Explorer-33 and 35 at 2–12 Å, in addition to the NRL measurements in the bands of 1–8 Å, 9–20 Å. The hardening ratio defined by $F(0\text{--}3\ \text{Å})/F(1\text{--}8\ \text{Å})$ was obtained additionally from VLF effects in a way described later in Section 10.1.1. He obtained ψ values at 70 km, using q_X profiles from X-ray spectral distributions so developed, and the maximum lowering of reflection height Δh (km) obtained from 18 VLF paths using frequencies in the range 13.6 to 60 kHz over different propagation paths and sec χ values ranging from 1.06 to 3.1. The ψ values below 70 km were found to be proportional to the total number density of the atmospheric constituents and are given in Figure 92. Here again the values are lower than the quietday values defined by Reid's curve (suitably corrected).

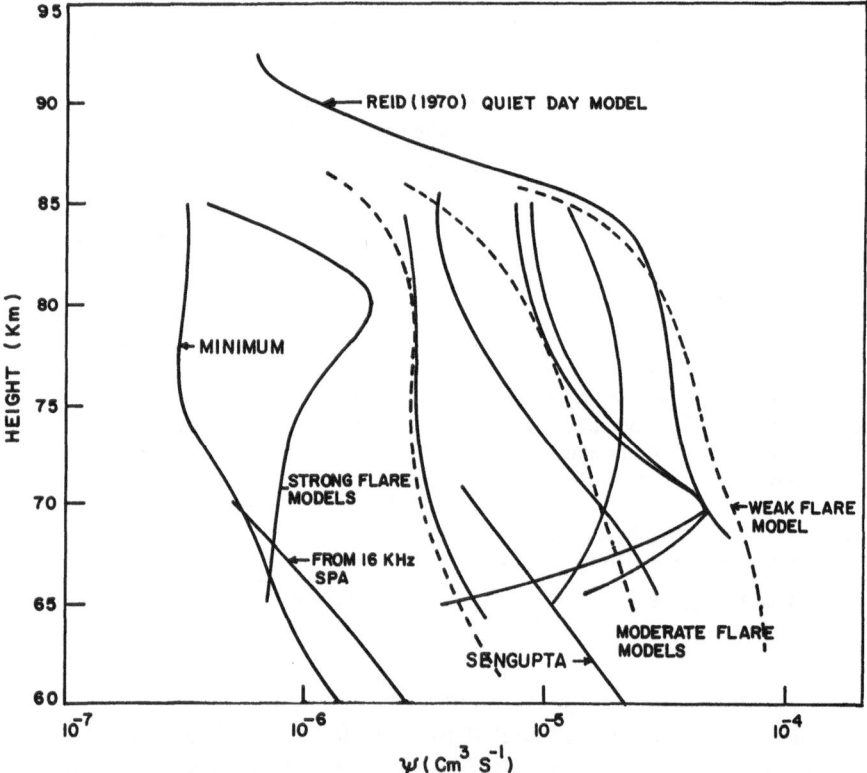

Fig. 92.   Profiles of q/N_e^2 for weak, moderate and large flare events compared with the quiet day situation given by Reid (1970a). ψ-profile deduced by Sengupta (1971) for the large flare event of July 8, 1968 is also shown, as well as a minimum distribution for conditions where the water cluster ions have completely disappeared.

The different values are summarised in Table XV (for 80 km) and in Table XVI (for the negative ion regions of 70 and 80 km). The drastic decrease in ψ is clearly evident in all cases.

(e) *Work of Ananthakrishnan et al. (1973)*

Low values of the D-region recombination coefficient were also derived by Ananthakrishnan *et al.* (1973) from simultaneous use of VLF phase and amplitude measurements made in Sao-Paulo, Brazil, on five frequencies during a solar flare that occurred on 22 January, 1972. The values were 5×10^{-7} and 2×10^{-6} cm^3 s^{-1} at 70 and 60 km respectively.

8.2.2. AVERAGE PROFILES FOR WEAK, MODERATE AND STRONG FLARE CONDITIONS

Although Montbriand and Belrose (1972) did not find any evidence for the decrease in ψ to be larger for large flares than for small flares, Figure 90 of Deshpande and Mitra (1972b) is quite convincing. It is important to note the following:

(i) The pre-flare value of ψ drops at first quite rapidly for increasing flux values $F(1-8 \text{ Å})$ but for large values of $F(1-8 \text{ Å})$ the variation slows down to a near-constant value. At both 70 and 80 km levels, this occurs for $F(1-8 \text{ Å}) \geqslant 10^{-1}$ erg cm^{-2} s^{-1}.

(ii) The decrease at 80 km from its pre-flare value to its minimum for very large flares is almost two orders of magnitude from around 10^{-5} cm^3 s^{-1} to about 2×10^{-7} cm^3 s^{-1}. The decrease at 70 km is smaller, from about 3×10^{-5} to about 3×10^{-6} cm^3 s^{-1} an order of magnitude.

(iii) At 80 km, the lowest value reached during large flare events is of the same order as the dissociative recombination coefficients of the simple molecular ions O_2^+ and NO^+ – the sort of values one encounters under normal conditions above 90 km.

On the basis of the results presented above, Mitra and Rowe (1972) constructed ψ-profiles to represent weak, moderate, and strong flare conditions. These are reproduced in Figure 92, along with the distribution obtained by Sengupta, and the minimum distribution expected when cluster ions have completely disappeared, and the positive ions consist only of NO^+ and O_2^+ (see also Chapter 9).

SIDs AS A TOOL FOR THE STUDY OF AERONOMY
AND ION CHEMISTRY

SIDs have been used to a considerable advantage in resolving some of the major problems in D-region aeronomy and ion and neutral chemistry, but have not been used so effectively for the E and F-regions.

In the lower ionosphere, the flare ionization suddenly changes the source of ionization from photoionization of NO by $L\alpha$ to photoionization of O_2 and N_2 by X-rays. The abundance of O_2^+ ions upsets the chemical scheme operating in the D-region and will, therefore, affect its ion composition and consequently the effective recombination rate. In the E and F-regions also some imbalance is expected – in the E-region due to larger contribution of ionization from X-rays in the 20–100 Å band, and in the F-region due to different distribution in the EUV radiation. There is, in addition, an increase in electron temperature in the F–region, causing, in very large flares, appreciable thermal expansion of the layer and a sudden surge of upward drift velocity.

9.1. Photoionization Rates During Solar Flares

Under normal conditions the main photoionization source below about 65 km is galactic cosmic rays, that below 85 km is principally photoionization of nitric oxide by $L\alpha$, and the X-ray contribution begins to be effective only above 85 km (Figure 93). The first and the third processes produce initially both O_2^+ and N_2^+ ions, but N_2^+ ions are soon converted into O_2^+ ions; the net result is, therefore, production of O_2^+ ions. The second process produces NO^+ ions.

The principal change during a solar flare is not merely an increase in the total photoionization rate, but also a relative imbalance between $q(O_2^+)$ and $q(NO^+)$. The dominance of NO^+ in the region between 65 and 85 km is gradually reduced, and may completely vanish. Two examples are shown in Figures 94a, b. Figure 94a gives the production rate profiles during normal conditions and during the peak of a flare event (1733 UT) occurring on 21 October 1968. At 1733 UT, $q(O_2^+, N_2^+)$ is dominant in the *entire* lower ionosphere. Figure 94b shows how $q(O_2^+)$ gradually takes over from $q(NO^+)$ at one selected height (80 km) for the same flare.

Around 100 km only enhancement in the soft X-rays above 20 Å affects photoionization. At higher heights any change in the photoionization rate comes from the EUV radiation. If the fluxes in the different EUV bands (Table III) are not enhanced uniformly, there is a change in the ionization rate profile. The relative emphasis in the production of atomic and molecular ions also changes. If one uses the enhancements in EUV radiation as given in Table V, then the resulting changes in q will be as given

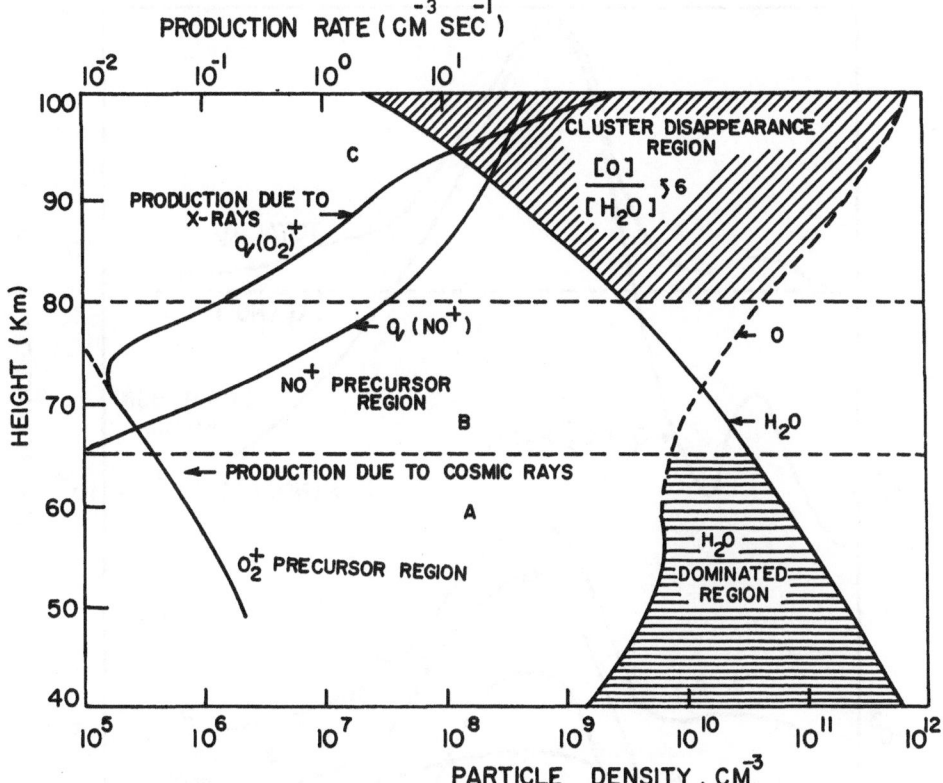

Fig. 93.   Relative magnitudes of $q(O_2^+)$ and $q(NO^+)$ in a normal atmosphere.

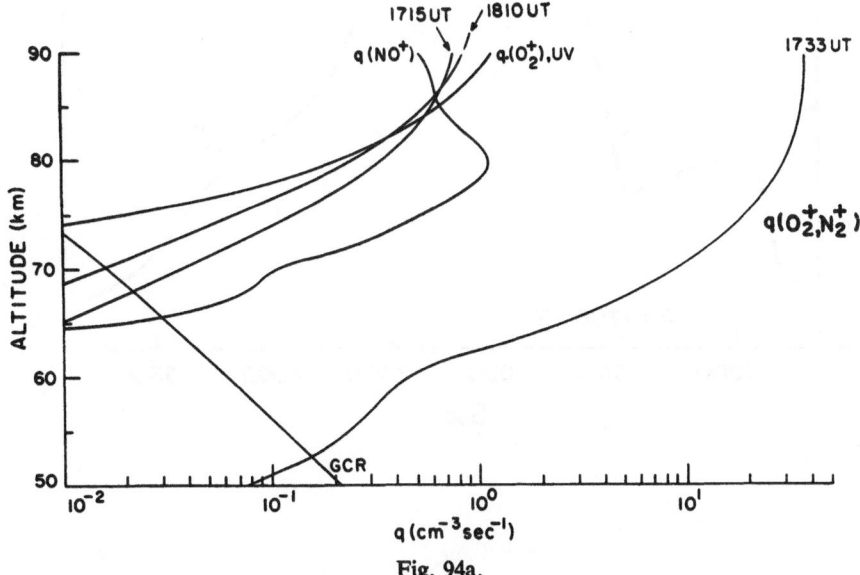

Fig. 94a.

Fig. 94a–b.   Gradual shift in the relative importance of O_2^+ and NO^+ production rates as the flare progresses. Curves in (a) give the example of October 21, 1968 flare, after Rowe (1972) showing the dominance of $q(O_2^+)$. Curves in (b) show how $q(O_2^+)$ takes over from $q(NO^+)$ at 80 km.

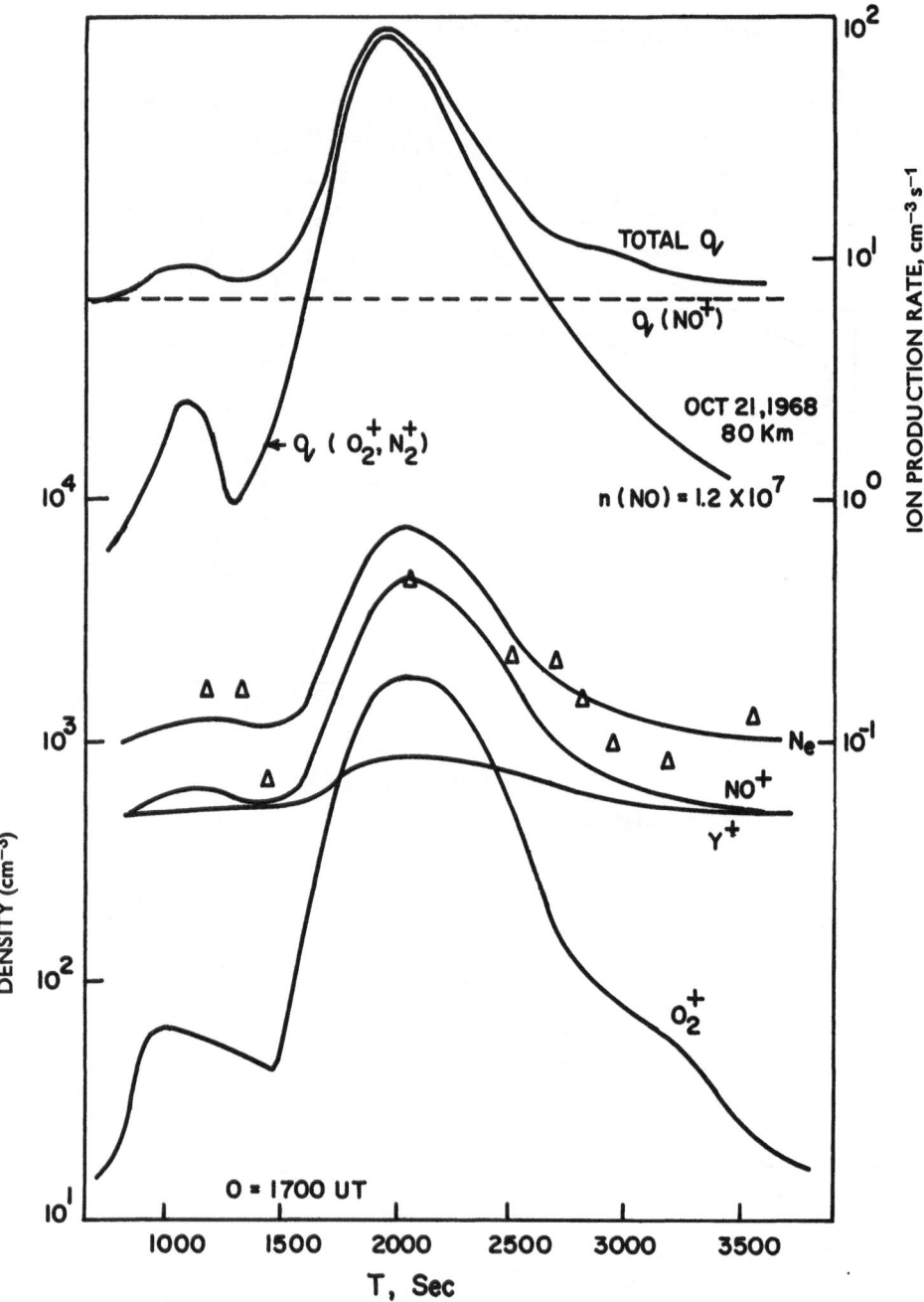

Fig. 94b.

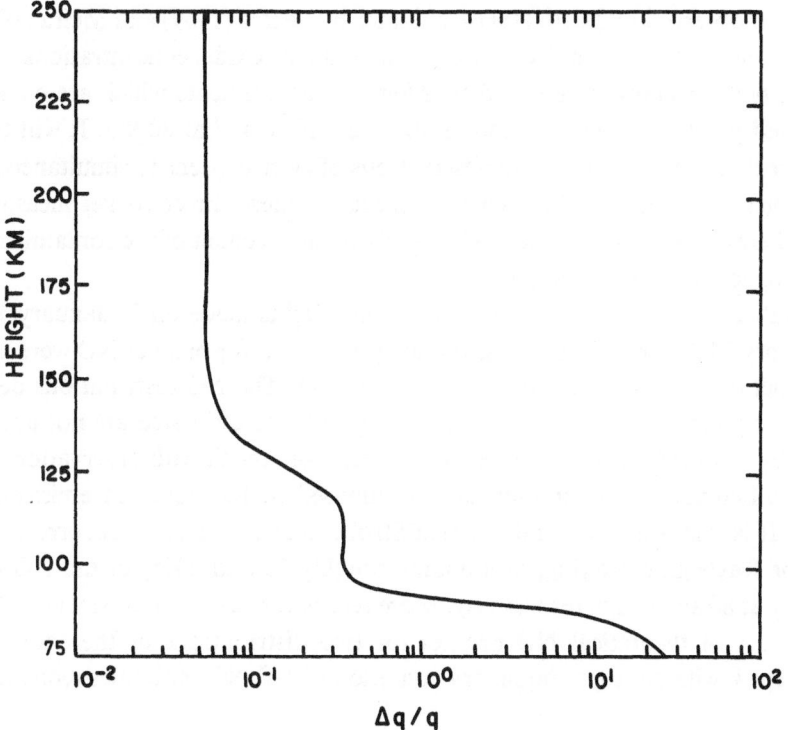

Fig. 95. Changes in the electron production rate at different levels in the ionosphere based on the changes in EUV emission lines observed by Hall and Hinteregger (1969) and given in Table V.

in Figure 95. The largest changes occur, of course, at heights below 100 km – changes above 150 km are very small, around 0.05 for $\Delta q/q$.

The ionization continuity equations during flares are given by:

$$(1 + \lambda) \frac{dN_e}{dt} + N_e \frac{d\lambda}{dt} = (q + \Delta q) - \alpha(h, q) N_e^2 \quad D\text{-region} \qquad (76a)$$

$$\frac{dN_e}{dt} = (q + \Delta q) - \alpha(h) N_e^2 \quad E\text{–}F\text{-region}. \qquad (76b)$$

9.2. Minor Neutral Constituents in the Mesosphere

In the mesosphere the minor neutral constituents play a very crucial role both during quiet and disturbed times. Those known to be important include O, O_3, NO, H_2O and CO_2. No reliable direct observations exist for O, H_2O and CO_2 at mesospheric levels. For O and H_2O measurements are available above 90 km. For CO_2 the usually accepted mixing ratio of 3×10^{-4} is generally used, but no direct observation exists at mesospheric levels.

9.2.1. NITRIC OXIDE IN MESOSPHERE

For nitric oxide, the ultraviolet dayglow due to NO bands in the region 2000–3000 Å have been observed by Barth (1966) and later by Pearce (1969) and Meira (1971). The severe controversy raised by Pearce's large nitric oxide concentrations has been largely, but not entirely, removed by Meira's measurements which are considerably lower and gives NO concentration of about 2×10^7 cm^{-3} at 80 km. It will be shown in the next section that new flare observations of electron density simultaneously with X-ray measurements can be used as a check on these direct rocket measurements, where there is always the problem of properly taking account of the contamination due to the Rayleigh scattered radiation.

Meira's estimates were based on two rocket flights made on 31 January 1969 and 6 February 1969 in which the emission rate profiles (1, 0) γ band of NO were measured as a function of height over the range 60–115 km. The NO distributions derived by Meira are given in Figure 96. That the NO profiles so estimated are not unique, was shown by Strobel (1972) on theoretical grounds from the fit with observational data of a very different NO distribution and normalised to the observed emission rate at 90 km. It is important to point out that Strobel made a substantial correction to the data for Rayleigh scattering, of amounts roughly 20 and 400% of the NO emission intensity at 80 and 60 km respectively. More recently, Brasseur and Nicolet (1973) have given a set of theoretical NO profiles for two distributions of the eddy diffusion coefficient k with an upper boundary condition of 10^8 NO molecules (consistent with

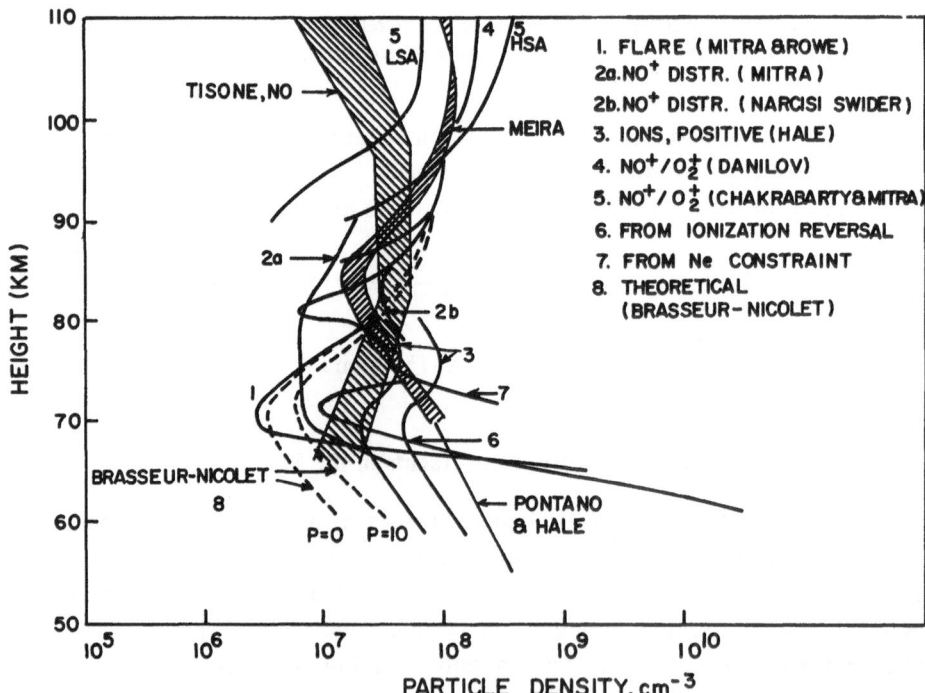

Fig. 96.   Nitric oxide profiles from rocket measurements and ionospheric considerations. The profile derived by using flaretime ionization profiles is curve 1. The basis for the other profiles as well as the theoretical profiles of Brasseur and Nicolet (dotted curves) is described in text
(after Mitra and Rowe, 1973).

Meira's) at 100 km, and *including* the effect of photo-dissociation. This leads to a rapid decrease of the nitric oxide concentration in the mesosphere. The results of additional production of atomic nitrogen (Brasseur and Nicolet use $P(N) = 1$ and $10 \, \mathrm{cm}^{-3} \, \mathrm{s}^{-1}$) have strong effects on NO distribution for k_{min} distribution but relatively little effects for k_{max} distribution for which transport effects are more dominant. The dotted curves in Figure 96 give Brasseur and Nicolet's theoretical curves for k_{max} profiles for $P(N) = 1$ and $10 \, \mathrm{cm}^{-3} \, \mathrm{s}^{-1}$ respectively.

It is important to note that not only do these theoretical profiles give NO concentrations significantly lower than Meira's below 80 km, the shapes are also different. Brasseur and Nicolet's distributions for k_{max} have minima around 70 km, considerably lower than the 85 km given by Meira.

In addition several ionospheric estimates are also available. These include the use of information on the level of ionization reversal with solar activity, the gradual loss of solar control in the diurnal variation of electron density at levels below 70 km (Mitra, 1969), and the magnitude and changes in the ratio of the two molecular ions [NO$^+$]/[O$_2^+$] (Mitra and Rowe, 1973), as well as the use of positive ions (Hale, 1973). It is important here to mention that although ionospheric estimates are necessarily approximate, it is almost always possible to define *an upper limit*, and that estimates of such upper limit have in the past acted as a restraint on hasty interpretations of rocket observations of $\gamma \, (1, 0)$ bands. Profiles obtained from different ionospheric approaches are summarised in Figure 96, along with the γ-band profiles of Meira (1971) and Tisone (1973) and the theoretical profiles of Brasseur and Nicolet. A rule-of-thumb value is $1 \times 10^7 \, \mathrm{cm}^{-3}$ all the way from 70 to 90 km.

Above 90 km, where the theoretical curves accept Meira's distributions, ionospheric estimates are in agreement with γ-band measurements, except that the latter fall somewhat intermediate between the low and high solar activity values. An increase in NO concentration from low to high solar activity by a factor of 4 above this height is indicated. The evidence of quiet-day midlatitude profiles between 70–80 km requiring a concentration of $5 \times 10^8 \, \mathrm{cm}^{-3}$ at 70 km is rather strong. A minimum around 70 km is also indicated by these measurements.

An intriguing measurement has been provided by Zipf *et al.* (1970) who observed at 120 km an unusually large NO concentration of 3.8×10^{10} (larger than O$_2$ concentration of 2.3×10^{10}) in a bright auroral arc above Ft. Churchill.

In the current theories of nitric oxide production, NO is formed in the *E-region* from dissociative recombination of NO$^+$ as follows:

$$\mathrm{NO^+} + e \rightarrow \mathrm{N(^2D)} + \mathrm{O}$$
$$\mathrm{N(^2D)} + \mathrm{O_2} \rightarrow \mathrm{NO} + \mathrm{O}. \tag{77}$$

Nitric oxide so produced is then moved downwards to mesospheric heights. Thus mesospheric nitric oxide has an *ionospheric* origin.

Keneshea *et al.* (1970) calculated changes in NO concentration for the situation for which $\mathrm{N(^2D)}$ principally (50%) from $\mathrm{NO^+} + e \rightarrow \mathrm{N(^2D)} + \mathrm{O}$. Their results are that at 140 km, NO concentration is the same at day and night ($\sim 10^7 \, \mathrm{cm}^{-3}$) while N concentration decreases from 10^7 at day to 2×10^2 cm^{-3} at night. At 130 km, variation in NO concentration is similar although the actual concentration is somewhat lower. The minimum in N concentration at night is about 2×10^2 cm^{-3}. At 160 km, a decrease from day to night is observed – from 3×10^6 cm^{-3} to 3×10^5 cm^{-3} at night, while [N] is now stable at $(1–5) \times 10^7$ cm^{-3}. Also at this height [NO] < [N].

9.2.2. OZONE IN MESOSPHERE

Another important constituent is ozone. Ozone plays an important role in negative ion concentration. Ozone measurements using ultraviolet absorption are essentially limited to a height of about 65 km, above which reported values must be considered suspect. Several measurements are now available, conducted mostly with rocket borne instruments, giving the ozone distribution below this height during day as well as at night. Evans *et al.* (1968) point out that the measurements by Johnson *et al.* (1954), often considered the most satisfactory measurements give an excellent fit, between 40–70 km, to the expression:

$$[O_3] = 8.0 \times 10^{10} \exp\left[- (h - 50)/4.53 \right] \text{ molecules cm}^{-3}. \tag{78}$$

In some Japanese experiments, a daytime maximum has also been observed at mesospheric heights. Typical daytime values are 5×10^{12} cm^{-3} at 40 km, 5×10^{10} cm^{-3} at 50 km, 8×10^9 cm^{-3} at 60 km and 6×10^8 cm^{-3} at 70 km (Figure 97).

Above 70 km a promising technique appears to be the measurement of $O_2(^1\Delta_g)$, which is produced by photolysis of ozone and has been found to be quite abundant in the mesosphere (Evans *et al.*, 1968). At those heights quenching is negligible and the concentration of $O_2(^1\Delta_g)$ is just the product of production rate and the radiative life time. On this consideration, Evans *et al.* (1968) obtain the following important equation, using a radiative lifetime of 3600 s:

$$[O_2, {}^1\Delta_g] = 34 [O_3]. \tag{79}$$

Ozone concentration derived from Equation (79) is larger than that would be obtained by extrapolating Equation (78) and may well be a result of increased atomic oxygen concentration because of downward diffusion of atomic oxygen.

There is no diurnal variation in ozone at 50 km (as one would expect from the large characteristic time at this altitude), but nighttime enhancements appear at greater heights in some of the rocket flights, although it is not so apparent in the theoretical oxygen-hydrogen calculations. There is a drastic decrease in atomic oxygen concentration after sunrise at altitudes below 80 km; this has a pronounced effect in mesospheric chemistry.

9.2.3. WATER VAPOUR AND OTHER HYDROGEN COMPOUNDS

The chemistry of water vapour and other hydrogen compounds are inextricably mixed with the chemistry of O and O_3, and the distributions of these constituents can be obtained theoretically in the same calculations. Hesstvedt (1969) finds the mixing ratio for water vapour to be 5×10^{-6} at 65 km, 4×10^{-6} at 80 km and 1×10^{-6} at

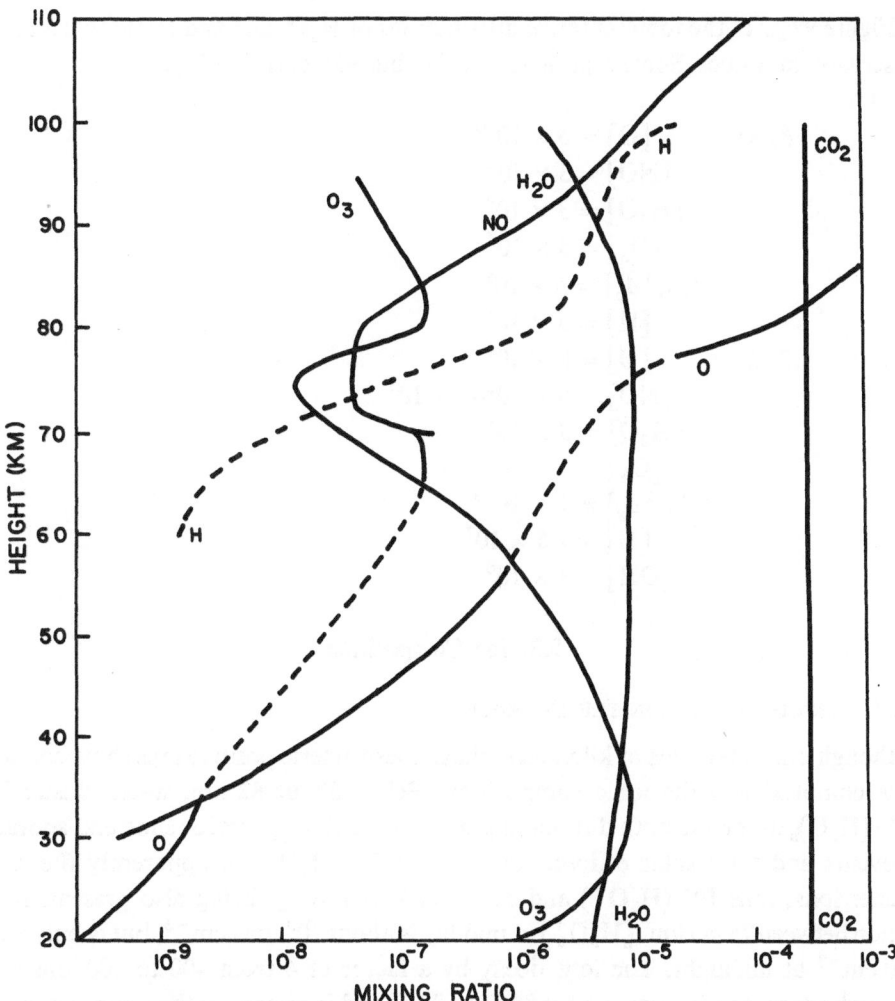

Fig. 97. Representative altitude distributions in the mesosphere of some of the principal minor constituents: NO, O_3, H_2O, H, O and CO_2.

95 km. With these mixing ratios, H_2O concentrations are large enough to play significant role in mesospheric ion chemistry.

Since water vapour is dissociated above about 70 km, and a substantial amount of atomic hydrogen produced escapes from the atmosphere, the 'sink-over-source' situation prevails, and there is a continuous transport of water vapour through the stratosphere and the mesosphere. The time constant for upward mixing is about one day and is about the same as the dissociation time constant. As a result, water vapour reaches a low concentration, about one scale height above the level of unit optical depth (~ 75 km). On the other hand, if eddy diffusion were to increase by a factor of 2, water vapour would be expected to extend to two scale heights above that level. This possibility may have important consequences for the hydration of the D-region positive and negative ions.

Figure 97 gives the representative distributions of H_2O, OH and H along with some observational values. Some representative daytime values (in cm^{-3}) are:

$$
\begin{aligned}
80\ km \qquad &[O] = 5 \times 10^{10} \\
&[NO] = 3 \times 10^7 \\
&[H_2O] = 3 \times 10^9 \\
&[O_3] = 6 \times 10^7 \\
&[O_2, {}^1\!\Delta_g] = 5 \times 10^9 \\
&[H] = 3 \times 10^8 \\
70\ km \qquad &[O] = 8 \times 10^9 \\
&[NO] = 5 \times 10^6\text{--}1 \times 10^7 \\
&[H_2O] = 3 \times 10^{10} \\
&[O_3] = 5 \times 10^7 \\
&[O_2, {}^1\!\Delta_g] = 1 \times 10^{10} \\
&[H] = 1.5 \times 10^7 \\
&[OH] = 3 \times 10^6 .
\end{aligned}
$$

9.3. Ion Composition

9.3.1. ION COMPOSITION IN THE D-REGION

Although only a few tens of kilometers thick, mesospheric positive ions show considerable complexity in the ionic composition. Below about 82 km, water cluster ions $H^+ \cdot (H_2O)_n$ dominate both day and night as well as during meteor showers, sporadic-E events and total solar eclipse. Ions $37^+ (H_3O^+ \cdot H_2O)$ are apparently the major cluster ions; ions $19^+ (H_3O^+)$ and $55^+ ((H_3O^+) \cdot (H_2O)_2)$ being also present. In the region between 75–85 km, $[H_5O_2^+]$ at midday is about 10^3 ions cm^{-3}, but is only about 100 cm^{-3} at midnight. The ions decay by a factor of 4 from 400 to 100 ions cm^{-3} through sunset as χ increases from 88.5° to 98.6° and increases by the same factor from 30 to 100 ions cm^{-3} at sunrise as χ decreases from 102.5° to 90°. Metal ion layers have been observed near 95 km, composed mainly of magnesium and iron, and about 5–10 km in the half-width, and a narrower layer at a somewhat higher level composed mainly of silicon, magnesium, or iron. NO^+ and O_2^+ ions are the dominant ions from 90 km to about 200 km, where $[O^+]$ equals $([NO^+] + [O_2^+])$. When the layer is metallic, a decrease in the density of molecular ions is observed. The stronger the metallic ion concentration the greater is the decrease. Molecular nitrogen ions, 28^+, are a minor ion constituent.

A striking feature in the mesosphere is the level of sudden disappearance of cluster ions. This occurs at around 82 km during day, but only at 86 km at night. There is a sharp ledge of electron density around this level, the depth of which is probably even less than 1 km. During a Polar Cap Absorption event this level moves downwards to below 70 km or, in other words, clusters vanish in the region above 72 km. Figure 98 gives an idealised version of this phenomenon. As we shall see later similar condition is expected during a solar flare.

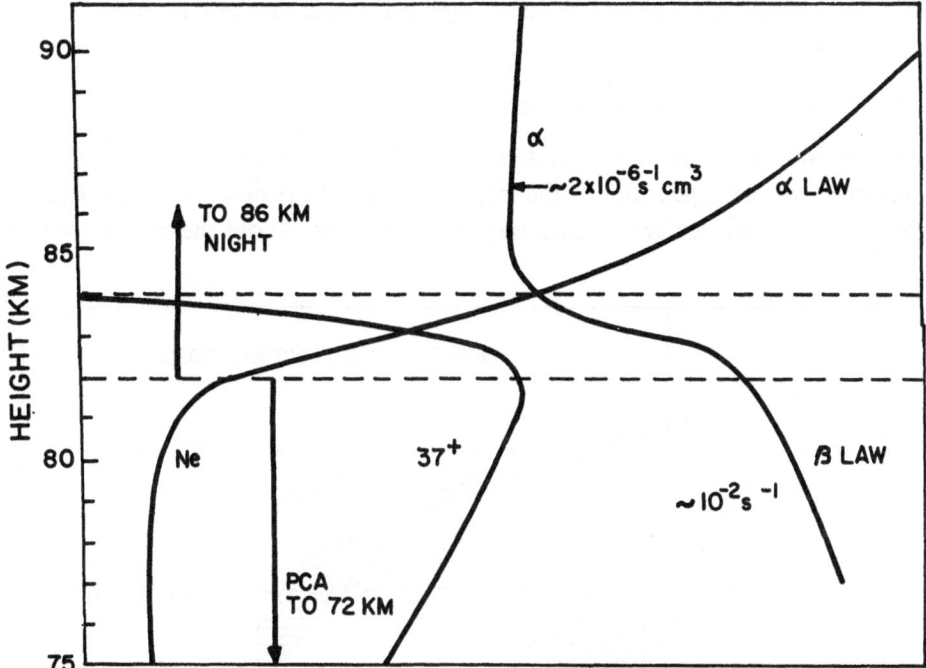

Fig. 98.   Ledge in electron density and level of disappearance of water cluster ions in the mesosphere.

Measurements of the total negative ion density are usually accomplished by measuring separately the positive ion density and the electron density and calculating the difference.

Measurements have been made of the negative ion composition at 70 km at night and during an eclipse (Narcisi *et al.*, 1971; Arnold *et al.*, 1971; Narcisi *et al.*, 1972d) but not during daytime full Sun conditions. The ions tentatively found include O^-, O_2^-, Cl^-, NO_2^-, NO_3^-, CO_3^-, HCO_3^-, CO_4^-, $NO_3^- \cdot (H_2O)$ and $CO_3^- \cdot (H_2O)$. Not all have been observed on the same flight or by the same experimenters, nor have these been the same in the German and American experiments. An intriguing feature is that while in Narcisi's experiments, the dominant negative ions are NO_3^- and its hydrates (upto sixth hydrate has been observed), in the German experiments the dominant ions are CO_3^-.

The quantity λ is often used in *D*-region work as a measure of the importance of negative ions. Figure 99 shows some estimates of λ; these include the results of measurements of positive ions and electrons with a certification rocket prior to the PCA event of November 2, 1969 and those during the PCA daytime conditions (Ulwick, 1972) – the two results are within a factor of two even though the production rates changed by an order of magnitude; use of Hale's average positive ion density (Hale, 1973) with summer upon average N_e profiles obtained with cross-modulation experiments with Bain-Harrison profile (Bain and Harrison, 1972) and Illinois active Sun profile (Mechtly *et al.*, 1972). The shaded portion is the range of λ values that we believe to be representative, although values higher than these have sometimes been suggested.

Table XVII gives an outline of major *D*-region features during day, night, eclipse, PCA and flares.

9.3.2. ION COMPOSITION IN *E*- AND *F*-REGIONS

Molecular ions of oxygen (32^+) and nitric oxide (30^+) dominate the *E* and lower *F*1-region. Atomic oxygen (16^+) is the major ion in the *F*2 region. *F*1 is the transistion region where the molecular ions give way to atomic ions. Above the cross over level,

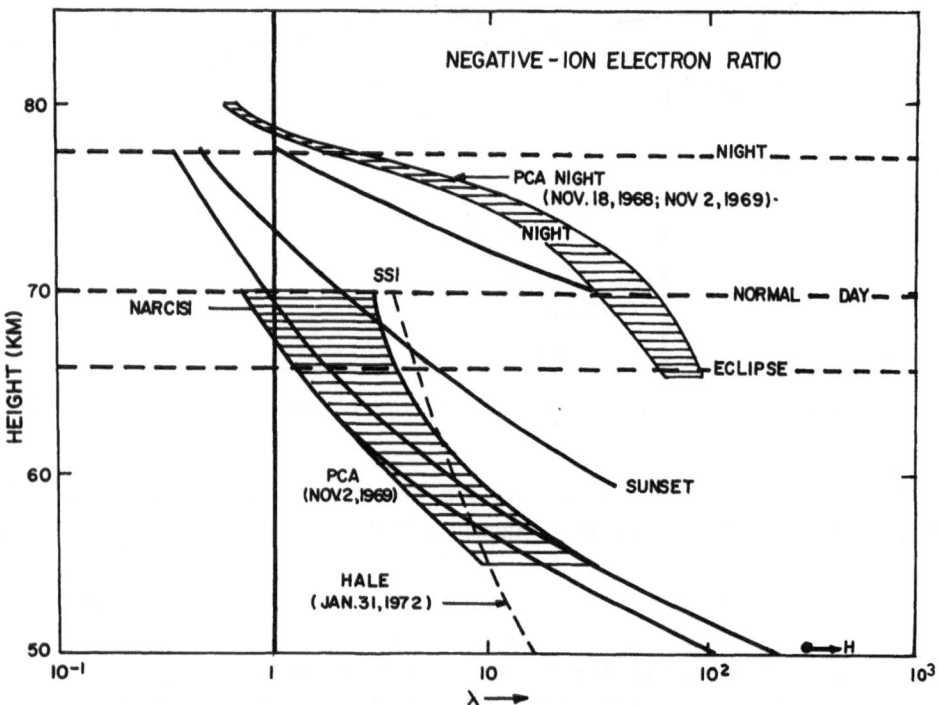

Fig. 99.   Altitude distribution of negative ion to electron concentration ratio (λ) in the lower iono-sphere for day and night conditions, and during PCAs (after Rowe *et al.*, 1974).
(The dashed horizontal lines denote heights where $\lambda=1$.)

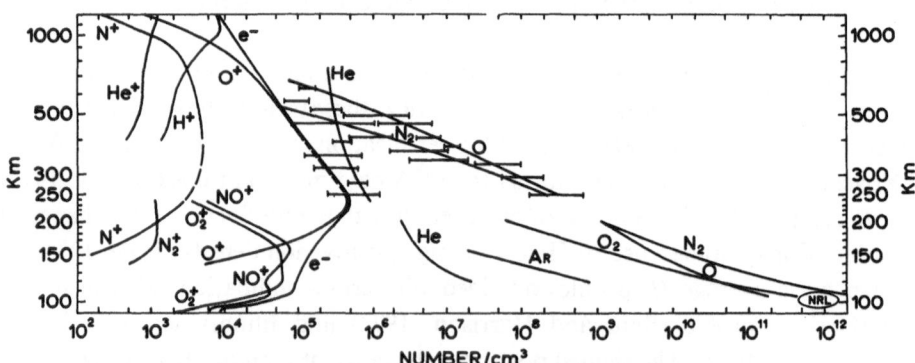

Fig. 100.   Ion composition in the *E* and *F*-regions (after Johnson, 1969).

the molecular ions 32^+ and 30^+ decrease in number density with a scale height similar to those of O_2 and N_2. Atomic and molecular nitrogen ions, (14^+) and (28^+) are always minor constituents in the ionosphere (Figure 100).

At sunset the E and lower $F1$-regions have decayed an order of magnitude. Molecular oxygen (32^+) has decayed two orders of magnitude in the E-region. Molecular nitrogen (28^+) has disappeared. Atomic oxygen (16^+) remains the dominant $F2$-region ion but has been consumed by ion-chemical reactions in the E and lower $F1$-region. The most surprising feature of this sunset ionosphere is the nitric oxide (30^+) molecular oxygen (32^+) relative density reversal above 175 km compared to the daytime value.

At night nitric oxide (30^+) dominates the E, $F1$ and the first 20 km of the $F2$-region. Molecular oxygen (32^+) remains constant at 100 ions cm^{-3} from 130 to 190 km, then increases in density to become the major molecular ion at 220 km. There is an increase in both NO^+ and O_2^+ densities above 190 km. At the cross over point the $NO^+ - O_2^+$ reversal seen in the sunset ionosphere is again observed.

In the topside atomic oxygen, nitrogen, helium and hydrogen are the major ions. The relative proportion of hydrogen to helium ions is seen to be a function of solar activity. During the minimum solar activity, H^+ density was larger than He^+ ions, whereas the opposite situation is observed when the activity is larger.

Thin layers of ionization, often of anomously high density, are observed in the E-region by radio techniques and rocket borne experiments (Smith and Matsushita, 1962). It has been suggested that these layers, in the mid-latitude, are formed by windshear actions on the ionosphere and the accumulation at the shear level of ions with the smallest recombination coefficients, i.e. atomic species. This has been confirmed by experiments in which trimethyl aluminum (TMA) was released from gun-launched projectiles with simultaneous operation of a ground based ionosonde.

A major clue to the mid latitude sporadic E problem was the measurements by Narcisi *et al.* (1969) of metallic ions within a sporadic E layer in which metallic ions of magnesium, sodium, silicon and possibly iron were the only species detected within the layer. The normal molecular ions of NO^+ and O_2^+ were not detected in the blanketing layer, although they existed both above and below it. They are rapidly lost by dissociative recombination in the dense electron layer whereas the metallic ions being atomic are removed by a slower process.

9.3.3. ION CHEMISTRY

The different types of reactions that occur in the ionosphere are listed in Table XVIII. The reactions written from left to right are, in general, exothermic. In the normal atmosphere and in a highly ionized atmosphere which has come to thermal equilibrium and is rather cool, the endothermic reactions are usually unimportant. In addition to ionization, excitation and dissociation created by the radiation and high-energy particles, high temperature may cause collisional excitation, dissociation and ionization. These are the reverses of the reactions in Table XVIII.

In the ambient temperature region, the processes of greatest importance are those

TABLE XVII

Important D-region features

(Most values refer to $\chi = 60°$ and midlatitudes)

Feature	Day	Night	Eclipse	PCA	Flares
1. Level of N_e ledge (Z_0)	82–85 km	—	82.5 km (1970 eclipse)	No evidence of ledge	—
$\left(\dfrac{dN_e}{dh}\right)$ at ledge	$(4\text{–}10) \times 10^2$ cm^{-3} km^{-1}		86.8 km (1966 eclipse)		
2. N_e at 95 km	$1.33 \times 10^4\ (R=11)^a$ cm^{-3}	—	—	—	—
	$2.73 \times 10\ (R=96)^a$ cm^{-3}	—	—	—	—
3. Cluster disappearance level (Z_1)	~82 km	86 km		<72 km (day)	Expected as in PCA
		86 km (twilight)			
4. Disappearance depth	~2 km (or less)	—	—	—	—
5. N_e at 80 km	600–1300 (Av: 900) cm^{-3}	—	—	—	—
N_e at 70 km	200–500 (Av: 300) cm^{-3}				
6. β below Z_0	Ranges between $(1\text{–}2) \times 10^{-2}$ s^{-1} down to 70 km, nearly constant. βN_e in this region gives q. Also old controversy that NO^+ production rate is larger than loss is removed.	—	8×10^{-3} down to $(Z_0 - 10)$ km. Nearly constant. Evidence of slight decrease from full-Sun value	—	—
7. Positive Ion Composition at and immediately below Z_0	Hydrates dominant. Hydrate population determined by Kebarle formula. $[\Sigma H^+\cdot(H_2O)_n]/[NO^+] = 0.1\text{–}0.8$	—	Clusters remain essentially unaffected, but NO^+ disappears quickly	Clusters disappear down to 72 km during day time and down to 77 km at night	Probably similar to PCA

[a] Follows solar activity. Empirical relationship given by Mechtly et al. (1972) is: $N_{95}(\text{cm}^{-3}) = 178\ R + 10800$.

Table XVII (Continued)

Feature	Day	Night	Eclipse	PCA	Flares
8. $\lambda = 1$ level	70 km	80 km (indicates atomic oxygen diurnal variation at least upto 80 km)	—	—	—
9. λ vs h	$\sim 2 \times 10^2$ at 50 km; ~ 5–10 at 60 km; ~ 1–3 at 70 km; $\ll 1$ at 75 km	30 at 70 km; 1 at 80 km	Evidence of formation of negative ions as high as 88 km. Decay of positive ions (a factor of 4 from full-Sun to totality) much less than in N_e	—	—
10. Solar Activity Reversal level	60–65 km	—	—	—	—
11. Approximate values of $(N_e)_h/(N_e)_l$ (active Sun to quiet Sun)	0.3–0.6 at 55–60 km; ~ 1 at 63–64 km; ~ 2 at 70 km; 2–4 bet. 75–90 km	—	—	—	—
12. Levels of little or no diurnal variation	$h < 65$ km	—	—	—	—
13. ψ below ledge	$(1$–$5) \times 10^{-5}$ cm^3 s^{-1} (nearly constant down to 70 km)	Increases rapidly below 80 km to 10^{-4} cm^3 s^{-1} near 75 km	—	Drastic decrease at least down to 70 km and probably also below.	Decreases by a factor of 5 at 80 km; also decreases at 70 km by smaller amount
ψ above ledge	2×10^{-6} cm^3 s^{-1}	—	Not affected	Not affected	Not affected

Table XVII (Continued)

Feature	Day	Night	Eclipse	PCA	Flares
14. Negative ions		Below 80 km, principally heavy ions of the type $NO_3^- \cdot (H_2O)_n$ with $n = 0$–5, CO_3^- dominant in Arnold and Krankowsky's measurements, with sharp decrease above 78 km – above negative ions mostly O_2^- and O^-.	A layer of large negative cluster ions near 88 km	Day: O_2^- major ions between 70–90 km, with $NO_3^- \cdot (H_2O)_n$ and CO_4^- in substantial amount. No information on O^-. Night: O^- major ions between 75–94 km.	
15. Region below 70 km	Possible existence of a 'C' layer. N_e at 60 km = 40–80 cm^{-3} (av. = 50) Other view: N_e at 60 km <10 cm^{-3}. Electron density profiles not related to changes in +ive ions. Dependent on T, but not N_e (Hale, personal communication)			N_e large between 10^3 to 10^4 cm^{-3} for strong PCA; two to three orders of magnitude larger than quiet day values	For a strong flare N_e is between 10^2 to 10^3 cm^{-3} at 60 km.

by which free electrons are removed by binding the electrons to any other particle, atomic or molecular, charged or neutral. These are in general exothermic reactions.

Electrons may be removed by direct recombination with positive ions (type I in Table XVIII). Dissociative recombination is very rapid (10^{-7} to 10^{-5} cm^3 s^{-1} at room temperature) and prevails where molecular ions are present. Radiative deionization is important only in a highly dissociated atmosphere, while threebody recombination is important only in a dense atmosphere.

Electrons may also be removed by attachment to neutral molecules or atoms (type III in Table XVIII). The constituents O, O_2, O_3 and NO_2 are important in this type of reaction. Negative ions can then be removed by either two- or three body mutual neutralization (type II in Table XVIII), or destroyed by detachment which releases free electrons.

Although recombination and attachment are the two main mechanisms for the removal of free electrons, several other types of reactions are important in leading to these mechanisms. The rate at which electrons disappear by the above reactions depends on the specific ion or, in the case of attachment, on the specific neutral particle. These are controlled by the ion-neutral reactions (type IV in Table XVIII) and by the neutral-neutral reactions (type V in Table XVIII). This is most easily seen by considering the ions rather than the electrons.

The photoionization and photodissociation reactions require no kinetic energy and are of importance in any system with an appreciable time scale when any radiating source is effective, e.g., during the daylight hours when the solar radiation has large effects.

For positive and negative ions, the complex reaction schemes can be understood by arranging them in order of their respective electrochemical series. Positive ions are arranged in the order of *decreasing* ionization potential and negative ions in the order of *increasing* electron affinity. Positive or negative ion-neutral reactions proceed *down* the electrochemical series.

9.3.3.1. *Important Reaction Rates*

(a) *Dissociative recombination*. The principal mechanism for deionization of molecular ions is dissociative recombination:

$$AB^+ + e \rightarrow AB^* \text{ (unstable)} \rightarrow A + B \text{ kinetic energy},$$

where A and B may be atomic but are not necessarily so.

The recombination coefficient is large, 10^{-7} cm^3 s^{-1} or larger.

The principal dissociative reactions of interest are those of O_2^+, NO^+ and of the different types of water cluster ions. The relevant values in cm^3 s^{-1} are (Biondi *et al.*, 1972):

$$\alpha_D(19^+) = (1.3 \pm 0.6) \times 10^{-6}$$
$$\alpha_D(37^+) = (2.7 \pm 1) \times 10^{-6}$$
$$\alpha_D(55^+) = (4.6 \pm 1) \times 10^{-6}$$
$$\alpha_D(\gtrsim 100^+) \approx 10 \times 10^{-6}.$$

TABLE XVIII

Types of reaction

I. Ion-Electron Recombination – Ionization
  A. Radiative Recombination – Photoionization $X^+ + e = X + h\nu$
  B. Three-body Recombination – Collisional Ionization $X^+ + e + M = X + M$
  C. Dissociative Recombination – Associative Ionization $XY^+ + e = X + Y$
  D. Ion-Electron Recombination with Dissociation – Ionization with Association $X^+ + e + YZ = X + Y + Z$
  E. Ion-Electron Recombination with Rearrangement – Ionization with Rearrangement $WX + YZ^+ + e = WY + XZ$

II. Ion-Ion Recombination – Ion Pair Formation
  A. Mutual Neutralization – Ion Pair Formation $X^+ + Y^- = X + Y$
  B. Associative Ion-Ion Recombination – Dissociative Ion Pair Formation $X^+ + Y^- + M = XY + M$
  C. Ion-Ion Recombination with Rearrangement – Ion Pair Formation with Rearrangement $WX^+ + YZ^- = WY + XZ$

III. Attachment – Detachment
  A. Radiative Attachment – Photodetachment $X + e = X^- + h\nu$
  B. Three-Body Attachment – Collisional Detachment $X + e + M = X^- + M$
  C. Dissociative Attachment – Associative Detachment $XY + e = X^- + Y$
  D. Attachment with Rearrangement – Detachment with Rearrangement $WX + YZ + e = WY^- + XZ$

IV. Ion-Neutral Reactions
  A. Charge Transfer, Positive Ion $X^+ + Y \rightleftharpoons X + Y^+$
  B. Ion-Molecule Reactions, Positive Ion $WX^+ + YZ \rightleftharpoons WY^+ + XZ$
  C. Charge Transfer, Negative Ion $X^- + Y \rightleftharpoons X + Y^-$
  D. Ion-Molecule Reactions, Negative Ion $WX^- + YZ \rightleftharpoons WY^- + XZ$
  E. Radiative-Ion-Neutral Association – Ion Photodissociation, Positive Ion $X^+ + Y = XY^+ + h\nu$
  F. Three-Body Ion-Neutral Association – Collisional Ion Dissociation, Positive Ion $X^+ + Y + M = XY^+ + M$
  G. Radiative Ion-Neutral Association – Ion Photodissociation, Negative Ion $X^- + Y = XY^- + h\nu$
  H. Three-Body Ion-Neutral Association – Collisional Ion Dissociation, Negative Ion $X^- + Y + M = XY^- + M$

V. Neutral-Neutral Reactions
  A. Radiative Neutral Recombination – Neutral Photodissociation $X + Y = XY + h\nu$
  B. Three-Body Neutral Recombination – Neutral Collisional Dissociation $X + Y + M = XY + M$
  C. Rearrangement $WX + YZ = WY + XZ$

All values are normalised to $300\,\mathrm{K}$ (using $T^{-1/2}$ relationship).

(b) *Positive ion-molecule reactions.* This type of reactions is one of the most important in ionospheric chemistry. While interest has for long been focussed mainly on the major atmospheric constituents N_2, O_2 and O, interest has recently also been extended to the minor constituents O_3, NO, OH, CO_2 and water clusters (present in the D-region), He and H (present in the exosphere) and to metals (lower E-region). Another important area of interest is the role of excited states of reactants. In several instances it has been observed that the rate depends very strongly on the internal energy of one or both reactants. The processes involving positive ions are:

(a) *Charge Transfer*

$$A^+ + BC \rightarrow A + BC^+ .$$

(b) *Ion-Molecule Reactions (or Ion-Atom Interchange)*

$$A^+ + BC \rightarrow AB^+ + C .$$

The basic difference between the two is that in (a) the identity of the reactants is conserved, in (b) they are altered.

Some of the major reactions of interest in the ionosphere are the following:

$$
\begin{array}{ll}
O^+ + O_2 \rightarrow O_2^+ \ + O & (k_1) \\
O^+ + N_2 \rightarrow NO^+ + N & (k_2) \\
N_2^+ + O_2 \rightarrow N_2 \ + O_2^+ & (k_3) \\
N_2^+ + O \ \rightarrow NO^+ + N & (k_4) \\
O_2^+ + N_2 \rightarrow NO^+ + NO & (k_5) \\
O_2^+ + NO \rightarrow NO^+ + O_2 & (k_6) \\
He^+ + O_2 \rightarrow He + O + O^+ & (k_7)
\end{array}
$$

$$
He^+ + N_2
\begin{cases}
\nearrow He + N + N^+ & (k_a) \\
\searrow He + N_2^+ & (k_b)
\end{cases}
$$

$$NO^+ \xrightarrow{B} H^+ \cdot (H_2O)_n . \qquad (B)$$

Of these the first two control the F-region chemistry. While their reaction rates are well determined at $300\,\mathrm{K}$ (2×10^{-11} and $2 \times 10^{-12}\ \mathrm{cm^3\ s^{-1}}$), one needs to know their values at temperatures of the order of $1000\,\mathrm{K}$. Measurements at different temperatures are now available for both these processes, and it appears that while for the ($O^+ + O_2$) reaction an empirical relationship of the following nature is appropriate:

$$k_1 = 2 \times 10^{-11} (T/300)^{-0.5}\ \mathrm{cm^3\ s^{-1}},$$

for the ($O^+ + N_2$) process no clear-cut relationship can be established. Recent experimental observations seem to indicate that its rate process slowly decreases to a minimum at around $1100\,\mathrm{K}$ and then increases. With a value $2 \times 10^{-12}\ \mathrm{cm^3\ s^{-1}}$ at

300 K, the rate would decrease to a value between $(1-1.4) \times 10^{-12}$ cm^3 s^{-1} at F-region temperatures.

An important result reported by Schmeltekopf *et al.* (1967) is that in the presence of internally excited N$_2$ the rate is much faster. This has far-reaching consequences in the F-region where an appreciable population of N$_2$ in the vibrational state is not unlikely.

The (N$_2^+$ + O$_2$) process which almost completely depletes N$_2^+$ ions in the ionosphere even though its production is the largest, has a rate of about 1×10^{-10} cm^3 s^{-1} and is essentially independent of temperature over ion temperatures of 400–800 K. For the process (O$_2^+$ + N$_2$) only the upper limit of 1×10^{-15} cm^3 s^{-1} is known, but the rate cannot be much less, and is expected to be around 1×10^{-16} cm^3 s^{-1}. Because of its low value, its temperature dependence can be important. Since it involves the breaking of *two* bonds, it will involve an activation energy (an exponential temperature dependence) as neutral chemical reactions often do. The process (O$_2^+$ + NO) is also essentially temperature independent (over ion temperatures of 400–2000 K) and so is the important upper-F-region process He$^+$ + N$_2$. The branching ratio for the latter is N$^+ \sim 70\%$ and N$_2^+ \sim 30\%$.

NO$^+$ conversion to H$^+ \cdot$ (H$_2$O)$_n$ consists of several steps and is the most vital in the mesosphere. This is discussed in detail later.

(c) *Negative ion reactions.* The most important negative ion reaction is the three-body attachment:

$$e + O_2 + O_2 \rightarrow O_2^- + O_2$$

for which the rate is about 2×10^{-30} [O$_2$]^2 and is essentially independent of temperature. The process is 40 times more effective than

$$e + O_2 + N_2 \rightarrow O_2^- + N_2.$$

Of great interest is photodetachment given by:

$$X^- + hv \rightarrow X + e \tag{ϱ_i}$$

in which X may be an atom or a molecule.

Photodetachment coefficients suggested for different species of negative ions are indicated below:

$$\varrho(O^-) \ = 1.4 \, \text{s}^{-1}$$
$$\varrho(O_2^-) \ = 0.33 \, \text{s}^{-1}$$
$$\varrho(O_3^-) \ = 1.3 \, \text{s}^{-1}$$
$$\varrho(CO_3^-) = 0.04 \, \text{s}^{-1}$$
$$\varrho(NO_2^-) = 10^{-2\pm1} \, \text{s}^{-1}$$
$$\varrho(NO_3^-) = 1.5 \times 10^{-3} \, \text{s}^{-1}.$$

Other negative ion processes of interest are those of associative detachment, the most important being:

$$O_2^- + O \rightarrow O_3 + e$$

and the ion-ion recombination:

$$X^+ + Y^- \rightarrow X + Y \hspace{3cm} (\alpha_{imn})$$

in which X and Y may either be atomic or molecular. One or both of the neutrals may be in excited state. The rate depends strongly on the nature of the X^+ and Y^-, but is generally between $(1-10) \times 10^{-7}$ cm^3 s^{-1} at mesospheric temperatures. Any imbalance between positive and negative ion composition will have large influence on the ionization loss rates.

The negative-ion molecule reactions (2-body types) have mostly rates around a few times 10^{-10} cm^3 s^{-1}.

9.3.3.2. *Positive Ion Schemes*

Figures 101a–e and Figures 102 and 103 represent the reaction schemes for the positive ions NO$^+$, O$_2^+$, O$^+$, He$^+$ and H$^+$ and the hydrated positive ions respectively.

For the D region the important point to note is that we have two different types of positive molecular ions. One type is represented by a very large dissociative recombination coefficient; these are the hydrated ions with the coefficient in the neighbourhood of 10^{-5} cm^3 s^{-1}. The other type is represented by the coefficient more than order of magnitude smaller $(2-5 \times 10^{-7}$ cm^3 s$^{-1})$; these are the simple molecular ions NO$^+$ and O$_2^+$. The consequence of a mixture of these two types, as we have in the D region, not only affects the net magnitude of the effective recombination coefficient, but also determines whether the ionization loss process is governed by a quadratic law (Equation (40)) or a linear law (Equation (41)). This was first discussed by Haug and Landmark (1970) in which the more rapidly recombining ions were called X_2^+ and the more slowly recombining ions X_1^+. A further criterion was that the production of X_2^+ is proportional to the concentration of X_1^+ i.e. X_2^+ is an ion chemical product of X_1^+. In such a case, one gets:

$$q = N_e(1 + \lambda)(\alpha_1 N_e + B) \frac{\alpha_2 N_e}{\alpha_2 N_e + B},$$

where B is the rate at which X_2^+ is converted into X_1^+. For values of B in the range:

$$\alpha_1 N_e \ll B \ll \alpha_2 N_e$$

the following linear relationship is obtained:

$$q = (1 + \lambda) B N_e.$$

9.4. Information on Nitric Oxide from Simultaneous Measurements of Electron Density and Solar X-Rays During Flares

We have already seen that evidence exists for concluding that large [NO] values are not acceptable from even elementary considerations of SID ionization. The ease with which even a minor X-ray enhancement produces an SID does not allow an NO concentration much in excess of 10^7 cm^{-3} at 80 km.

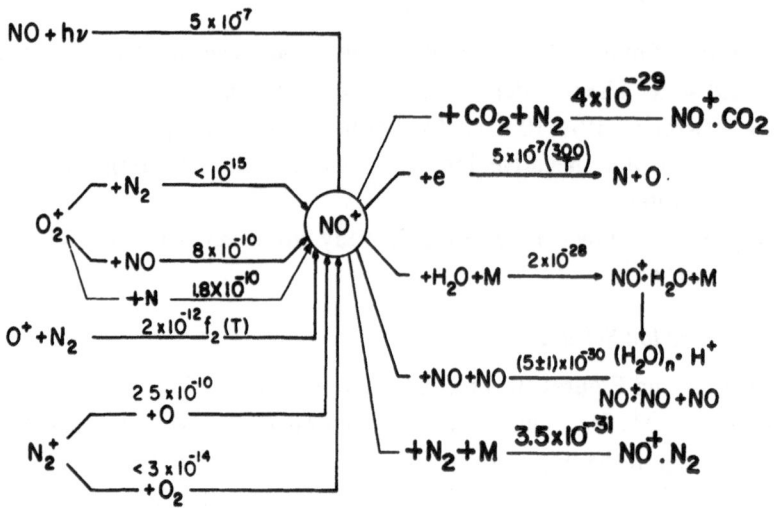

Fig. 101a.

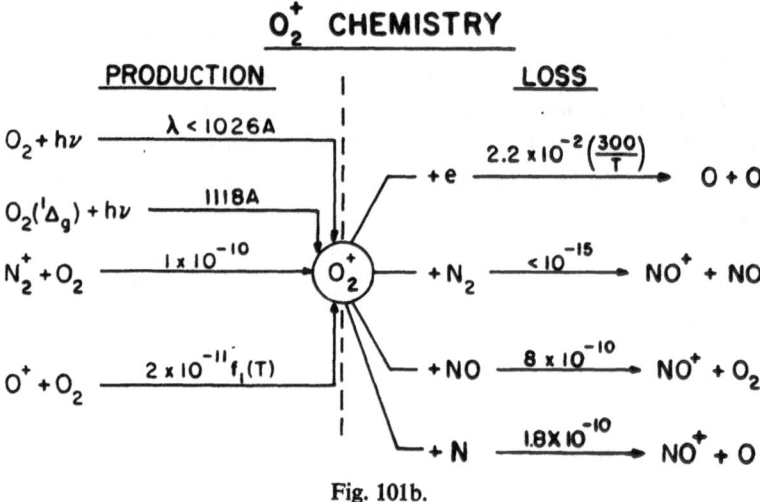

Fig. 101b.

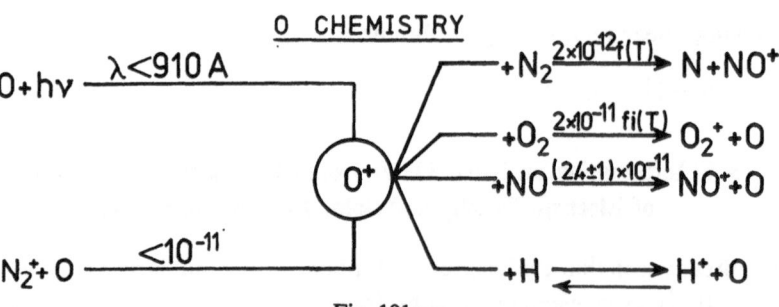

Fig. 101c.

Fig. 101a–e.   Reaction schemes for positive ions: (a) NO+, (b) O₂+, (c) O+, (d) He+, and (e) H+
(after Mitra, 1971).

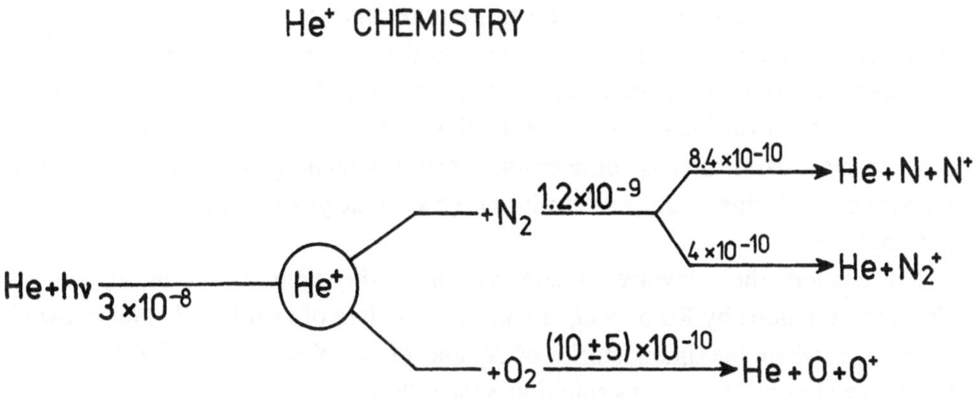

He⁺ CHEMISTRY

Fig. 101d.

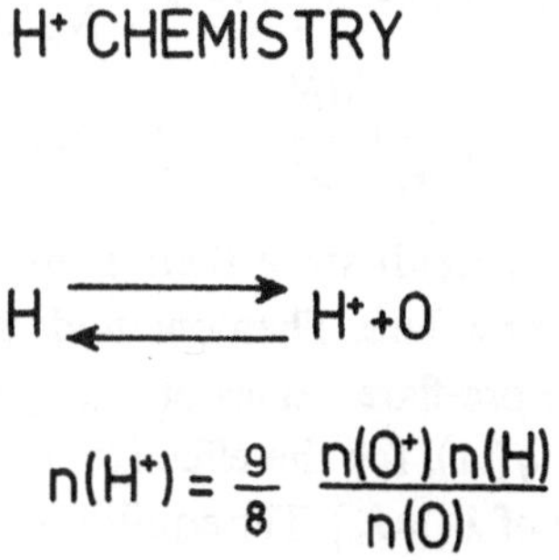

H⁺ CHEMISTRY

$$n(H^+) = \frac{9}{8} \frac{n(O^+)\, n(H)}{n(O)}$$

Fig. 101e.

A simple consideration is to relate observed $\Delta N_e/N_e$ at any level with the computed value of $\Delta q/q$. If one has the complete information on the ionizing spectrum (especially for wavelengths below 3 Å) so that Δq can be computed with some reliability, then one can use this relationship to obtain an upper limit for q either: (a) by comparing peak values of $\Delta q/q$ and $\Delta N_e/N_e$ or more properly (b) by comparing the entire time sequences of q and N_e. Let q be divided into two parts:

$$q_X + q(A)$$

in which q_X represents production due to X-rays and $q(A)$, production due to all sources including Lα.

In the first case one can write simply,

$$\frac{N_f^2}{N_n^2} = \frac{\psi_f}{\psi_n} \left[\frac{q_{X_f} + q(A)}{q_{X_n} + q(A)} \right] \tag{80}$$

in which, for most heights below 80 km, $q(A) \rightarrow q(L\alpha)$.

This gives an estimate of $q(A)$ and hence an upper limit of NO concentration if it is assumed that $\alpha_f \approx \alpha_n$. This procedure, however, has not been found desirable, firstly because, as we have seen, the effective recombination coefficient does not remain constant during a flare and its value has to be independently determined, and secondly the time lag between the maximum of N_e and that of q does not permit the use of Equation (80).

The second alternative of using the entire time curve, is, therefore, the only appropriate procedure. For this one must be able to derive the complete time curve of q_X. This depends not only on the availability of the X-ray flux as a function of time, but also its spectral distribution as well as the time variation of this spectral distribution. Since in most flares there is considerable spectral hardening during a flare, the exact amount of hardening must be determined following the procedures such as those given in Chapter 3.

The method, then, involves a comparison of the time curves of N_e and q_X. This has been done by Rowe *et al.* (1970) with the flare of October 21, 1968 shown in Figure 85a, where the time histories of N_e and of the X-ray production rate, q_X, at 80 km, are shown. The starting equation is the following:

$$\frac{d\lambda}{dt} + \lambda^2 [\alpha_i N_e] + \lambda \left[(\alpha_i + \alpha_D) N_e + \frac{1}{N_e} \frac{dN_e}{dt} \right] +$$

$$+ \left[\frac{1}{N_e} \left(\frac{dN_e}{dt} - q \right) + \alpha_D N_e \right] = 0. \tag{81}$$

One should retain the term $d\lambda/dt$ since there is no compelling physical reason why $d\lambda/dt$ should be zero during a flare. Then one finds that $q(A)$, the non X-ray production term, depends on the pre-flare values of α_D, α_i and λ. The Penn State approach was to assume values for $q(A)$, for the effective ion-ion recombination coefficient α_i and for the pre-flare value of $\lambda (=\lambda_n)$. The equilibrium equation (with $d\lambda/dt = dN_e/dt = = 0$) was then solved for α_D. The differential equation was then integrated numerically using the above values and using time histories of N_e and q_X as shown in Figure 85a. Rowe *et al.* found that $q(A) = 8$ cm^{-3} s^{-1} is the largest $q(A)$ which gives positive λ for all times. If $q(A)$ is reduced below 3 cm^{-3} s^{-1}, λ does not return to λ_n in the same time interval that N_e returns to normal. The conclusion is that $q(A) < 10$ cm^{-3} s^{-1} at 80 km if $\alpha_i = 1 \times 10^{-7}$ cm^3 s^{-1} and $\lambda_n < 8$; $q(A)$ is probably about 3 or 4 cm^{-3} s^{-1} if λ_n is about 5 (Baker, 1969). If all of the $q(A)$ at 80 km is due to Lα ionizing NO, then $[NO] = 3 \times 10^7$ cm^{-3} at 80 km.

Similar analysis was made for the flare of March 27, 1969 (1630–1700 UT). For this event the following values were obtained:

(a) $\alpha_i = 1 \times 10^{-6}$ cm^3 s^{-1}
         $\lambda_n = 5$, $q(A) = 4$ cm^{-3} s^{-1}
         $\lambda_n = 10$, $q(A) = 25$ cm^{-3} s^{-1}
(b) $\alpha_i = 1 \times 10^{-7}$ cm^3 s^{-1}
         $\lambda_n = 5$, $q(A)$ negative
         $\lambda_n = 10$, $q(A) = 3.5$ cm^{-3} s^{-1}.

A result of immediate interest is that, with variable λ, $q(A)$ could be as large as large as 8 cm^{-3} s^{-1}. This would correspond to an NO concentration of 2.4×10^7 cm^{-3}, whereas the constant loss rate analysis gives $[NO] = 3 \times 10^6$ cm^{-3}.

Similar analysis were made for other heights. Estimates of [NO] from the above considerations have resulted in the profile marked 1 in Figure 96 which is shown along with the previous ionospheric estimates and rocket results. At 80 km there appears to be a general convergence to about 1×10^7 cm^{-3}. Pearce's large [NO] values which created much controversy, are entirely inconsistent with flare values. We should also note that [NO] values deduced from flares are even somewhat lower than Meira's – a conclusion that is in agreement with other ionospheric approaches and the theoretical arguments of Brasseur and Nicolet.

Similar conclusions were also made by Somayajulu and Aikin (1970) from their rocket measurements of flare time electron density profiles (Figure 41).

9.5. Chemistry of D-Region During Flares

Two height regions have been distinguished – one in which positive ions dominate and the other in which in addition to positive ions negative ions play competitive or even dominant roles. For the first category Mitra and Rowe (1972) chose the height of 80 km; for the second category Mitra and Rowe chose 70 km, while Thomas et al. (1973) chose 72 and 64 km.

9.5.1. CHEMISTRY AT 80 km

At and around 80 km, the principal ions are $H_5O_2^+$ and NO^+; the total positive ion concentration at midday is of the order of 10^3 cm^{-3}. Negative ions, as indicated earlier, do not play any part.

Under quiet conditions, the main production source for electrons is $NO + L\alpha \rightarrow NO^+ + e$. Production of electrons from O_2 is very small since both the ionization of O_2 by X-rays and of $O_2(^1\Delta_g)$ by $\lambda \leqslant 1108$ Å are much smaller than the NO^+ production rate. At a solar zenith angle of 60° the O_2^+ production rate is about an order of magnitude less than the NO^+ production rate.

Thus, the precursor ion for clustering is likely to be NO^+. The hydrated positive ion chemistry, as currently understood, is shown in Figure 102.

The major difficulty with reactions currently observed in the laboratory concerns the mode of conversion of NO^+ ions to cluster ions. In the region between 70–85 km, production of NO^+ through photo-ionization of NO by $L\alpha$ far exceeds the combined production of O_2^+ through the photoionization of $O_2(^1\Delta_g)$ by the radiation 1027–1118 Å and of N_2 and O_2 by X-rays and galactic cosmic rays as shown in Figure 93. At the levels where O_2^+ production rate is dominant, there are two possibilities: (1) If atomic oxygen concentration is low, as at levels below 70 km, the conversion of O_2^+ into hydrates is quick and efficient, and (2) At levels where atomic oxygen concentration is large so that $10^{-9}[H_2O] \ll 3 \times 10^{-10}[O]$, or in other words $[O]/[H_2O] \gg 3$, the back reaction $O_4^+ + O \rightarrow O_2^+ + O_3$ is more efficient than the forward reaction leading to clustering.

If we define the critical level as one in which the back reaction $O_4^+ + O \rightarrow O_2^+ + O_3$ is as large as that for the forward reaction $O_4^+ + H_2O \rightarrow O_2^+$. $H_2O + O_2$, then for the concentrations given in Figure 93, condition 2 begins to be effective at and above 80 km, and condition 1 at all heights below about 78 km. However in this region for

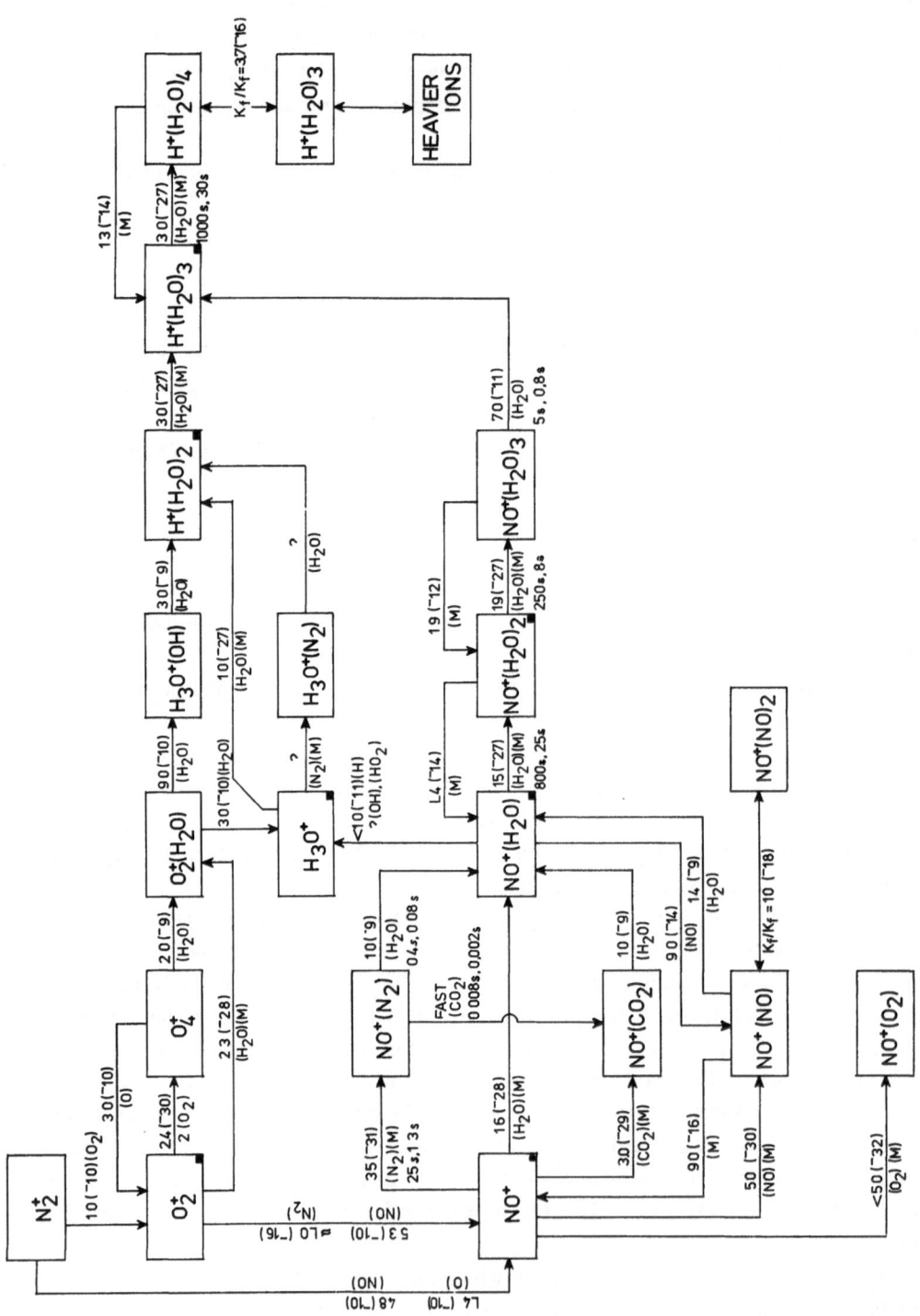

Fig. 102.   Positive ion chemistry involving water clusters in the mesosphere as currently understood. Reaction times are indicated on the basis of laboratory reaction rates and CIRA 1972 density models (after Rowe *et al.*, 1974).

heights down to somewhat below 70 km, there is insufficient production of O_2^+ ions, and consequently, hydration from O_2^+ ions is possible only at levels somewhat below 70 km. The sharp disappearance of cluster ions between 82–85 km comes, on this hypothesis, either from the sharp rise of atomic oxygen or from the predominance of O_2^+ production rate at this level, or to a combination of both. In Figure 93 we see three regions of interest. Below 65 km both the conditions for hydrates originating from O_2^+ are satisfied, $q(O_2^+) \gg q(NO^+)$ and $[O]/[H_2O] \ll 3$. This is the *only* region where O_2^+ is the precursor ion for hydration. In the second region (roughly between 65 and 85 km) although H_2O concentration is still larger than O concentration, $q(NO^+)$ now exceeds $q(O_2^+)$. Also NO^+ would be formed continuously by charge transfer from O_2^+. This is, therefore, the region where NO^+ is the precursor ion. Since $q(O_2^+)$ is not entirely negligible, O_2^+ clusters channel is not entirely non-existent, but its rate is determined by the ratio $[H_2O]/[O]$ and, in particular, whether or not there is a valley in atomic oxygen around 75 km. If, for example, one uses the atomic oxygen distribution shown in dotted curve in Figure 93, from the computations given by Park (1972), no effective hydration can occur from O_2^+ ions in region B. In region C the sharp rise in atomic oxygen concentration makes the condition for cluster disappearance $[O]/[H_2O] > 6$ effective, and consequently, even though the cluster formation rate may be large, there is a quick cut off of the clusters.

We now come to region B. This is the region which poses the maximum difficulty. In the chain of reactions that convert NO^+ to $H^+ \cdot (H_2O)_n$, NO^+ must undergo three hydrations before it can be converted to $H^+ \cdot (H_2O)_3$ and then to other types of water cluster ions, and in this chain dissociative recombination of the intermediate ions with electrons will greatly reduce the efficiency of the hydration process. Furthermore, as we will see later, the observed ratio of $[\Sigma H^+ \cdot (H_2O)_n]/[NO^+]$ indicates that the rate of conversion B in the channel $NO^+ \xrightarrow{B} H^+ \cdot (H_2O)_n$ has to be at least 10^{-2} s^{-1} for the entire region B, and consequently, none of the steps in the hydrations process can have a time constant more than 100 s. The characteristic times of reaction in the different steps are indicated in Figure 102. The hydration steps are too slow. It is clear that existing reactions are unable to reproduce the observed large densities of the hydrates in this region, and instead produce a larger abundance of NO^+ clusters than are observed. Suggestions have been made for making the clustering process speedier. Heimerl *et al.* (1972), for example, have suggested that NO^+ need only have one hydration, forming $NO^+ \cdot H_2O$ after which a re-arrangement reaction with a free radical such as OH or HO_2 may occur,

$$NO^+ \cdot H_2O + OH \rightarrow H_3O^+ + NO_2$$
$$NO^+ \cdot H_2O + HO_2 \rightarrow H_3O^+ + NO + O_2$$
$$\rightarrow H_3O^+ + NO_3.$$

While these reactions will provide necessary shortening of the clustering sequence, and will also give an additional mechanism for the rapid disappearance of water cluster ions above 83–85 km, as both OH and HO_2 concentrations decrease very rapidly above this height in all theoretical models, neither has been measured in the laboratory.

The reaction rates required for the characteristic time to exceed 100 s at 80 km are between 10^{-9} to 10^{-8} $cm^3 s^{-1}$ for both. While this sequence looks attractive, it is purely hypothetical.

It is clear that in the matter of modelling of the D-region, there is considerable danger in using complex reaction schemes which are either too slow or involve steps of a purely hypothetical nature. Nor, as we will see now, is it necessary.

Mitra and Rowe (1972) have described a simplified six-ion model of D-region ion chemistry, which takes the form shown in Figure 103. It offers certain advantages over complex schemes in constructing D-region profiles under quiet or disturbed conditions. Firstly, it eliminates some of the uncertain and hypothetical (and for electron density determination, unnecessary) steps in the reaction sequence; secondly, as the parameters involved are relatively fewer and are related explicitly to specific measurements on electron density, positive ion density, or positive and negative ion composition, a semiempirical approach is possible. Thirdly, any part of the simple scheme can be expanded when and where further steps are considered reliable. Some of the factors considered in the design of this model are outlined below.

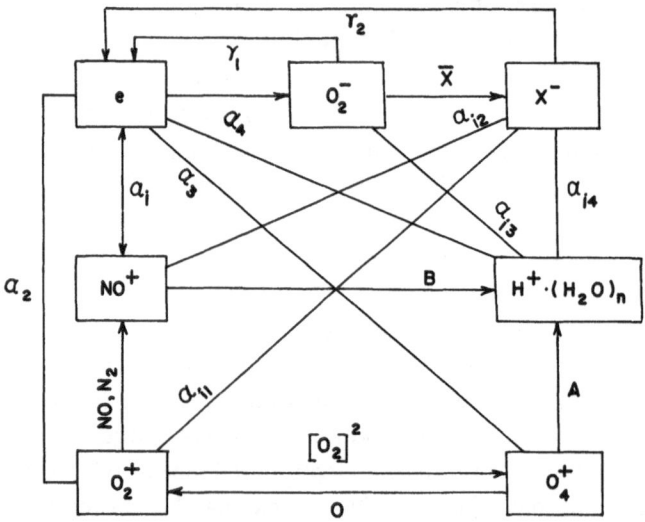

Fig. 103. Simplified ion chemical scheme for the lower ionosphere suggested by Mitra and Rowe (1972).

In the D-region N_2^+ converts rapidly to O_2^+. Thus, O_2^+ is considered to be the primary ion produced by radiation other than $L\alpha$. The primary ion produced by $L\alpha$ is NO^+.

The relative concentrations of the ions $H^+ \cdot (H_2O)_n$ for different values of n are not well determined because of experimental difficulties, for the heavier clusters tend to decompose to lighter ones by the shockwave of a passing rocket. Mitra and Rowe consequently lump the whole series $H^+ \cdot (H_2O)_n$ and call it Y^+. The individual concentrations of different clusters can, then, be obtained from the following equation:

$$\log\left(\frac{[H^+ \cdot (H_2O)_n]}{[H^+ \cdot (H_2O)_{n-1}][H_2O]}\right) = \frac{A \times 10^3}{T} - B,$$

where values of A and B are available from Kebarle *et al.* (1967) for $n=2-8$.

The effective rate of conversion of NO^+ into hydrated ions is called B (as in **Haug** and Landmark, 1970):

$$NO^+ \overset{B}{\rightarrow} H^+ \cdot (H_2O)_n.$$

The requirement on B so that the water clusters form a concentration of 50% of the total positive ions is:

$$B = 10^{-2} \, s^{-1}.$$

During the flare we suddenly have a large supply of O_2^+ ions produced both through direct photoionization of O_2 and through charge transfer from N_2^+. There is now a sudden shift of the clustering channel from $NO^+ \rightarrow$ clusters to the channel $O_2^+ \rightarrow O_4^+$, and from O_4^+ partly to hydrates and partly back to O_2^+ through back reaction involving atomic oxygen. With $[H_2O] = 10^9 \, cm^{-3}$ and $[O] = 10^{10} \, cm^{-3}$ there is an equal sharing of O_4^+ ions between the two channels. Thus, some of the ions are lost for clustering and go back into O_2^+ and NO^+. Thus we expect a decrease in the percentage of simple ions. Here $A = 1.0 \times 10^{-9}[H_2O]$ and $B = 10^{-2} \, s^{-1}$.

The concentrations of H_2O and O are quite critical also quite different in the various theoretical models currently in use. Mitra and Rowe have varied $[O]$ from 10^9 to $10^{11} \, cm^{-3}$ and $[H_2O]$ from 10^8 to $10^{10} \, cm^{-3}$. Different flare conditions were represented by using different values of $q_X = q(O_2^+)$. At 80 km, a range of 10–500 $cm^{-3} \, s^{-1}$ can be expected in q_X for different flares. The most remarkable consequence is the change in ion composition resulting as q_X was increased from its normal value of 0.5 $cm^{-3} \, s^{-1}$; this is shown in Figures 104a, b. In all cases, the percentage of hydrated ions decreases as q_X is increased. When q_X is 0.5 $cm^{-3} \, s^{-1}$ hydrated ions represent about 50% of the total; when q_X is 500 $cm^{-3} \, s^{-1}$, hydrates represent only about 5% of the total positive ion concentration. This reduces the effective recombination coefficient by about a factor of 5 since ψ for the hydrated ions at the mesospheric temperatures is around $10^{-5} \, cm^3 \, s^{-1}$, compared to $6 \times 10^{-7} \, cm^3 \, s^{-1}$ for NO^+ and $4 \times 10^{-7} \, cm^3 \, s^{-1}$ for O_2^+. This effect is indicated in Figure 104b. As one might expect, larger concentration of O makes the simple ions increase faster as q is increased; the opposite effect occurs with increase in $[H_2O]$. A noticeable effect occurs even with $[O]$ as low as $10^9 \, cm^{-3}$.

We also notice that for large values of q, the percentage of hydrates and the value of ψ tend to level off, in agreement with observations of Figure 90. Under these conditions, the positive ion chemistry reduces to the situation of simple molecular ion chemistry.

Danilov (personal communication) has pointed out that a decrease in the cluster content of the mesospheric ions can occur even if one considers only the $NO^+ \overset{B}{\rightarrow}$

clusters channel, simply because N_e is increased. In this case, the principal reactions are:

$$NO^+ \xrightarrow{B} H^+ \cdot (H_2O)_n$$

$$H^+ \cdot (H_2O)_n + e \xrightarrow{\alpha_D} \text{neutrals}.$$

The relative proportion of $[NO^+]$ and $[\Sigma H^+ \cdot (H_2O)_n]$ ions in this case is given by:

$$\frac{[\Sigma H^+ \cdot (H_2O)_n]}{[NO^+]} = \frac{B}{\alpha_D N_e}. \tag{82}$$

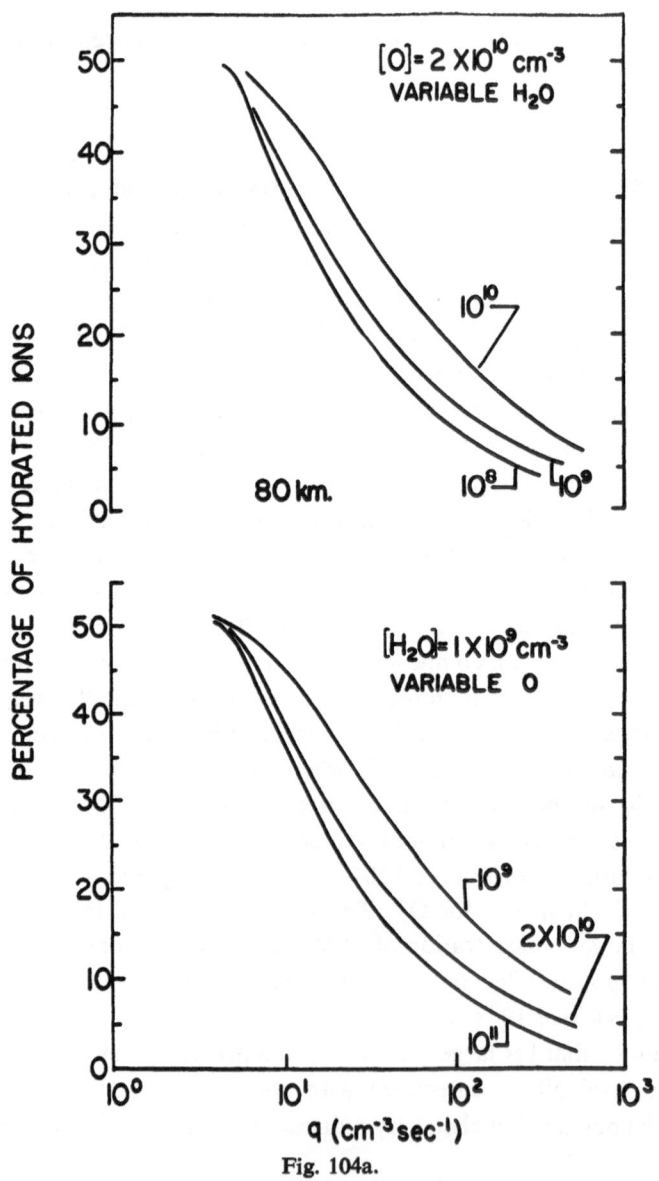

Fig. 104a.

Fig. 104a–b.   Decrease in water cluster ions and in the effective recombination coefficient for flares
of increasing intensity (after Mitra and Rowe, 1972).

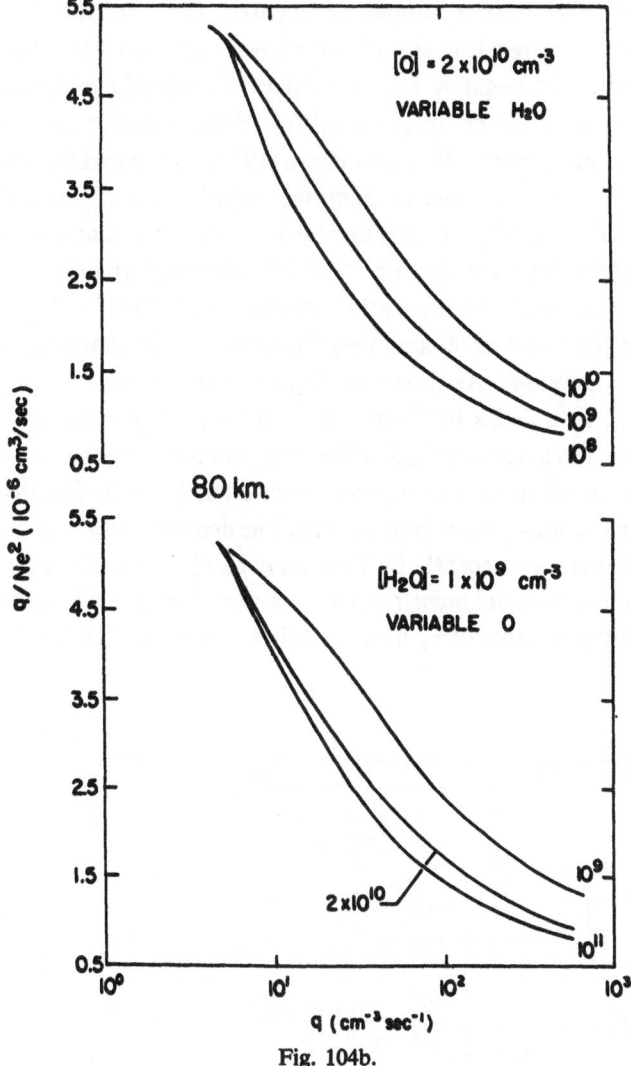

Fig. 104b.

As N_e increases, the cluster population decreases. This will occur during flares, during PCAs or under any conditions when N_e is suddenly and greatly increased.

To what extent the channel $O_2^+ \to O_4^+$ is important depends on the relative rates of the channels:

(a) $O_2^+ \to O_4^+$
(b) $O_2^+ \to NO^+$.

The first is given by $6 \times 10^{-30} [O_2]^2$ and the second by $8 \times 10^{-10} [NO]$. At 80 km, the respective rates are about 5×10^{-2} s^{-1} and 8×10^{-3} s^{-1}. Channel (a), therefore, is six times larger than channel (b). If this is so, Equation (82) is too simplistic.

9.5.2. NEGATIVE ION REGIONS (70 km AND BELOW)

At those levels where the negative ion composition is significant, we have an extremely complex situation. We have a number of negative ions reacting with each other, a number of positive ions reacting as outlined above, and reactions between the positive and negative ions. The negative ion chemistry as currently understood is given in Figure 105. In view of the complex chemistry of the negative ions and the various uncertainties that still persist, Mitra and Rowe (1972) considered the simplified scheme given in Figure 103, in which the unidentified negative ion X^- is a cluster ion and is presumably $NO_3^- \cdot (H_2O)_n$. There is no reason to suppose that the different mutual neutralization coefficients are the same, but for simplicity and since firm values were not at hand, they assumed that these all had values of 10^{-7} cm^3 s^{-1}.

In contrast to the result at 80 km, they found very little change in the positive ion composition as q increased. As shown in Figure 106 percentage of hydrates decreased from about 99% at $q_X = 2 \times 10^{-2}$ cm^{-3} s^{-1} to 94% at $q_X = 100$ cm^{-3} s^{-1}. On the other hand, there was a decrease in λ which was probably due partly to faster destruction of clusters through mutual neutralization and partly to the fact that negative ions were produced more slowly than the electrons. The decrease in ψ was of the order of 2.

Further work was done recently by Thomas *et al.* (1973) for the negative ion region by assigning different detachment rates to the negative ions CO_3^-, NO_2^- and NO_3^-, instead of the lumped value of γ_2 used by Mitra and Rowe (1972). The detachments

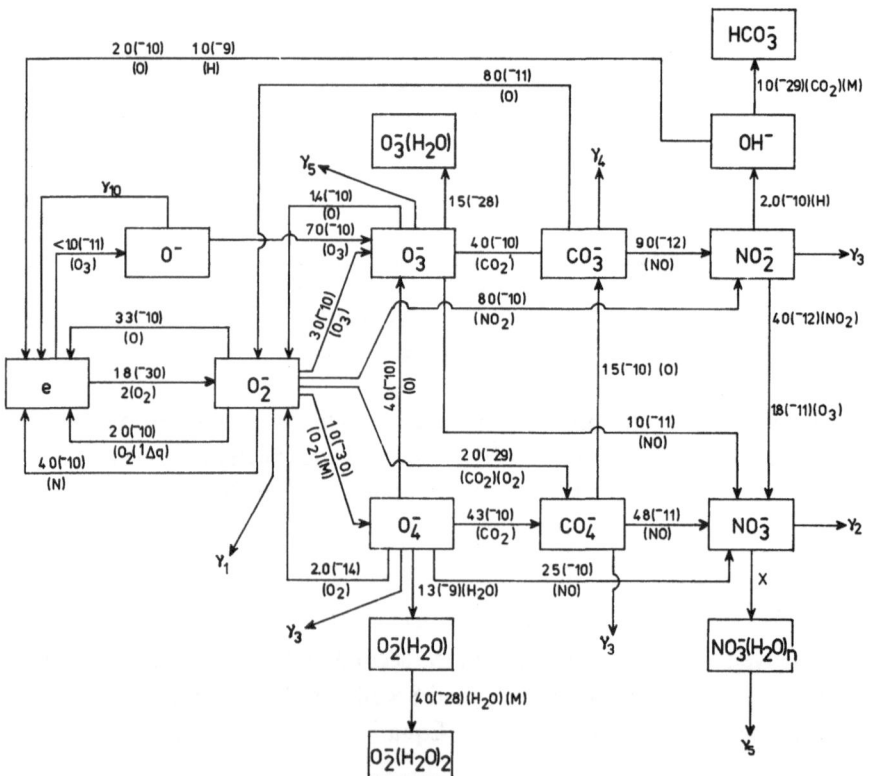

Fig. 105. Negative ion chemistry in the *D*-region as currently understood. The reaction time for each step is indicated (after Rowe *et al.*, 1974).

were, however, assumed to be entirely photodetachment, and the rates were taken to be those indicated in Section 9.3.3. The negative ion channels are principally two: one goes along the line $O_2^- \rightarrow O_3^- \rightarrow CO_3^- \rightarrow NO_2^- \rightarrow NO_3^-$; the other along the line $O_2^- \rightarrow O_4^-$ $\rightarrow CO_4^- \rightarrow NO_3^-$ hydrates. Two principal points of interest are:

(i) the role played by atomic oxygen;

(ii) the role played by detachment from all the intermediate and final negative ion products.

Atomic oxygen, by converting O_3^- and CO_3^- to O_2^-, reverses the trend of O_2^- conversion to heavier negative ions. The detachment from CO_3^-, NO_2^-, NO_3^- and presumably from cluster negative ions can also substantially reduce the formation of negative ions formation. Photodetachment of these ions are in the neighbourhood of 10^{-2} s^{-1}. Chakrabarty and Mitra (1974) from a study of eclipse time variations in D-region electron density also find $\gamma_2 \approx 10^{-2}$ s^{-1}. The principal detachments may, therefore, mostly be photodetachment. To what extent these detachment rates are changed during a flare are not known. The changes are not likely to be very large, since the photodetaching radiations are in the visible and in the infrared in which no appreciable variations during flares have yet been noted.

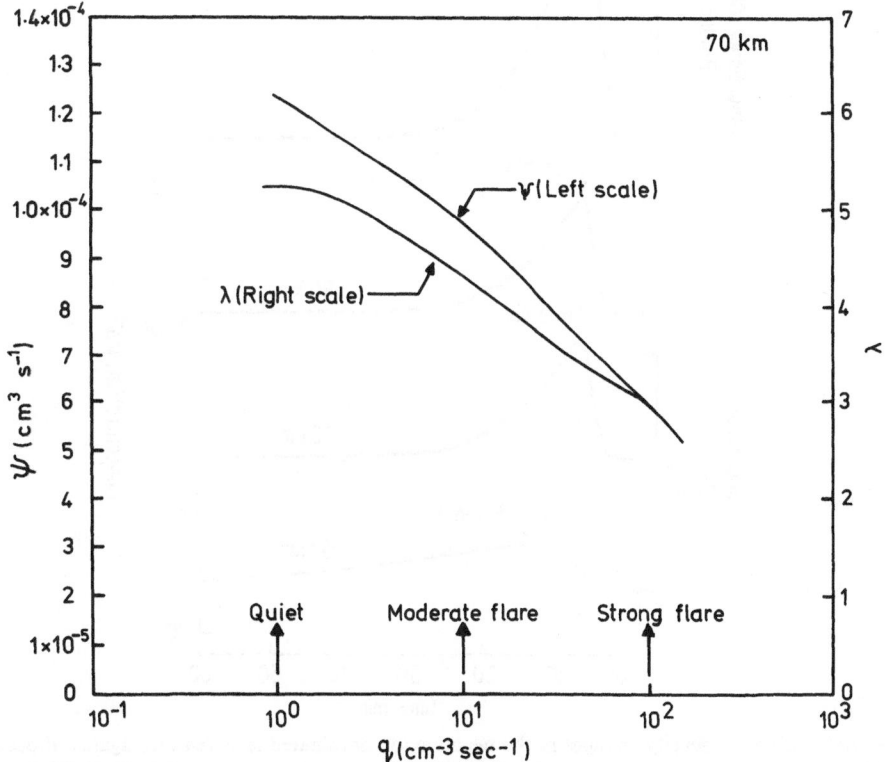

Fig. 106.   Variation of $\psi = q/N_e^2$ and λ as a function of X-ray flare magnitude for the height of 70 km (after Mitra and Rowe, 1972).

Thomas *et al.* assumed a production rate variation given by the following expression, over a 5-min interval during a flare:

$$q(t) = \left(1 + \frac{t^3}{25}\right) \tag{83}$$

and the decay in production given by an exponential term with a time constant 5 min:

$$q(t) = q_n(1 + 5 \exp[-0.2(t - 5)]), \tag{84}$$

where q_n represents the pre-flare value and t is measured in minutes.

Three heights were chosen by Thomas *et al.*: 64, 72 and 80 km of which the first was entirely negative ion-dominated. Thomas *et al.* confirmed that the quantity q/N_e^2 decreased, becoming minimum shortly after the corresponding maximum in N_e.

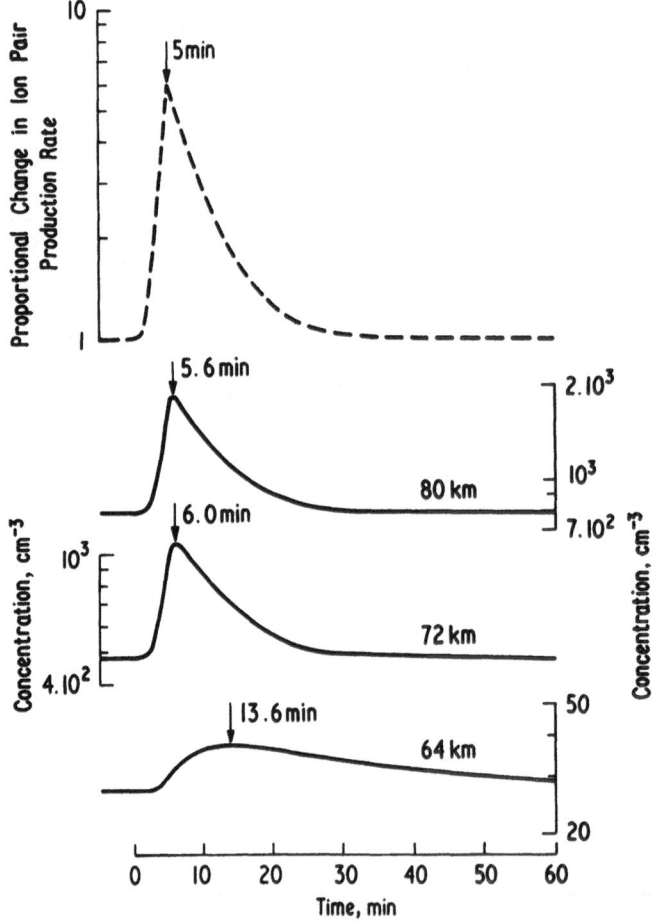

Fig. 107. Electron density changes in the negative ion dominated level (64 km) against those in the levels dominated by changes in positive ions (80 km) and at an intermediate level (72 km). The increasing time delay in N_e curve compared to the q-curve as one goes to the negative ion region may be noted (after Thomas *et al.*, 1973).

However, as long as the production rate was assumed to be enhanced by the same proportion and had an identical form at all heights, the delay between the peak of production and maximum electron concentration increased with decrease in height. Under such situation and in the rapidly changing conditions of a solar flare, they questioned whether the essentially equilibrium concept of an effective recombination coefficient as defined by the expression q/N_e^2 or $[(1+\lambda)(\alpha_D+\lambda\alpha_i)]$ is valid. In the case where photodetachments were included, with q reaching maximum after 5 min, the electron concentration maximised at 5.6 min and 6.0 min at 80 and 72 km, but considerably later at 13.6 min in the negative ion dominated height of 64 km (Figure 107). As a result of this delay, they found that q/N_e^2 increased initially and reached a maximum value prior to the peak in q. The magnitude and the subsequent decrease was determined both by the magnitude of the enhancement in N_e and its delay relative to the peak of production. It is important to note that the inclusion of photodetachment of

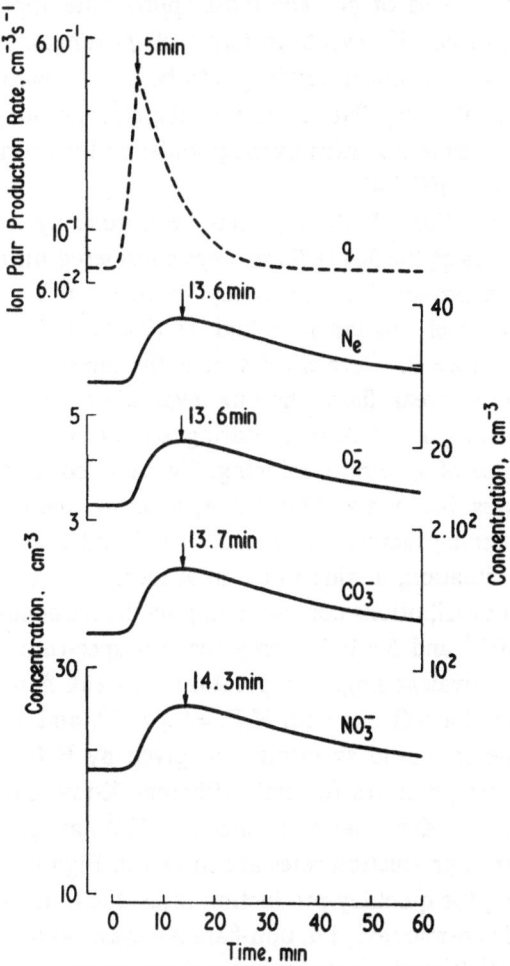

Fig. 108.   Delays in negative ions O_2^-, CO_3^- and NO_3^- in the presence of photo-detachment for an assumed variation of ion pair production rates (after Thomas *et al.*, 1973).

electron from CO_3^-, NO_2^- and particularly from NO_3^- retarded the decay. The delays in N_e and of the various negative ions were not the same. In Figure 108 the delays in N_e, O_2^- and CO_3^- were found to be about 8.6 min and for NO_3^- about 9.3 min. A relative shift in the negative ion composition can therefore shift the nature of the time curve.

9.5.3. MODELS OF FLARETIME IONIZATION PROFILES

Models of flaretime ionization profiles are somewhat more difficult to construct than the quietday profiles since one must take account of two different factors: (a) the flux and spectral characteristics of the X-ray event, and (b) the nature and magnitude of the variation of the effective recombination coefficient expected for the event. If one assumes, on the basis of Section 3.3 and that both are known to first degrees of approximation, then representative ionization profiles may be developed for different types of flares. The nature or class of a flare, as defined optically, or by radio means, or even by the magnitude of the X-ray flux, can be quite misleading; what we are basically concerned with is the value of q_X. The most appropriate approach for ionospheric effects is, therefore, to classify events in terms of q_X ranges. However, in practice, the basic objective for such model-building is to be able to predict enhancement profiles for specific types of X-ray flares, and it is, therefore, more practical to start with flare classifications in terms of X-ray flux magnitudes and the corresponding hardening ratio given by $F(0.5-3 \text{ Å})/F(3-8 \text{ Å})$.

The only model-building of this type has been done by Mitra and Deshpande (1972). Four categories of the X-ray flares were considered on the basis of the intensity in 0–8 Å band expressed for a greybody spectrum at $T = 2 \times 10^6$ K. For strong flares the 0–8 Å flux level was taken as high as $(2-5) \times 10^{-1} \text{ erg cm}^{-2} \text{ s}^{-1}$, and for moderate flares, about $2 \times 10^{-2} \text{ erg cm}^{-2} \text{ s}^{-1}$. In the threshold case flux was 2×10^{-3} erg cm^{-2} s^{-1} while for weak flares the flux level taken was $5 \times 10^{-4} \text{ erg cm}^{-2} \text{ s}^{-1}$. The large enhancements in 0–8 Å band during strong flares are not likely to occur without a high degree of spectral hardening. Two composite spectrum models with separate temperatures below and above 3 Å, based on the ratio of 0–3 Å to 0–8 Å flux levels (the hardening factor, H.F.) as 5×10^{-2} and 2×10^{-2} were used. Under moderate intensity situation, a wide range in spectral distribution is possible. They chose three spectral distributions corresponding to the hardening factor, H.F., values of 5×10^{-2}, 1.5×10^{-2} and 5×10^{-3}; only the soft spectrum (H.F. $= 5 \times 10^{-3}$) was represented by an equivalent single temperature. In weak flares, the spectral distributions taken included a soft one with H.F. $= 5 \times 10^{-3}$, and a harder spectrum with H.F. $= 2 \times 10^{-2}$. The threshold spectrum was given by H.F. $= 1.5 \times 10^{-2}$. The flux levels and spectral temperatures for eight different X-ray spectra belonging to the four broad categories of X-ray flares (extended to 20 Å) are given in Figure 109. The corresponding electron production rates are shown in Figure 110 for zenith angle of 50°. For comparison, the quietday production rates due to the usual sources are also shown; these include cosmic rays, Lα, non-flare X-rays as well as a contribution to the production from $O_2(^1\Delta_g)$ and are shown in the same figure.

Flare time production rates exceed the normal time rates of production from the sources during strong and moderate events; however, for a moderate flare with a soft spectrum, the rates of production q_X (flare) are comparable to those due to Lα. Under threshold condition, q_X (flare) is lower than q(Lα), but exceeds q_X (normal) at all heights. During weak flares q_X (flare), although much less than q(Lα), exceeds the q_X (normal) below 85 km; this being the effect of spectral hardening. Above 85 km flare time rates are lower than q_X (normal) because the latter refer to moderate and high solar activity period with 0–8 Å and 8–20 Å levels larger than for our weak flare models.

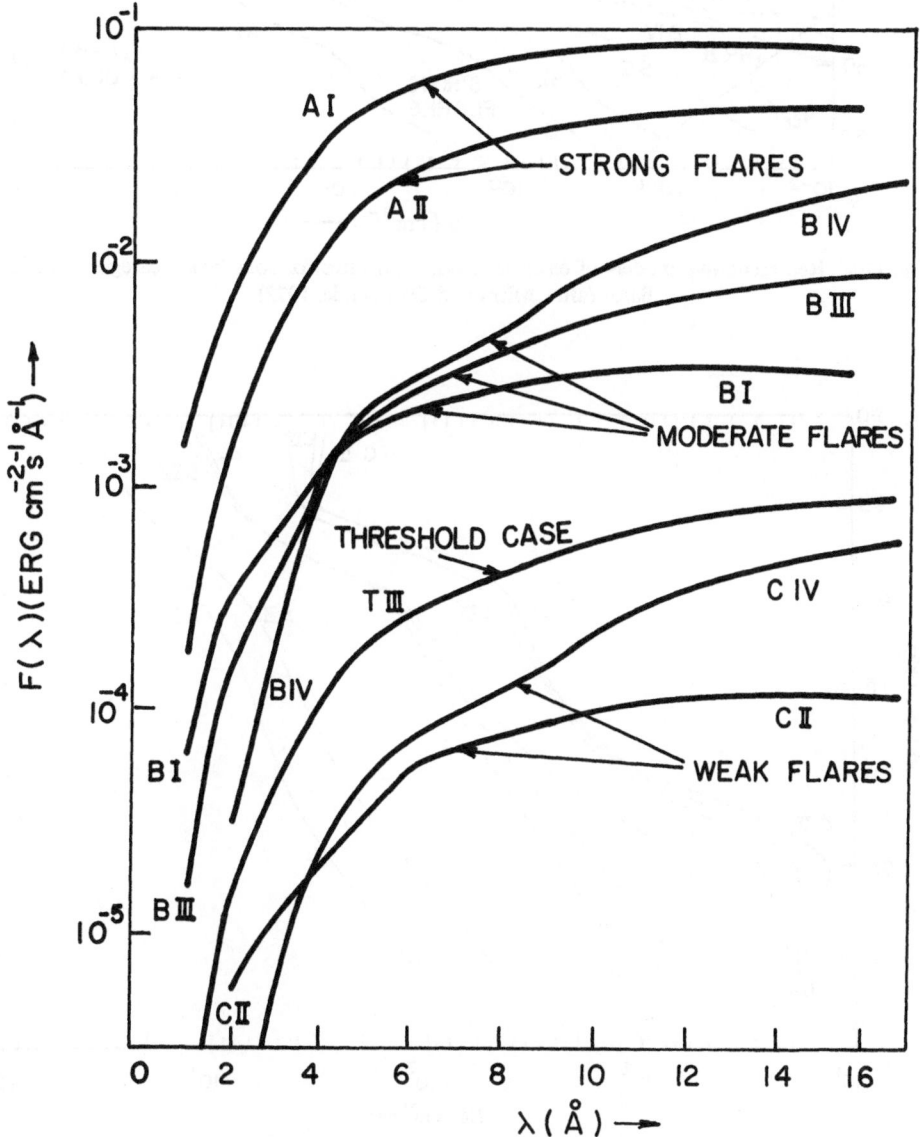

Fig. 109. Representative models of the X-ray spectra for four broad categories of X-ray flares (after Mitra and Deshpande, 1972).

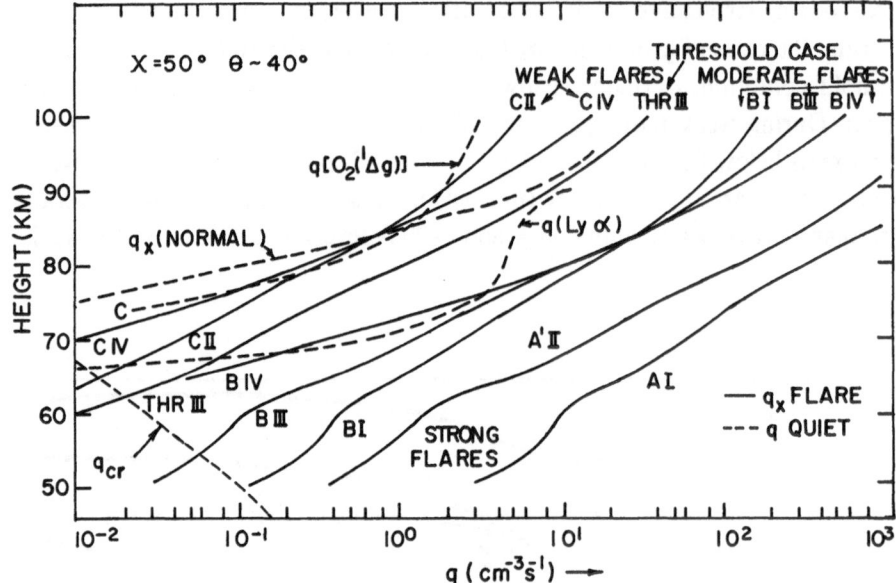

Fig. 110.   Representative models of electron production rates for four broad categories of X-ray
flares (after Mitra and Deshpande, 1972).

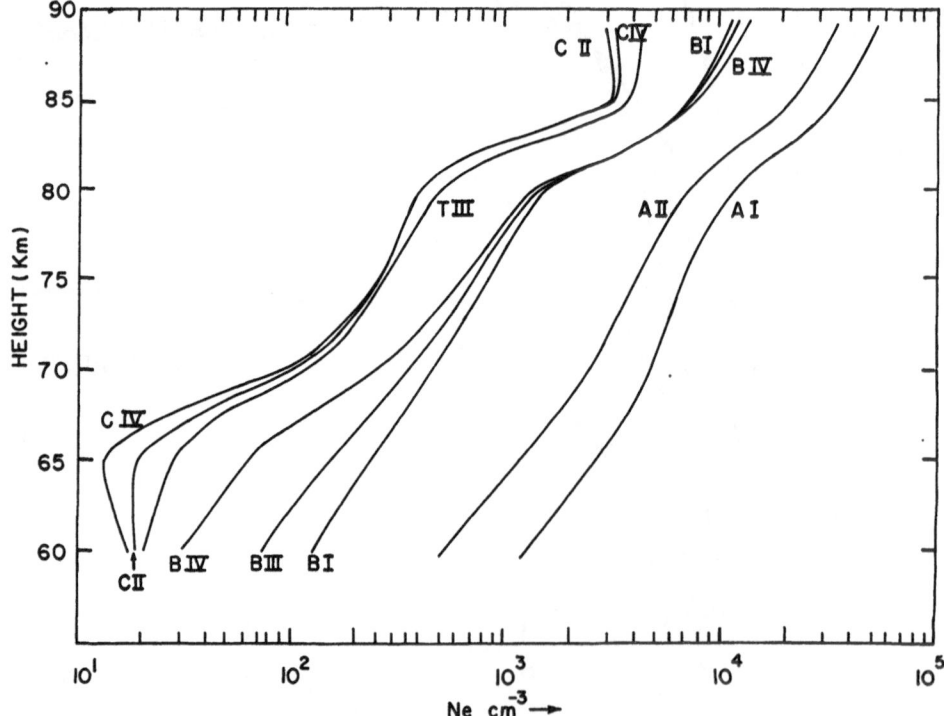

Fig. 111.   Representative models of electron density profiles for four broad categories of X-ray
flares (after Mitra and Deshpande, 1972).

The ionization profiles for strong, moderate and weak flares are shown in Figure 111. Electron concentrations at 80 km during strong, moderate and weak flares are about 1×10^4 cm^{-3}, 1.5×10^3 cm^{-3} and 5×10^2 cm^{-3} respectively. During strong flares concentrations at 60 km may be as large as 1×10^3 cm^{-3}. Effect of the variation in spectral composition is significant only below 80 km. Weak flares with flux levels below the 0–8 Å threshold value can only be seen below 70 km; above this level the ionization is indistinguishable from the quietday value. During strong flares, the production rates are large and for the associated low ψ profiles the electron concentrations become very high. On the contrary, during weak flares, the production rates are low and with high values of ψ, low electron concentrations result. A consequence of the variable ψ is thus to cause a larger variation in electron concentration than that caused by a change in q alone.

9.6. F-Region During Flares

Since F-region ionization changes (only a few per cent) are considerably smaller than in the D-region, they are more difficult to investigate. Although several techniques (Sections 2.8, 2.11, 2.12) have been used, they are not all reliable for the quantitative determination of ionization profile changes. Perhaps the most reliable profiles are those given by the incoherent scatter experiment. Only a few events have, however, been reported so far. We will see that ionization profiles deduced from SFDs and from ionograms tend to overestimate $\Delta N_e/N_e$.

9.6.1. Ionization Changes

From the Incoherent Scatter observations reported in Section 2.12, it is clear that there are large enhancements ($\sim 100\%$ in the E-region) decreasing progressively upwards but with discernible effects up to 300 km. Profiles of $\Delta N_e/N_e$ for the two flare events of May 21 and May 23, 1967 observed at Arecibo are given in Figure 112, along with the profiles obtained at Malvern, England for the flares of 11 December 1968, February 20, 1970, and 19 August 1970, and the profile obtained at Millstone Hill for the flare of August 7, 1972.

Ionization profiles have also been obtained at levels above 100 km from:
(a) Specially run ionosondes,
(b) Multifrequency SFD analysis.

From ionograms taken at Kodaikanal at 15-min intervals during the proton event of July 7, 1966, Bhattacharyya and Balakrishnan (1967), as we have seen before, have derived changes of electron density at six levels in the 160–260 km range. These changes (shown in Figure 73b) show an increase in electron density at all the heights plotted, beginning at nearly the same time as the optical flare. There were two maxima – the authors attributed the first (occurring at about 0630 IST) to the solar XUV radiation and the second (occurring immediately after 0700 UT) to collisional ionization by high-energy particles from the Sun. There was a progressive delay in the time

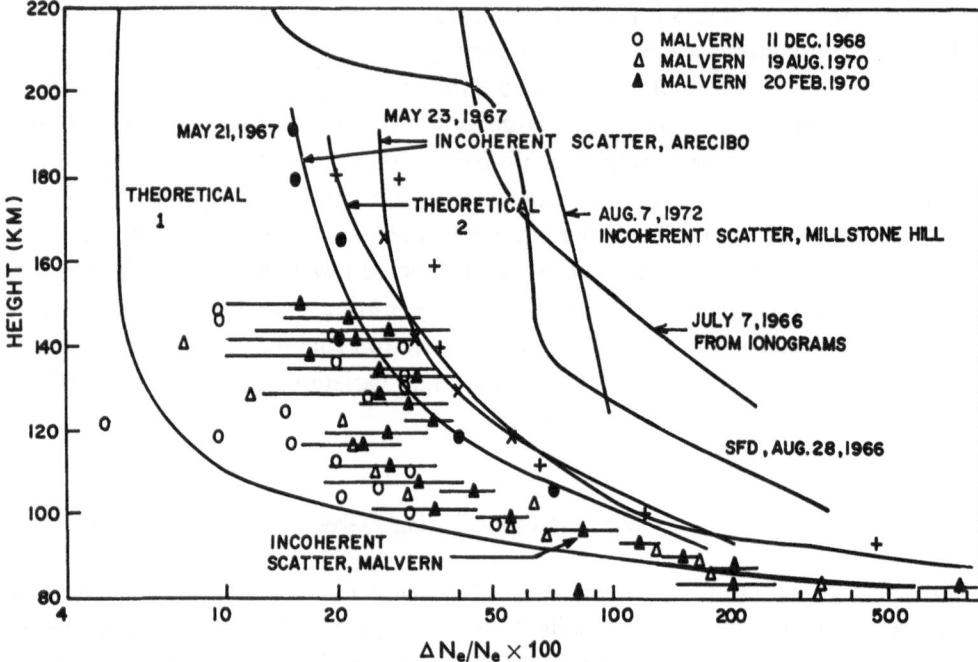

Fig. 112.   Profiles of $\Delta N_e/N_e$ in the E and F-regions (i) incoherent scatter measurements at Arecibo for the flares of May 21, and 23, 1967 (after Thome and Wagner, 1971), (ii) incoherent scatter measurements at Malvern, U.K. for the flares of December 11, 1968, February 20, 1970, and August 19, 1970 (after Taylor, 1974), (iii) multifrequency SFD measurements (after Donnelly, 1968d), (iv) incoherent scatter measurements at Millstone Hill, U.S.A. for August 7, 1972 flare (after Mendillo and Evans, 1974). Note also the two theoretical curves: theoretical curve 1 is based on Hall-Hinteregger flux of Table V; theoretical curve 2 is based on flux values given in text.

of maximum ionization as one goes to higher levels, and the changes in ionization were not unlike those obtained with incoherent scatter technique. Observations of this kind are, however, very rare.

On the other hand, many profiles have been derived by Donnelly and his colleagues from observations of SFDs. Deduction of ionization increases from SFDs is somewhat involved. The procedure used by Donnelly (1968d) is as follows:

(i) Firstly, the pre-flare electron density profile was obtained along the oblique SFD propagation path using available ionograms at the appropriate locations.

(ii) Secondly, the height profile of dN_e/dt at beginning of the flare was adjusted until the corresponding frequency deviations, computed from the generalized equation derived by Bennett (1967) (Equation (8a)) with a ray tracing program developed by Jones (1966), varied with transmission frequency in agreement with the observations. The electron density profile was approximated by the preflare profiles. Since a unique solution for dN_e/dt was not available, only those solutions that were consistent with expected changes in EUV radiation were considered; solutions having sharp discontinuities were discarded.

(iii) Thirdly, the time dependence of the oblique path SFD data was used as a first order estimate of the time dependence of dN_e/dt at all heights to estimate $dN_e(h)/dt$ at later times.

(iv) Fourthly, the first estimates of $dN_e(h, t)/dt$ were integrated with time to determine the first estimate of $\Delta N_e(h, t)$. This was added to the pre-flare electron density profile $N_n(h)$ to correct for the electron density enhancement.

(v) Finally, the frequency deviations were recomputed with the revised electron density profiles, and $dN_e(h)/dt$ was adjusted to make the computed frequency deviations again fit the observations. The entire process was repeated until no corrections were necessary.

Ionization profiles were obtained by Donnelly using the above procedure for several large flare events, one of which is discussed here. This is the proton flare of 1522 UT, August 28, 1966, for which observations of SFDs were available over six frequencies – two on an oblique path (11.1 and 8.9 MHz and 1288 km ground range) and four at vertical incidence (5.054, 4.0, 3.3 and 2.108 MHz). The ionization profiles finally selected, are given in Figure 113. At 300 km, Donnelly estimated the increase in ionization to be less than 1%.

A comparison of the profiles given in Figure 112 indicates that while there is simi-

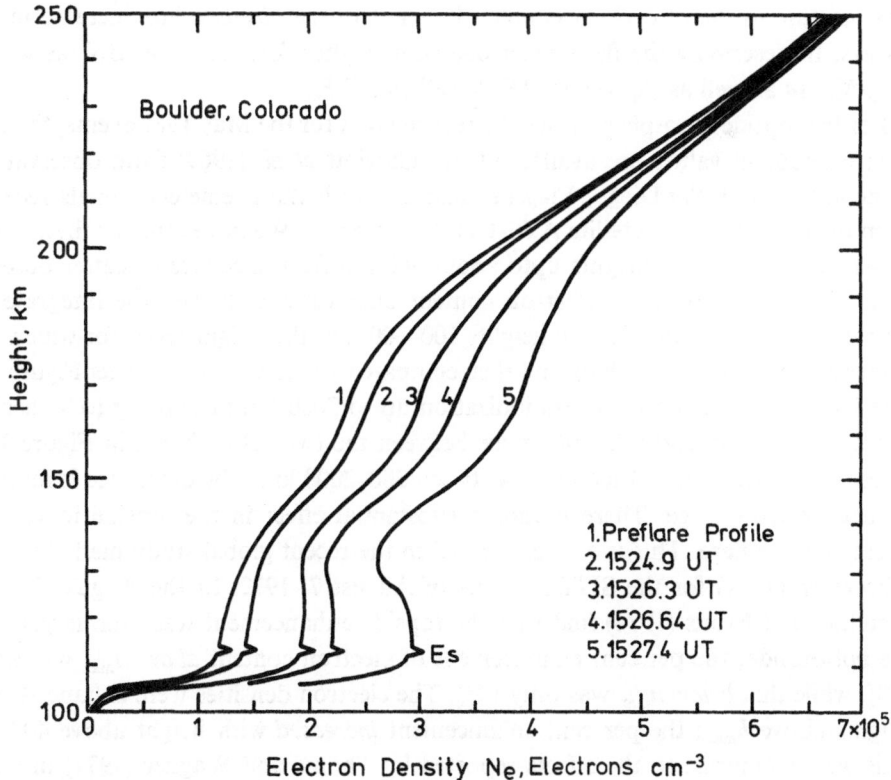

Fig. 113.   Electron density profiles obtained by Donnelly during the growth and decay of the large flare event of August 28, 1966 from SFD measurements over six frequencies (after Donnelly, 1968d).

larity in the $\Delta N_e/N_e$ profiles obtained from the incoherent scatter measurements, (at Arecibo and at Malvern), those obtained from SFD analysis or from ionograms are on the high side. However, values as high as these (or higher) have also been seen. An example is the incoherent scatter results at Millstone Hill during August 7, 1972 event. In this case (see Figure 112 and Table XIX), the per cent ionization enhancement was 60% at 200 km, 20% at 300 km and about 12.5% at 525 km. It should be recognized, however, that the ionization profiles derived from SFDs and from incoherent scatter are not strictly comparable, since the SFD profiles refer to the impulsive phase of the flare whereas the incoherent scatter results are good mainly during the slow phase of the flare.

There are two major points to note. One is that the ionization change is appreciable even at heights well beyond the maximum of the $F2$-region. The second is the very sharp rise in $\Delta N_e/N_e$ below 100 km. This is remarkable even when we remember that the flare time changes in the soft X-rays are considerably larger than in the ultra-violet radiation. A compounding factor is likely to be the change in the chemistry that occurs at 85 km and below decreasing the effective recombination coefficient and, therefore, increasing the enhancement of electron density at these heights (Taylor, 1974). A further point, often missed, is that while the fractional change in the soft X-rays during a flare is considerably larger than that in the ultraviolet, the *absolute* flux increase in the ultraviolet is much larger than the *absolute* flux increase in soft X-rays. In describing the flare effect one should, therefore, consider ΔN_e as well as $\Delta N_e/N_e$, Δq as well as $\Delta q/q_n$ and ΔF as well as $\Delta F/F_n$.

For the topside ionosphere, it will be recalled that for the May 1967 events, the total electron content values are available from Garriott *et al.* (1969) from observations made at Stanford, SanDiego, Flagstaff and Ely with the telemetery signals received from the geostationary satellite ATS-1 at 137.35 MHz. We have thus for these events $N_e - t$ curves at specific heights upto about 300 km from incoherent scatter observations and $\int N_e \, dh$ from the electron content observations. If we now integrate the incoherent scatter data over the heights 100–300 km (the height levels for which ΔN_e values are reliably measurable), and then compare with TEC values, we get Figure 114. The latter, we may recall, covers ionization up to 2000 km (according to Titheridge 1972), and consequently the difference between the two (also shown in Figure 114), gives the change of electron content between 300–2000 km. The difference is found to be unexpectedly large. There is thus a substantial effect in the topside ionosphere during a solar flare. This is also confirmed in the recent global study made by Mendillo *et al.* (1974) for the SITEC events of August 7, 1972. In the August 7 event, Mendillo and Evans (1974) find that the topside enhancement was even larger than the bottomside; the per cent enhancement of electron content *above* h_{max} was about 40%, while that *below* h_{max} was only 14%. The electron densities were enhanced at all heights above h_{max}; the per cent enhancement *increased* with height above 400 km. This was in contrast to the effects reported by Thome and Wagner (1971) in which only small *negative* changes were observed in the topside ionosphere, but in agreement with the observations that TEC results gave twice as much change in N_T than the

TABLE XIX

Comparison of flaretime changes in F-region for different flares

are	Technique	$(\Delta N_e/N_e) \times 100$ (approximate values)				ΔN_T (el m^{-2})	Other changes	Reference
		90–130 km	200 km	300 km	Topside			
ly 7, 66	Ionograms	200 (130 km)	45	—	—	—	—	Bhattacharyya and Balakrishnan (1967)
ug. 28, 66	SFD	120 (125 km)	35	—	—	—	—	Donnelly (1968d)
ay 23, 67	Incoherent scatter	40 (125 km)	15	5	3–10 (Negative effect)	2.4×10^{16} ($\approx 5\%$)	Shortlived T_e/T_i change of 15%	Thome and Wagner (1971)
	Beacon satellite	—	—	—	—	4.4×10^{16} (2.2×10^{16} topside)	—	Garriott et al. (1967)
b. 20, 70 c. 11, 68 ig. 19, 70	} Incoherent scatter	~ 1000 (80 km and below) 100 (90 km) 30 ($100 < h < 130$ km)	—	—	—	—	—	Taylor and Watkins (1970) Taylor (private communication, 1974)
ig. 7, 72	Incoherent scatter	100 (125 km)	60	20	12.5 (525 km)	3.8×10^{16} ($\approx 30\%$) 13.9% below h_{max} 42.3% above h_{max}	T_i nearly constant. $\Delta T_e = 400$ K in topside. Dramatic upward surge in drift	Mendillo and Evans (1974)
	Satellite Beacon	—	—	—	—	1.8 to 8.6×10^{16} (15–30% range) (17 stations)	TEC rise times of about 10 min with excellent time correlation with solar burst at 35000 MHz	Mendillo et al. (1974)

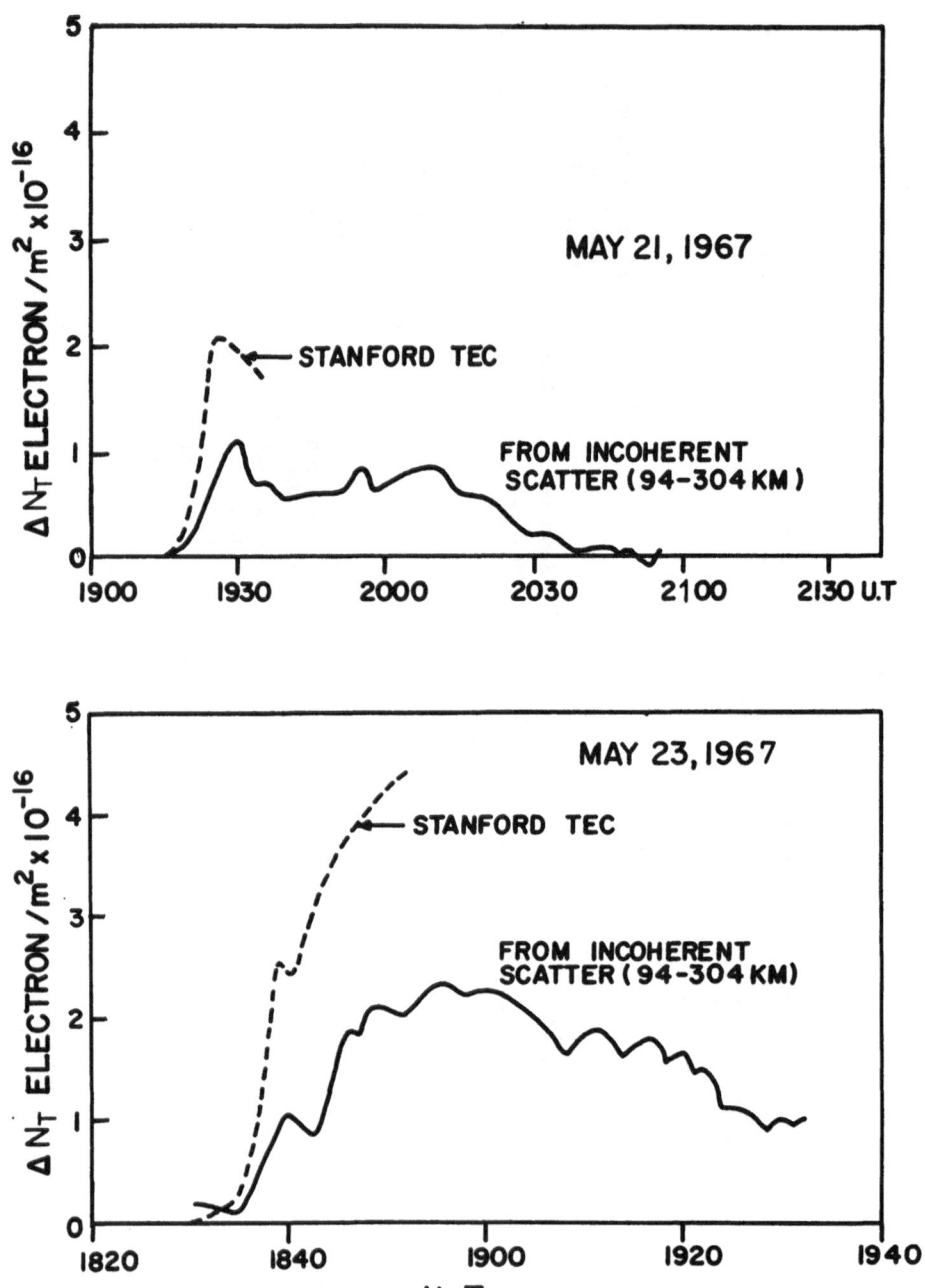

Fig. 114.   Comparison of $\int N_e \, dh$ obtained with satellite beacon experiments with values of $\int N_e \, dh$ for the range 100–304 km obtained from incoherent scatter observations for the flares of May 21 and 23, 1967.

integrated content obtained from incoherent scatter at Arecibo upto the height of 304 km, for the events of May, 1967.

A result of far-reaching significance is the increase in electron temperature observed by Mendillo and Evans during the August 7, 1972 flare at all altitudes examined (of about 400 K in the topside, with little change in the ion temperature), and a simultaneous dramatic upward surge in which the drifts increased linearly with height over height 300–750 km (Figure 115). In the May 23, 1967 event also a shortlived T_e/T_i change of 15% was observed.

If we now consider the physics of these changes, it is not difficult to estimate the EUV distribution necessary to produce the observed electron density enhancements in

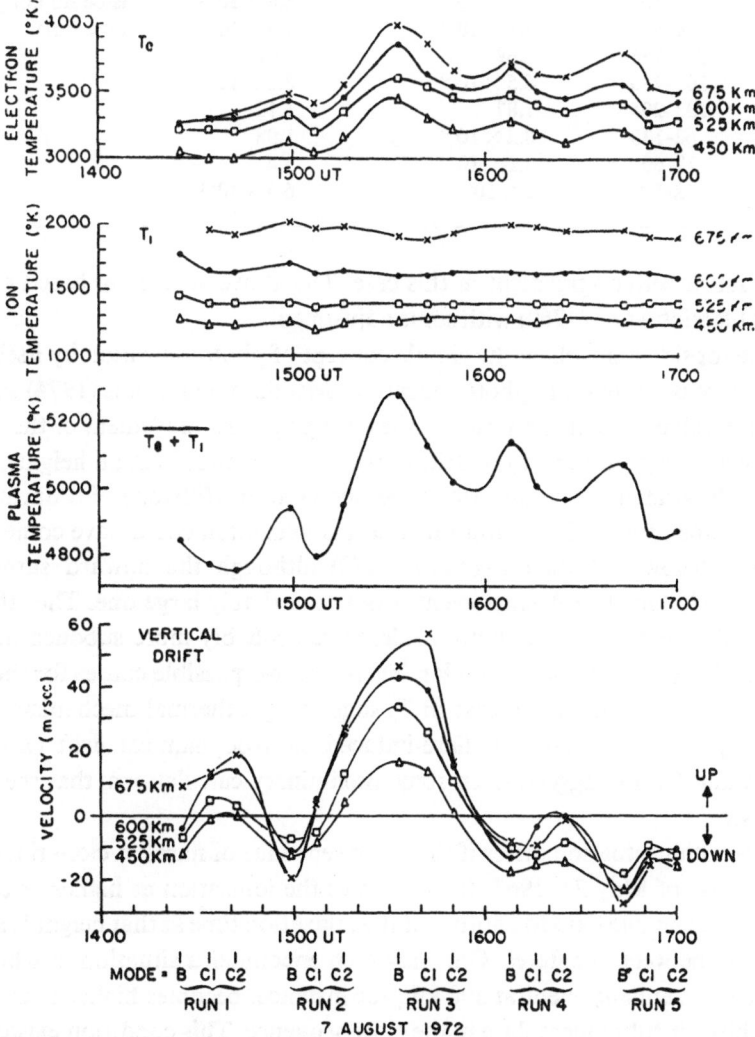

Fig. 115. Changes in electron temperature and in vertical drifts during the outstanding flare of August 7, 1972, obtained with the incoherent scatter equipment at Millstone Hill, U.S.A. (after Mendillo and Evans, 1974).

the bottomside ionosphere. The details of how this is done are discussed in the next chapter. It is appropriate, however, to point out the two theoretical curves marked 'Theoretical 1' and 'Theoretical 2' in Figure 112. Curve marked '1' is the profile one would get from the enhancements observed by Hall and Hinteregger (Chapter 3, Table V) in various EUV lines near the maximum of a 3⁻ flare starting at 0022 UT on 22 March 1967. Note that the observed EUV enhancements are insufficient to give the observed $\Delta N_e/N_e$ changes. The curve marked '2' has been obtained on the following basis (flux in erg cm^{-2} s^{-1}, wavelength in Å):

Wavelength	Flux before flare	Flux at peak	Remarks
1216	4.80	5.80	The different wavelengths
911–1027	6.2×10^{-1}	6.8×10^{-1}	listed do not peak at the
796–911	7.1×10^{-1}	7.8×10^{-1}	same time.
280–796	2.69	3.01	
205–280	6.5×10^{-1}	8.1×10^{-1}	
138–205	1.87	4.12	
62–138	8.2×10^{-1}	1.98	
12–60	4.1×10^{-1}	3.4	
2–12	2×10^{-3}	6.5×10^{-1}	

There is much better agreement in this case. The above values are based on the estimates of Garriott *et al.* (1969) with some adjustments.

For the topside ionosphere the simple concept of photoionization by itself does not suffice. There is not enough photoionization. Mendillo and Evans (1974) suggest that the topside enhancement is a result of the strong upward movement of the ionization. Under normal circumstances the drift is negative downwards at all heights, with little or no height gradient. The upward surge observed at Millstone Hill during the flare will cause enhancement in the topside, but this ionization could have come from both above and below 375 km (August 7, 1972) although the upward surge was not seen below 375 km. The August event was a remarkably large one. The effects, judging from the changes in the bottomside, were probably more subdued in the other cases, but the trends should be similar. There are two possible causes for the observed upward drift: (i) the large increase in T_e suggesting a thermal mechanism, similar to the existing at sunrise and (ii) flare-induced electrodynamical drift ($\vec{E} \times \vec{B}$ drifts). Mendillo and Evans suggest, from some preliminary calculations, that the first effect is adequate.

An interesting situation arises if there is a sequence of flares at close time intervals, as in the case of May 23, 1967. In such cases the ionization at higher levels goes on building up from one flare to another, if the relaxation time at that height is larger than the interval between the flares. One may then encounter a situation in which the per cent ionization enhancement at 300 km, for instance, becomes higher than at 150 km for the third or subsequent flare in the close sequence. This condition existed for May 23, 1967 flare, and is shown in Figure 116. It will be seen that during the first of the three closely spaced flares the computed per cent ionization enhancement at 300 km was considerably lower than at 140 km as expected. At the end of the first flare the 140 km ionization rapidly decays, but that at 300 km decays far more slowly and was substantially above normal by the time the second of the three flares starts. During the third flare 300 km ionization enhancement was comparable to or larger than the enhancement at 140 km. It is unfortunate that the incoherent scatter observations did not continue till the third flare: a comparison of the observed values would have thrown light on the loss processes at different heights during flares, for on these the nature of the cumulative effect depends. The curves shown in Figure 116 are computed theoretically on the following basis:

(i) EUV flux enhancements are given as in table above.

(ii) Time variations in the EUV flux follow centimeter radio bursts.

(iii) E and F-region recombination coefficients are obtained from laboratory reaction rate data (such as those given by Mitra and Banerjee 1971).

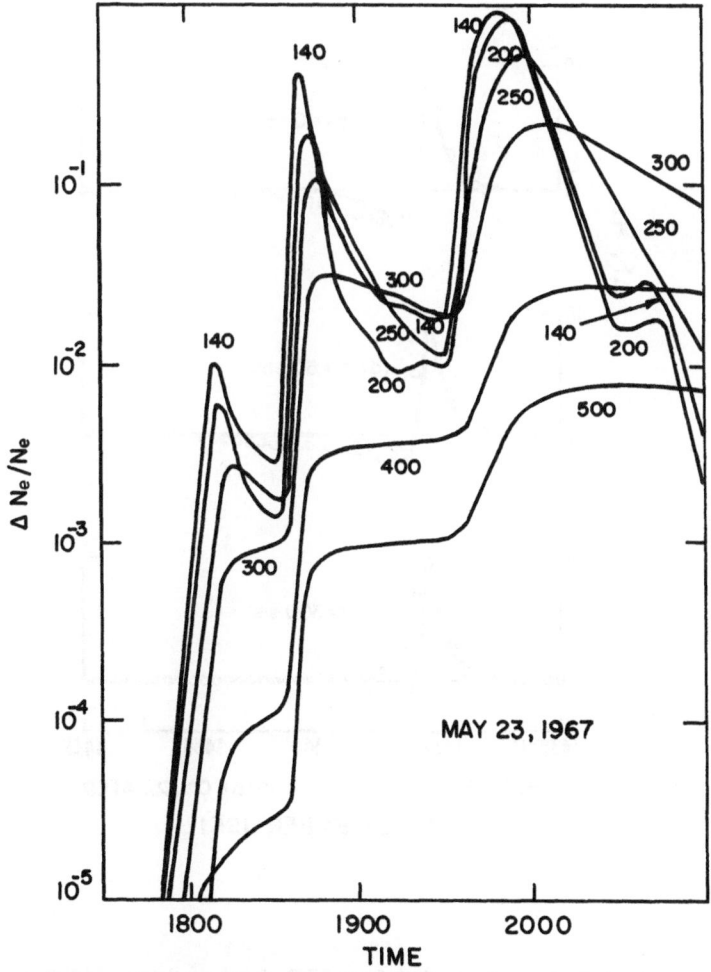

Fig. 116. Cumulative retention of charge enhancements in the F-region relative to lower levels (after Mitra, 1973).

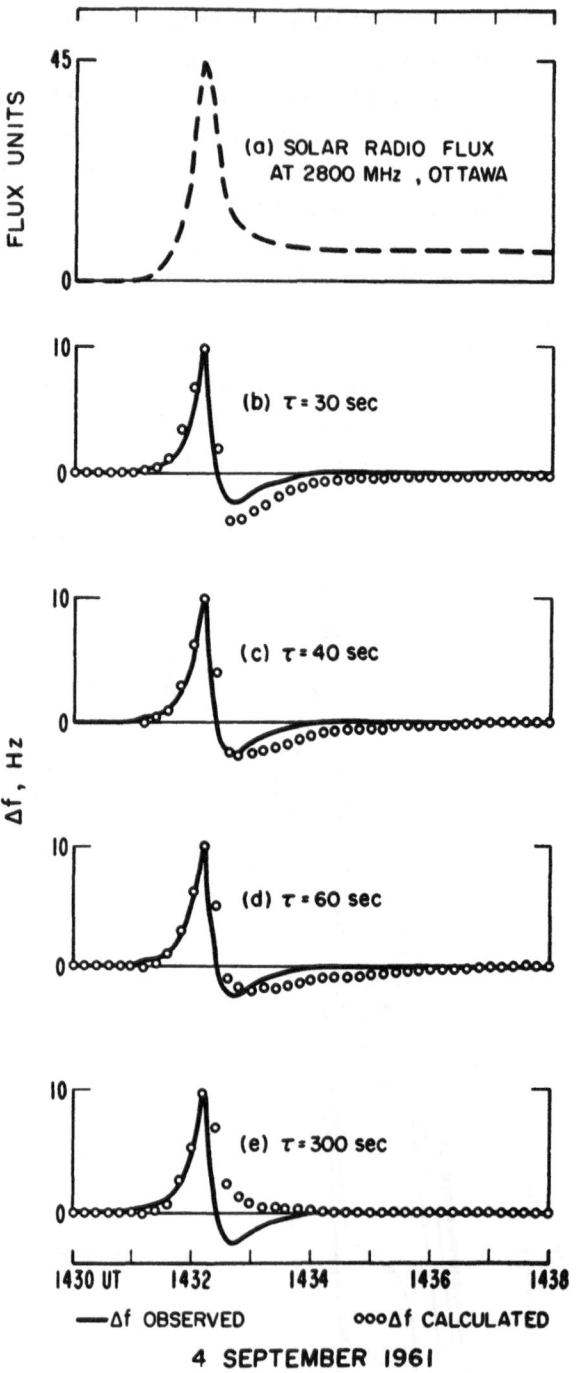

Fig. 117.   Determination of relaxation time from SFDs from synthesis of SFDs using solar radio
noise flux at 2800 MHz (after Baker *et al.*, 1968).

9.6.2. Effective loss rate

SFD data were used by Baker *et al.* (1968) to estimate the effective electron loss time constant (τ) and recombination coefficient in the E and F1-regions by two methods: (1) for SFD's with a pronounced negative decay stage, (implying that the ionizing radiation decayed rapidly), the time dependence of the decay of negative frequency deviation late in the event was used to estimate α, on the assumption that radiation had come back to normal by that time; and (2) the time dependence of 2800 MHz radio bursts was used to estimate the time curve of the ionizing radiation responsible for SFD's which was then used to compute the corresponding frequency deviation by varying the electron loss time constant until the best fit with the SFD observations was achieved.

In the first case the decay is given by:

$$\Delta f = \Delta f_R \exp\left(-\frac{t - t_R}{\tau}\right), \tag{85}$$

where Δf_R is the negative frequency shift at some reference time t_R, and a plot $\log \Delta f$ vs time gives a straight line for which the slope gives $(-1/\tau)$ and hence the relaxation time. An example of the second case is given in Figure 117. The best fit in this case was around $\tau = 40$ s.

The average results for the upper E-region and lower F1-region were $\tau = 35$ s and $\alpha_{\text{eff}} = 1.3 \times 10^{-7}$ cm^3 s^{-1}. This is in agreement with quietday value at these levels, there is thus no indication of any change in α in E or F-region heights.

Ogawa (1970), on the other hand, from an examination of two SFDs recorded over a 360 km path over Japan on 5 MHz found somewhat smaller values of α using the decay of Δf. His values were 9.4×10^{-8} cm^3 s^{-1} and 5.8×10^{-8} cm^3 s^{-1}. The two events were chosen with care: one was a single impulsive type event that occurred on October 2, 1965, and the other a slightly more complicated event on October 4, 1966.

Donnelly also examined the electron loss rate by using the time dependences of the OSO-3 measurements of He II 303.8 Å, O v 629.7 Å, H Lγ 972.5 Å or C III 977 Å to compute the time dependence of the frequency deviations. The effective recombination coefficient was varied until the theoretical and observational deviations agreed, especially for the rise, peak, and subsequent dip to near zero, rather than for the negative decay stage. Donnelly derived values of 2×10^{-7} cm^3 s^{-1} at 100 km, 1×10^{-7} at 160 km, 5×10^{-8} at 210 km and 1×10^{-8} at 235 km, for α_{eff}. However, the above α_{eff} model could be increased or decreased by about 40% before the time dependence of the computed frequency deviation differed significantly from the SFD observations. The SFD data involved were not sensitive to α_{eff} values above 200 km.

CHAPTER 10

SIDs AS FLARE RADIATION MONITORING TECHNIQUES

SIDs can, in several ways, supplement information on enhanced XUV flux during solar flares obtained directly with satellite-borne detectors. For some spectral ranges, the information provided by the SIDs can be unique. These include:

(i) Spectral information below 3 Å available from VLF propagation data and from multifrequency riometer observations, especially for periods *before* direct satellite-borne detector measurements in this region became available.

(ii) In the near and far ultraviolet, where flaretime variations are available for only a few flare events (Section 3.2), F-region SFDs have been used to estimate EUV enhancements in different wavelength bands.

The position is, however, changing very fast. There is now nearly 100% coverage of the solar XUV emissions with several satellites (e.g. Solrad, Vela, Atmospheric Explorer). Although flare spectra vary from flare to flare, and with time during a flare, there is a distinct pattern to these variations which is a complicated function of the wavelength but a simpler function of the flare source temperature. Since SIDs provide primarily the wavelength information; their usefulness is best realised when these are combined with direct satellite measurements.

10.1. Information on the X-Ray Region

10.1.1. VLF OBSERVATIONS

The procedure for converting the observed changes in amplitude and phase to those in the ionizing radiation below 3 Å was given by Barletti and Tagliaferri (1969). The core problem is to trace back the steps in the block diagram given in Figure 118. Here F_0 is the ionizing radiation under normal conditions, $F(\lambda, t)$ is the solar XUV flux at the top of the atmosphere during flare, $h_r(t)$ is a reference altitude, $E(t)/E_0$ is the ratio between received VLF field intensities in flare $[E(t)]$ and pre-flare $[E_0]$ conditions, $\Delta\phi(t)$ is, as before, the variation in phase in VLF and $\beta_r(t)$ is the conductivity gradient.

The observational parameters are $E(t)/E_0$ and $\Delta\phi(t)$, and the problem is not to

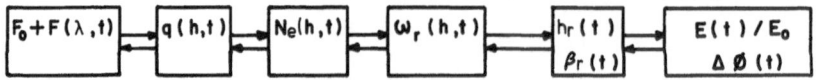

Fig. 118.   Steps for deriving spectral information below 3 Å from VLF phase and amplitude measurements (after Barletti and Tagliaferri, 1969).

derive the electron density profile as in Section 7.2, but to obtain $F(\lambda, t)$ for an assumed shape of the conductivity parameter ω_r. The basic equations used are those involving the following:

(i) Conductivity parameter $\omega_r(t)$ given by Equation (73) (Section 7.2)

(ii) Phase and amplitude variation parameters given by Equation (71) (Section 7.2) and the equation:

$$\left(\frac{E(t)}{E_0}\right) = \frac{h_r(0)}{h_r(t)} \, 10^{(d/20)(A_0 - A(t))},$$

A is the attenuation in dB/1000 km, respectively.

(iii) Production rate parameter given by Equation (86).

(iv) Ionization continuity equation given by Equation (38).

(v) Assumption concerning the normal situation defined by h_n and β_n; one may use the values given by Wait and Spies (1964) (70 km and 0.3 respectively) or any modified values.

One may write the equation for electron production rate in the form:

$$q(h, t) = \int_{0.3}^{3} F(\lambda, t) \, q'(h, \lambda) \, d\lambda + q_n(h), \tag{86}$$

where q_n is initial production rate corresponding to preflare radiation and q' is the electron production rate per photon impinging at the top of the atmosphere. On dividing the interval 0.3–3 Å in n parts,

$$q(h, t) = \frac{3 - 0.3}{n} \sum_{i=1}^{n} F_i(t) \, q_i'(h) + q_n(h). \tag{87}$$

Writing Equation (87) for n altitudes and solving for the $F(t)$ values, it is possible in principle to get the spectral photon flux at the top of the atmosphere. In practice, this approach is highly sensitive to the errors in the input data and is not workable. Therefore, another approach was used by assuming a power law distribution in the radiation spectrum between 0.3 and 3 Å (Equation (16)). With such an assumption, it is possible to give Equation (87) a new form with only two unknown quantities: the spectral index m and the flux $F(\lambda_1)$ at 0.3 Å (in erg cm^{-2} s^{-1} Å^{-1}) as follows:

$$\frac{F(\lambda_1) \lambda_1 \Delta\lambda}{hc} \left[q_1' + q_2' \left(\frac{\lambda_2}{\lambda_1}\right)^{m+1} + \cdots + q_n' \left(\frac{\lambda_n}{\lambda_1}\right)^{m+1} \right] = q(h) - q_n(h), \tag{88}$$

where h is the Planck's constant;

$$\Delta\lambda = \frac{3 - 0.3}{n},$$

or

$$p_m(h) F(\lambda_1) = q(h) - q_n(h). \tag{89}$$

In the narrow height range corresponding to the VLF reflection, both $\ln p_m(h)$ and $\ln[q(h)-q_n(h)]$ can be approximated by a straight line. It is, therefore, possible to find out the value $\bar{m}$ of m for which the slope of the function $\ln p_m(h)$ best approximates that of $\ln[q(h)-q_n(h)]$.

Now, from (89), $F(\lambda_1)$ can be obtained, so that

$$F(0.3\text{–}3) = \int_{0.3}^{3} F(\lambda)\, d\lambda = \frac{0.3}{m+1}\, F(\lambda_1)\left[10^{m+1}-1\right]. \tag{90}$$

Application of this procedure to six events in 1966 have been given by Barletti and Tagliaferri and are indicated in Figure 119. The estimated 0.3–3 Å flux for the events are compared with observations obtained with Solrad-8 satellite.

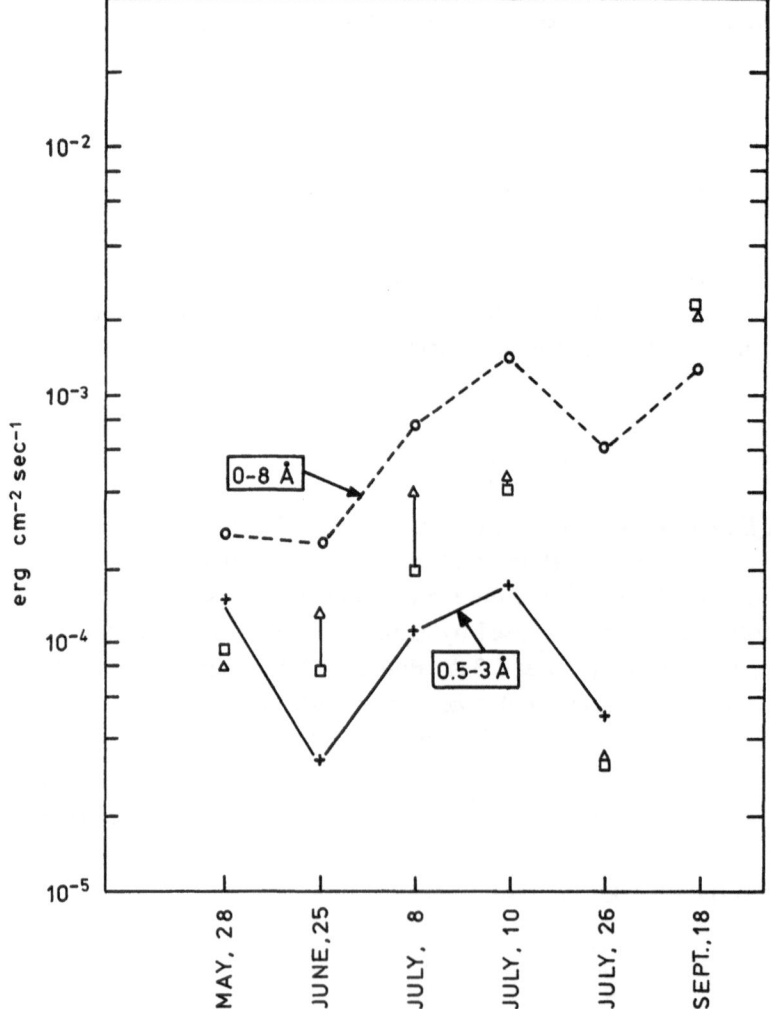

Fig. 119. Solar 0.3–3 Å flux deduced from VLF data by Barletti and Tagliaferri (1969) for several flare events in 1966, along with Solrad-8 measurements.

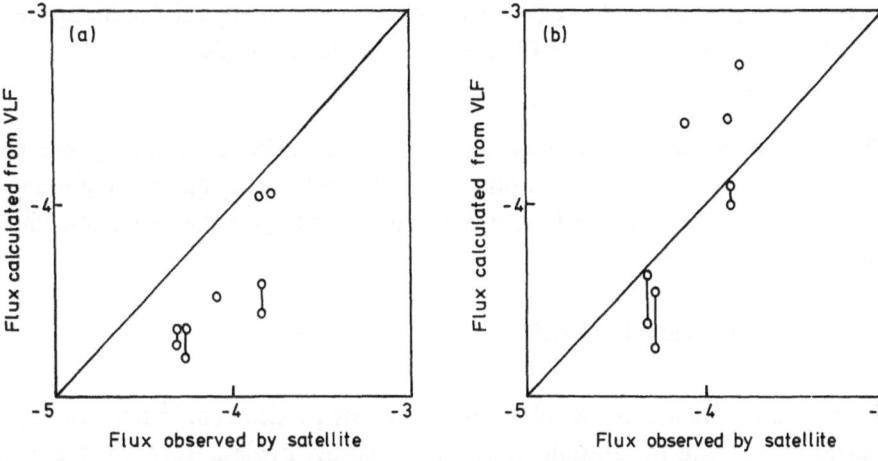

Fig. 120. Solar 0.3–3 Å flux deduced from VLF propagation data plotted against direct measurements made by Solrad-8 satellite under two different assumptions: (a) for $dN_e/dt=0$, $d\lambda/dt=0$; (b) for a rectangular burst and a complete solution of the time curve (after Barletti and Tagliaferri, 1972).

Further improvement in the estimate of the ionizing radiation can be effected by removing the assumption of equilibrium situation and the constraint of an unchanged loss rate. Neither assumption, as we have seen, is correct. The first can be removed by assuming a time function for the ionizing radiation. This may either be Mitra's function given by Equation (45) or, as in a recent work by Barletti and Tagliaferri (1972), by a rectangular burst lasting as long as the growth time (t_g) of the VLF event. In the latter case, Barletti and Tagliaferri, performing numerical integrations of the continuity equations for N_e and N^+ according to Runge-Kutta method by means of a high speed computer, find an improvement in flux estimate as shown in Figure 120. In this (a) gives the flux deduced under the equilibrium condition and unchanged loss rate; (b) gives the fluxes deduced when these assumptions are removed. Spectral information can be obtained when the value of the production rate q is derived at two different altitudes: from these one gets $F(\lambda_1)$ and m in Equation (90).

Sengupta (1971), however, feels that while the flux values and indices thus obtained give an approximate estimate of the 0.3–3.0 Å flux, the method of calculation is somewhat artificial because at 3.0 Å, thermal emission from the flaring region may be more important than the non-thermal emission obeying power law.

Taubenheim et al. (1970, 1974), Sengupta (1971), and earlier in a very simplified form Chilton et al. (1965) have given a different method for estimating the spectral characteristics or the hardness factor of flare emission in terms of an effective wavelength λ_{eff} from the observed VLF effects of the flare. It has been shown that the ionization due to radiation in a narrow spectral band within a small range on height below the height of peak ionization production can be approximated by ionization due to an equivalent monochromatic flux with an 'effective wavelength' λ_{eff}, dependent upon the spectral distribution and intensity F_{eff} proportional to the total flux in the wavelength range. It is assumed that we are concerned with ionization at a level where the pre-flare ionization can be neglected (an assumption usually valid at the appreciably lowered level of reflection at VLF) so that, from Equation (29)

$$q(h) = F_{\text{eff}} n(h) \, \sigma_{\text{eff}} \eta_{\text{eff}} \exp\left[-\xi(h) \, \sigma_{\text{eff}}\right], \tag{91}$$

where F_{eff} is in erg cm^{-2} s^{-1} at wavelength λ_{eff} incident above the atmosphere, $n(h)$, is the total number density of the ionizable constituents, being air in the present case, σ_{eff} is the mean absorption coefficient of air at λ_{eff} and η_{eff} is the ionization efficiency at the wavelength, and:

$$\xi(h) = \sec\chi \int_{h}^{\infty} n(h)\,\mathrm{d}h = \sec\chi n(h)\, H_0(h) \tag{92}$$

Physically, $\xi(h)$ stands for the total number of air particles cm^{-2} that the radiation traverses in reaching the altitude h, and the quantity $\xi(h)\sigma_{\text{eff}} = \sigma_d(h)$ is the effective optical depth at the height. For the VLF flare effects we are concerned with altitudes below 70 km, and the ionization equation is given by (from Equation (40)):

$$N_e^2(h) = q(h)/\psi(h) = \frac{F_{\text{eff}} n(h) \, \eta_{\text{eff}} \, \sigma_{\text{eff}} \exp\left[-\xi(h)\,\sigma_{\text{eff}}\right]}{\psi(h)}$$

$$= \text{constant} \times F_{\text{eff}} \exp\left[-\xi(h)\,\sigma_{\text{eff}}\right] \tag{93}$$

if one assumes $n(h)/\psi(h)$ remains approximately constant. Sengupta (1971) and Taubenheim *et al.* (1974) found $\psi(h)$ to be in fact proportional to $n(h)$, so that this assumption is probably valid.

If quasi-equilibrium condition holds during the effect or at least $\mathrm{d}N_e(h)/\mathrm{d}t$ has the same value at the reflection heights as F_{eff} changes (both questionable assumptions), and that the flare emission spectral distribution does not change considerably during the process, then

$$\frac{F_{\text{eff}}(t) \exp\left[-\xi(h_f)\,\sigma_{\text{eff}}\right]}{v^2(h_f)} = \text{constant} \tag{94}$$

if VLF reflection takes place at a constant value of the conductivity parameter ω_r. Here h_f is the reflection height at time t. Assuming an exponential $v(h)$ profile given by:

$$v(h) = 6 \times 10^6 \exp\left[-0.13(h - 70)\right].$$

Sengupta derived:

$$F_{\text{eff}}(t) = \text{constant} \times \exp\left[\xi(h_f)\sigma_{\text{eff}} - 0.26(h_f - h_n)\right], \tag{95}$$

where h_n is the reflection height at $t = 0$.

If the reflection height changes from h_1 to h_2 as F_{eff} changes from $F_{\text{eff}}(t_1)$ to $F_{\text{eff}}(t_2)$, assuming that the spectral distribution of the flare emission does not change considerably during the process we have:

$$\ln F_{\text{eff}}(t_2) - \ln F_{\text{eff}}(t_1) = \Delta \ln F_{\text{eff}} = \sigma_{\text{eff}}[\xi(h_2) - \xi(h_1)] + 0.26(h_1 - h_2)$$

or

$$\sigma_{\text{eff}} = \frac{\Delta \ln F_{\text{eff}} - 0.26(h_1 - h_2)}{\xi(h_2) - \xi(h_1)}. \tag{96}$$

Since it is assumed that F_{eff} is proportional to $F(1-8 \text{ Å})$ for a constant spectral distribution, F_{eff} and $\Delta \ln F_{eff}$ can be replaced by $F(1-8 \text{ Å})$ and $\Delta \ln F(1-8\text{Å})$ respectively in Equations (95) and (96). One can then derive the mean σ_{eff} (corresponding to λ_{eff}) by plotting $\xi(h_f)$ against $[\ln F(1-8 \text{ Å}) t - 0.26(h_f - h_n)]$. The above procedure is valid if VLF reflection occurs at constant ω_r. If, on the other hand, it occurs at constant electron density, one should use the equation:

$$\sigma_{eff} = \frac{\ln F (1-8 \text{ Å})}{\xi(h_2) - \xi(h_1)}. \tag{97}$$

Thus the effective wavelength can be obtained from observed changes in 1–8 Å flux and corresponding changes in reflection height given by the associated SPA time profile. It must, however, be noted that in such computations the finite timelag between the change in the ionizing flux and the resulting change in electron density need to be taken into account. Sengupta does this by using a constant time-lag of 1 min for heights below 70 km. One may improve upon this approach by using values of the type given in Figure 86.

By applying this procedure to the large flare occurring on July 8, 1968, beginning at 1707 UT flare and the precursor flares (Figure 76a) of which the one starting at 1630 UT alone produced ionospheric effects, Sengupta obtained the following results:

	λ_{eff}	h_n	F_{eff}	$(F_{eff} \times 100)$
	(Å)	(km)	(erg cm^{-2} s^{-1})	$F(2-12 \text{ Å})$
Flare 1	1.75	66.2	5×10^{-5}	0.6
Flare 2	1.45	54	4×10^{-2}	10

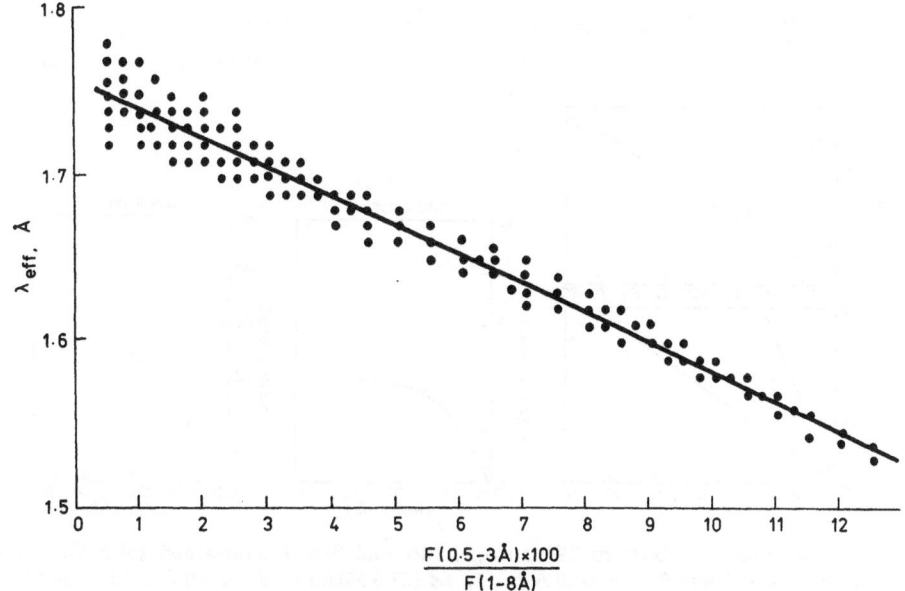

Fig. 121.   Relationship between λ_{eff} and hardness factor H.F. (after Sengupta, 1971).

One observes that Flare 2 with lower λ_{eff} value and F_{eff} =10% of the measured 2–12 Å flux had a much harder spectrum than Flare 1.

As mentioned before λ_{eff} is a measure of the hardness of the spectrum. Its relationship with the hardness factor H.F. (0–3 Å/1–8 Å) as obtained by Sengupta for 120 flare events is given in Figure 121. One finds that λ_{eff} is linearly related with H.F. This figure gives for λ_{eff} = 1.40 Å a value of H.F. = 18% which is somewhat larger than but in close agreement with the calculation given above. One thus finds a combination of SPAs and observations of solar X-ray flux in 1–8 Å channel may quite conveniently be used to determine the hardness of the flare X-ray spectrum.

10.1.2. HF ABSORPTION PHENOMENA INCLUDING SCNA OBSERVATIONS

We are now concerned with somewhat higher levels in the D-region (70–100 km) with a peak around 80–85 km than those involved in VLF propagation, and consequently softer X-ray are involved. Schwentek *et al.* (1970) have examined 60 SWFs of varying strengths received at Lindau recorded on frequencies of 2.50, 2.614, 2.775 and 6.09

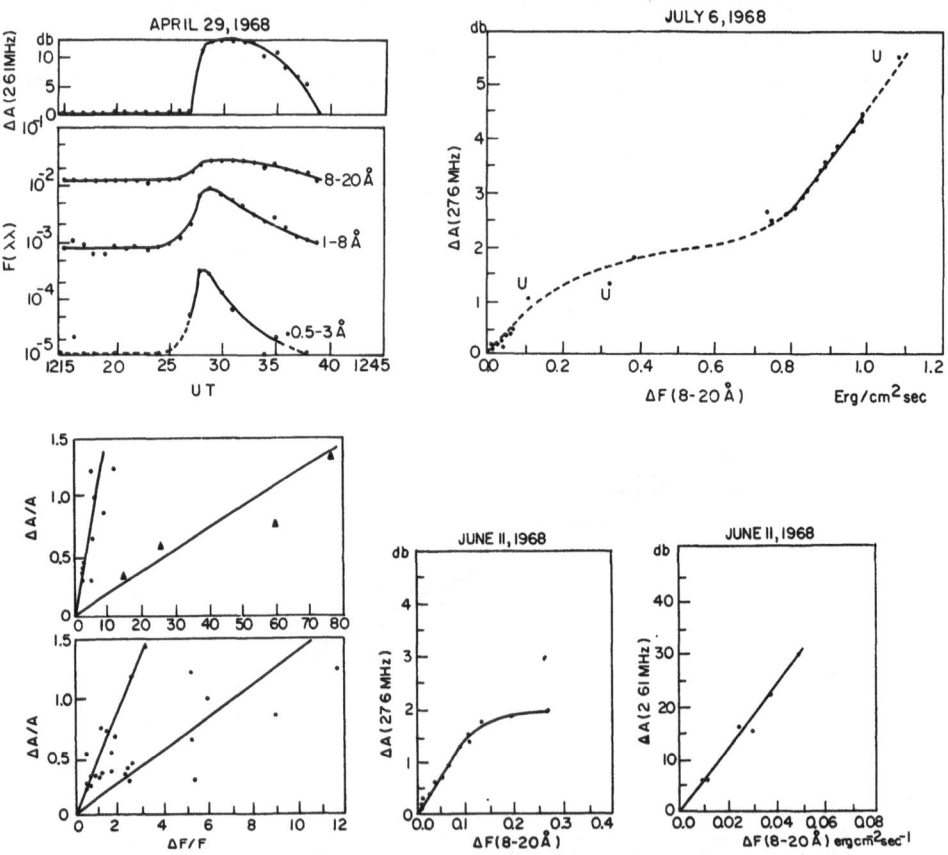

Fig. 122. Relationships between SWFs magnitudes and fine structures and solar X-ray fluxes: fadeout shape and X-ray flux in 8–20 Å band; ΔA (27.6 MHz) vs ΔF (8–20 Å); ΔA (2.61 MHz) vs ΔF (8–20 Å) and $\Delta A/A$ vs $\Delta F/F$ (Schwentek *et al.*, 1970).

MHz (with transmitter-receiver distance of 300 km) and a riometer operating on 27.6 MHz with a corner reflector directed towards the pole star.

Simultaneous observations were available during these events of solar X-ray flux (in minute intervals) over the wavelength ranges 0.5–3 Å, 1–8 Å, 8–20 Å and 44–60 Å from OGO-4 and Solrad-9 satellites. They found that:

(a) the shape of the fadeout is determined mainly by the wavelengths 8–20 Å (Figure 122).

(b) the sudden and slow SWFs have their counterparts in X-rays, and small variations in 8–20 Å flux can be correlated to those in signal strength on 2.61 MHz.

(c) Superimposed on the above is a kind of fine structure in the decrease of signal strengths occurring in the very beginning of some fadeouts that are connected with 0.5–3 Å and 1–8 Å flux variations.

Schwentek *et al.* then established empirical relationships between ΔA (27.6 MHz) and ΔF(8–20 Å) which can be used either for obtaining the time curve of the X-ray flux during a flare event (Figure 122 shows the remarkable correlation between ΔF(8–20 Å) and ΔA for large events), or as a monitor of the peak X-ray flux in the range 8–20 Å for different events. For the latter it is more appropriate to use $\Delta A/A$ instead of ΔA, as previously noted. In Figure 122 we show the results given by Schwentek *et al.* relating $\Delta A/A$ with $\Delta F/F$(8–20 Å) for absorption measurements made on 2.61 MHz. Only relatively small effects can be studied (upto 35 dB), since larger events cause total fadeout. A complicating factor in this representation is that there are three clearly separated groups of points. All these groups can be described by:

$$\frac{\Delta A\left(t_m\right)}{A} = \text{constant}\ \frac{\Delta F\left(t_m\right)}{F}. \tag{98}$$

For the first group (15 points; $\Delta F/F < 3.5$) the constant is 0.45; for the second group (10 points; $2 < \Delta F/F < 12$) constant is 0.13; for the third group (4 points; $13 < \Delta F/F < 80$) the constant is 0.018. The difference in the three groups is probably an indication of the effect that there is a change in the effective loss rate as the flare intensity increases.

The entire time curve of the total ionizing radiation can be obtained rather simply, if it is assumed that the peak absorption region has not been displaced greatly during a flare. In that case, the ionizing radiation $F(\Delta\lambda)$ is given by:

$$\frac{q}{q_n} \doteq \frac{F\left(\Delta\lambda\right)}{F_n} = \frac{\tau_0}{2} \frac{\mathrm{d}}{\mathrm{d}t}\left(\frac{A}{A_n}\right) + \left(\frac{\alpha}{\alpha_n}\right)\left(\frac{A}{A_n}\right)^2 ,$$

where τ_0, the preflare 'relaxation time', can be obtained in the way outlined in Section 5.3. The value of $2\alpha N_n (= 1/\tau_0)$ is in the neighbourhood of $4 \times 10^{-4}\ \mathrm{s}^{-1}$ (Mitra *et al.*, 1966; Appleton and Piggott, 1954). The time curve of q/q_n has been obtained by several workers; for example by Taubenheim (1962) assuming $\alpha = \alpha_n$. Nestorov (1971) and Taubenheim (1962) find that the time variations of the ionizing radiation so obtained are sometimes complex (and correspond to a complex type of radio burst) and sometimes simple. The common feature is a very sharp peak of ionizing radiation

('flash phase'), which mostly coincides with the peak of the radio burst, but sometimes shows delay upto 3 min. There is, however, considerable uncertainty in the
method, since the value of αN_n is doubtful, $\alpha \neq \alpha_n$, and the level of peak absorption may
move considerably during a flare: both uncertainties can be reduced by obtaining
αN_n from the decay curve of the same flare and checking on the relative constancy of
the absorption level by plotting $(1/A)(dA/dt)$ vs time during the decay portion
(see Chapter 8). Somewhat more reliable values can be obtained if $N_e - t$ values are
available for specific height (as with partial reflection and cross-modulation observations) and the complete continuity equation (76a) is used in conjunction with an ion
chemical scheme of the type of Figure 103. These provide $q(t)$ curves which can then
be connected to $\lambda - t$ curves through a set of equations relating $q_1, q_2, ..., q_n$ corresponding to heights $h_1, h_2, ..., h_n$ with fluxes of ionizing radiations at wavelengths
$\lambda_1, \lambda_2, ..., \lambda_n$ as given on p. 230.

A more valuable piece of information comes from multifrequency absorption
measurements, which allows identification of events in which there is severe spectral
hardening. For such events there is a substantial lowering of the peak absorption level
to heights where ω is no longer larger than v, as is usually the case, and the ratio of
absorption at two frequencies (A_1/A_2) shows a deviation from the pre-flare ratio of
ω_2^2/ω_1^2. According to the Appleton-Hartree equation,

$$\frac{A_1}{A_2} = \frac{\omega_2^2 + v^2}{\omega_1^2 + v^2} \simeq \frac{\omega_2^2}{\omega_1^2}. \tag{99}$$

If the spectrum is appreciably hardened the ionization to levels where $v^2 \gg \omega^2$, one gets

$$\frac{A_1}{A_2} \to 1.$$

In more elaborate calculations, where one must use the complete equation:

$$\frac{A(f_{e1})}{A(f_{e2})} = \frac{\int_{h_0}^{\infty} k_0(f_{e1}, h) N_e(h)\, dh}{\int_{h_0}^{\infty} k_0(f_{e2}, h) N_e(h)\, dh} = \frac{K_1}{K_2} = f(h_0)$$

and use generalised Sen-Wyller (1960) magneto-ionic theory. Then $A(f_{e1})/A(f_{e2})$ is
given by the absorption integral $K_1/K_2 (= f(h_0))$. If we now refer to Figure 78 we see
that the ratio falls rapidly with decreasing height as with spectrum hardening the
radiation enters deeply into the lower ionosphere. The first examples of this kind given
by Morriss (1960) are given in Figure 24. Other examples identified at Delhi with
simultaneous observations at 20 and 30 MHz are given in Figure 25 (Mitra et al.,
1966). These give events of unusually large spectral hardening. From the derived
ratio of $A(f_{e1})/A(f_{e2})$ one may obtain h_0 using Figure 78 and consequently the
hardening ratio H.F.

10.2. Information on EUV Flux from F-Region SIDs

The F-region SID effects are principally caused by flare radiation in the 100–1030 Å band but at E-region heights X-ray contribution from 10–100 Å band can be competitive. Since the time histories in the EUV region are at present available for only a few events, and these are known to differ from one event to another (Kreplin et al., 1969). It is important to examine to what extent F-region SIDs can provide information on this radiation.

Detailed studies in this line are available from: (a) ionogram analysis (Bhattacharyya and Balakrishnan, 1967), (b) SFD analysis (Donnelly, 1968–71, 1973), and (c) incoherent scatter (Garriott et al., 1969).

The starting point, as in the D-region, is the continuity equation. In this, the major differences from the D-region are the following:

(i) The negative ions are no longer important, and, therefore, the terms containing λ do not appear in the equation.

(ii) In addition to the terms defining production or destruction of ionization, we have also to consider the movement of ionization. The terms 'photochemical' and 'transport' serve as convenient labels for these two categories.

If the transport processes result in a net drift velocity v, then the electron continuity equation takes the form:

$$\frac{\mathrm{d}N_e}{\mathrm{d}t} = q - L - \mathrm{div}\,(N_e v),$$

where L is the loss term and $\mathrm{div}\,(N_e v)$ represents the effect due to the drift term.

The relative importance of the photochemical and transport terms varies with height; photochemical processes dominate below about 250 km during daytime; and the transport term above this level. The loss term L has a complex behaviour in the E and the F region. While between 100 and 150 km the effective recombination coefficient is determined by the dissociative recombination of the two principal molecular ions NO^+ and O_2^+, at higher levels the dominating O^+ ions go through the following chain of processes:

$$O^+ + O_2 \xrightarrow{k_1} O_2^+ + O$$

$$O^+ + N_2 \xrightarrow{k_2} NO^+ + O$$

$$O_2^+ + e \xrightarrow{\alpha_D(O_2{}^+)} O' + O''$$

$$NO^+ + e \xrightarrow{\alpha_D(NO^+)} N' + O'$$

Under these conditions, the effective recombination coefficient α_{eff} for quasi-equilibrium conditions, is given by:

$$\alpha_{\mathrm{eff}} = \frac{(k_1\,[O_2] + k_2\,[N_2])\bar{\alpha}_D}{(k_1\,[O_2] + k_2\,[N_2]) + \bar{\alpha}_D N_e} = \frac{\beta\bar{\alpha}_D}{\beta + \bar{\alpha}_D N_e},$$

where

$$\beta = k_1 [O_2] + k_2 [N_2]$$

and

$$\bar{\alpha}_D = \frac{\alpha_D(O_2^+)[O_2^+] + \alpha_D(NO^+)[NO^+]}{[O_2^+] + [NO^+]}.$$

The transport term, which begins to become important for daytime conditions above about 250 km, is difficult to take into account in a proper way especially during the solar flares. As we have seen earlier, there is a substantial change in the gradient of the drift velocity at the time of the solar flare. When information on the profile of the vertical velocity is available, it is possible to take into account all these effects appropriately but in most cases such information is lacking. Fortunately, much of the information on the EUV radiation during the flare can be obtained from the changes in ionization below 250 km for which the effect of the transport term has no appreciable effect and can be ignored.

A major source of uncertainty in these flux determination, common to all F-region SID techniques, is the question of choosing the appropriate recombination coefficient profile. One may obtain this profile in two ways:

(i) From the use of laboratory reaction rates for k_1 and k_2, an atmospheric model appropriate for the local time and solar activity and latitude and the use of preflare electron density profile; or

(ii) From merely deriving the preflare profile q/N_e^2, using a standard EUV flux spectrum such as that given by Donnelly and Pope (1973) and the preflare electron density profile. The problem is that the two do not always agree. If the discrepancy is small, a minor adjustment in k_2, in the flux model or in the atmospheric density model may be in order. Where the discrepancy is large, one must choose which of the two profiles one should use. The possibilities of error in the second procedure is generally larger, because of the uncertainties in estimates of q, and it is, in general, more advisable to make adjustments in (i) than to accept (ii) in full.

In Figure 123, we show the kind of differences in $\alpha - h$ profiles that may arise. Two typical cases are shown. These refer to preflare conditions of May 21 and May 23, 1967 appropriate for Arecibo. The thick solid curve is an average 'reference' profile obtained by Chakrabarty (1974) for $k_1 = 5 \times 10^{-13}$ cm^3 s^{-1}, $k_2 = 1 \times 10^{-9}$ $T^{-0.7}$ cm^3 s^{-1}, a solar flux spectrum which is twice that of Hinteregger below 350 Å and 3 times that of Hinteregger above 350 Å and the atmosphere model given by CIRA 1972. The difference between these at certain heights amount to an order of magnitude; the errors caused by the uncertainties in α in the estimates of flaretime q are thus also about an order of magnitude.

There is an additional error factor coming from the estimates of flaretime q. It is that in the region between approximately 140 to 180 km, where the recombination coefficient is changing from a quadratic law (αN_e^2) to a linear law (βN_e), α is itself a function of N_e. At heights below 140 km, where O_2 and N_2 densities are large,

$$(k_1 [O_2] + k_2 [N_2]) \gg \bar{\alpha}_D N_e$$

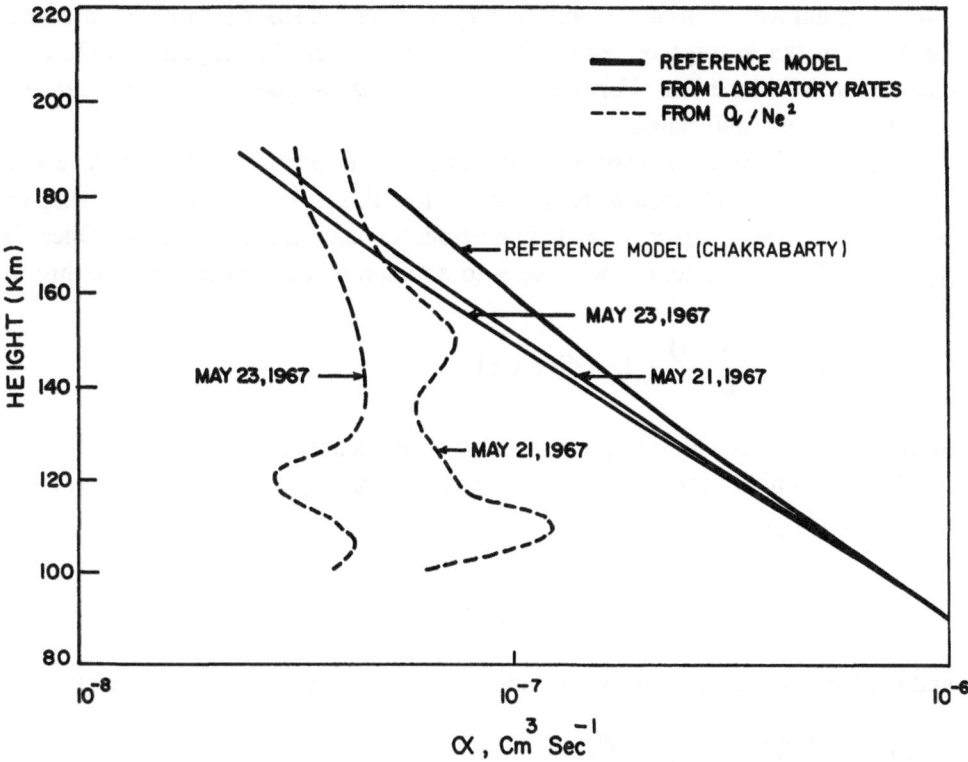

Fig. 123.   Differences in the *F*-region recombination coefficient profiles. The thick solid line gives the 'reference' model of Chakrabarty (1974). The thin solid line curves are those obtained by using laboratory reaction rate data appropriate for the preflare times of the events of May 21 and 23, 1967. The dotted curves are those derived from q/N_e^2 at these times. The differences are considerable especially at lower heights.

and

$$\alpha_{\text{eff}} = \bar{\alpha}_D .$$

At heights above 200 km

$$k_1 [O_2] + k_2 [N_2] \ll \bar{\alpha}_D N_e$$

and

$$\alpha_{\text{eff}} = \frac{k_1 [O_2] + k_2 [N_2]}{N_e} = \frac{\beta}{N_e} .$$

In the intermediate region where the entire equation must be used, an enhancement in N_e reduces the value of α_{eff}. If N_e is enhanced by 20%, the reduction amounts to somewhat less than 20% depending on the relative importance of the terms $(k_1 [O_2] + k_2 [N_2])$ and $\bar{\alpha}_D N_e$.

The actual amount of reduction is given by the term $[\bar{\alpha}_D \Delta N_e / (\beta + \bar{\alpha}_D N_e)]$ in the equation:

$$\alpha_f = \alpha_n \left(1 - \frac{\bar{\alpha}_D \Delta N_e}{\beta + \bar{\alpha}_D N_e} \right).$$

The maximum reduction can amount to $\Delta N_e/N_e$. Such reductions have been calculated for Malvern (an average of several flare cases) and for the August 7, 1972 event recorded at Millstone Hill. Fortunately, this effect is confined to a limited range of heights and is also only marginal.

The primary step for estimating enhancements in the ultraviolet flux during a solar flare is to obtain an electron density profile. How this is done for different SIDs has been discussed in Chapter 9, with indications of inaccuracies involved. Once the $N_e - h$ profile is available, the next step is to obtain a production rate profile through the equation:

$$q = \frac{dN_e(h, t)}{dt} + L + \text{div}(N_e v)$$

in which, for most calculations, the transport term is ignored.

A further simplification occurs in the F region where $\Delta N_e/N_e \ll 1$ and $\Delta q/q \ll 1$. In this region, therefore,

$$\frac{d(\Delta N_e)}{dt} = \Delta q - 2\alpha N_0 \Delta N_e - \alpha(\Delta N_e)^2$$

and writing $\tau_0 = (2\alpha N_0)^{-1}$ and neglecting the second order term $\alpha(\Delta N_e)^2$ we have:

$$\frac{d(\Delta N_e)}{dt} = \Delta q - \frac{\Delta N_e}{\tau_0}.$$

This equation is used for the F-region for the SFD for which the frequency deviation is given by $d(\Delta N_e)/dt$.

From the $q - h$ profile, the energy spectrum of the ionizing radiation can be deduced in the following way:

Suppose $q_1, q_2, ..., q_m$ are the rates of ion production at heights $h_1, h_2, ..., h_m$; and the fluxes of ionizing radiation at wavelengths $\lambda_1, \lambda_2, ..., \lambda_n$ are $F_1, F_2, ..., F_n$; then we can form the following equations:

$$q_1 = a_{11}F_1 + a_{12}F_2 + a_{13}F_3 + \cdots + a_{1n}F_n$$
$$q_2 = a_{21}F_1 + a_{22}F_2 + a_{23}F_3 + \cdots + a_{2n}F_n$$
$$\vdots$$
$$q_m = a_{m1}F_1 + a_{m2}F_2 + a_{m3}F_3 + \cdots + a_{mn}F_n$$

in which

$$a_{mn} = \eta_n A_n n_m \bar{e}^{\tau_{mn}}.$$

The above equations can be solved if the electron density changes are available at sufficiently close intervals of time.

10.2.1. EUV FROM IONOGRAMS DURING FLARES

One of the simplest and most direct way (but one that is very difficult to record) is to obtain ionograms with vertical incidence ionosondes during a flare at times when the

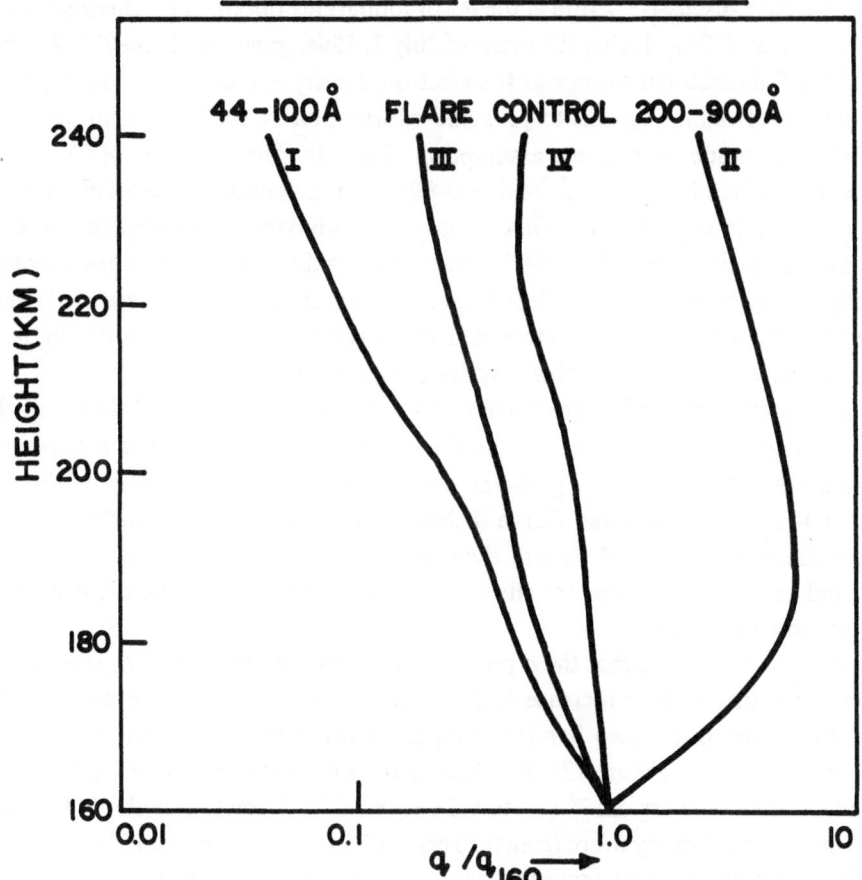

Fig. 124. Height distribution of electron production rates due to X-rays (44–100 Å) and UV (200–900 Å) during the flare of July 7, 1966, compared to those during control conditions (after Bhattacharyya and Balakrishnan, 1967).

fadeout effects are not very severe. This can occur at sunrise and sunset times. One such case that we have discussed before is the set of ionograms obtained by Bhattacharyya and Balakrishnan (1967) during the event of July 7, 1966, given in Figure 73b. Bhatta-charyya and Balakrishnan examined the electron density enhancements during these events obtained from ionograms to get some information on the changes in the XUV radiation. They made an extreme assumption. They divided the entire XUV region into two broad bands: 44–100 Å and 200–900 Å. The ionograms available during the flare were at 15 min intervals. Bhattacharyya and Balakrishnan used the electron densities at six levels in the 160 to 260 km range during these times. For normal condi-tions, mean electron densities for these heights for four adjacent days were used. They assumed that the time of peak electron density at 160 km is consistent with the peak intensity of the ionizing agency. This, however, is not entirely correct.

The production rate profiles for these two bands during the control days and the flare time are given in Figure 124. The production rates were normalised with respect to the values at 160 km. Curve I gives the production rate profile if whole of the flux lies within 44 to 100 Å band and Curve II that due to radiation in 200 to 900 Å band. Against these, Curves III and IV are their estimates, from the flaretime ionization profiles and assumed loss rates, for the actual production rates during the flare peak and during the control days.

The prominent feature is that the q-profile was pulled towards the X-ray side during the flare, indicating greater increase in the X-radiation than in the ultraviolet. The average flux values in the two bands during the control period and during the flare period are also given in Figure 124. It is seen that while in the 200 to 900 Å band, the flux changed from 0.09 erg cm^{-2} s^{-1} to 0.11 erg cm^{-2} s^{-1} and that in the 44 to 100 Å band changed considerably more from 0.25 erg cm^{-2} s^{-1} to 1.10 erg cm^{-2} s^{-1}.

It will be seen in the next sections that these values are quantitatively incorrect. However, the conclusion that the X-ray radiation is increased in the 44 to 100 Å by a larger percentage than in the ultraviolet is basically correct. Since 44–100 Å flux is dominated by slow coronal flare emission, while the impulsive emission is strong in 200–900 Å range, the low time resolution in these observations have essentially emphasised the slow components in the 44–100 Å range.

10.2.2. SFDs AS ULTRAVIOLET MONITOR

The Sudden Frequency Deviations have, however, been used with considerably more success. Much of the work with SFDs in this line has been done by Donnelly and his colleagues (1968–71, 1973) in Boulder, where SFD measurements at a number of frequencies have been in progress for many years. Depending on the geometry and the frequency, the SFD paths may involve: (i) the F-region and the upper E-region or (ii) the bottom of the E layer or sporadic E. We are concerned here principally with the former category. Furthermore, since 1 to 10 Å flux changes are usually much slower than the impulsive EUV burst, 1 to 10 Å portion is usually removed in the analysis of the F-region SFDs. The broad radiation band on which the SFD provides information is thus from 10 to 1030 Å. The upper limit is due to the fact that 1027 Å is the longest wavelength that will give efficient photoionization for any of the major constituents of the upper atmosphere.

Donnelly (1970) has introduced a sensitivity parameter S_λ for the SFDs to the incident solar ultraviolet radiation. S_λ is given by:

$$S_\lambda = \Delta f(f_v)/R_t R_\chi \Delta F(\lambda)\, \mathrm{d}\lambda,$$

where R_t gives the time dependence of the flux and R_χ gives the the solar zenith angle dependence, and f_v, as before, is the equivalent vertical incidence frequency of the SFD probing radio wave. S_λ varies with wavelength, with solar zenith angle, and with ionospheric and atmospheric properties. As mentioned before, the sensitivity drops off abruptly at about 1030 Å, and there are similar decreases near 900 Å and 800 Å because O and N_2 cease to be ionized around these wavelengths. There is also a rapid drop in S_λ for radiation below 10 Å since such radiation ionize principally the D-region. The net result is that there is a broadband of 10 to 1030 Å over which S_λ has roughly the same value. In many ways it is very much like a very broad-band satellite detector. In its relationships to solar zenith angle, it has a dependence like the aspect angle dependence of a satellite detector. The observed frequency deviation is linearly related to 10–1030 Å flux, upto enhancements of about 1 erg cm^{-2} s^{-1}.

Donnelly has pointed out another limitation in the detection capability of the SFD. The time constant τ determines how impulsive the ultraviolet burst has to be in order to be detected by the SFD system. Figure 125a explains this. In this the solid line gives

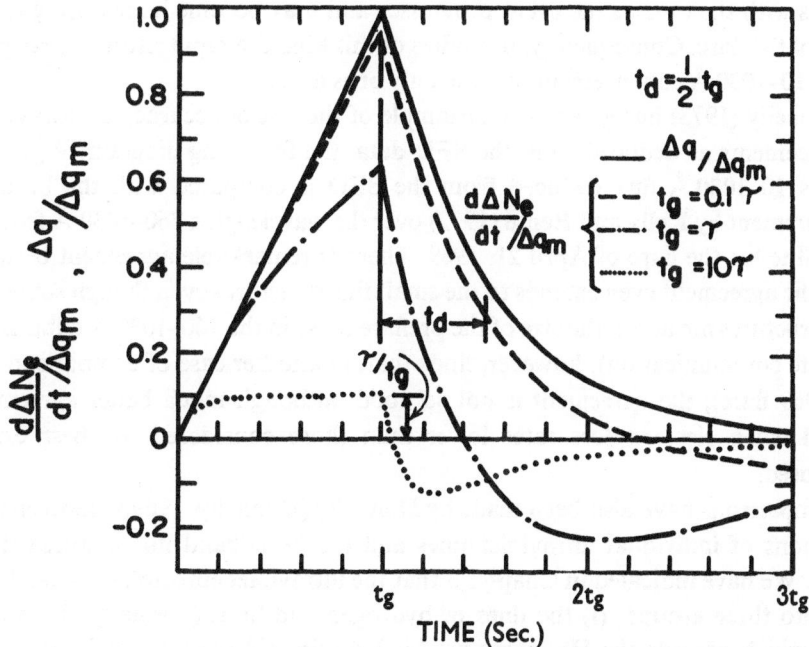

Fig. 125a.   Diagram indicating the sensitiveness of SFDs to the impulsiveness of EUV bursts (after Donnelly, 1970).

the time change of $\Delta q/\Delta q_m$ from a hypothetical ultraviolet flare event in which the rise time to the maximum is t_g and the decay time $t_d = \frac{1}{2}t_g$. The other groups give the time dependence of $d(\Delta N_e)/dt$ (and hence the SFD) normalised with Δq_m for three cases of impulsiveness: in one the rise time is 1/10th of the ionospheric time constant; in the second the two are equal, and in the third the rise time is ten times the time constant. When the ultraviolet burst is fast with respect to τ, then $d(\Delta N_e)/dt$ and Δf have essentially the same time variation as the ultraviolet burst, except late in the decay stage. When the ultraviolet event has rise and decay times comparable to τ, then again Δf has a time dependence similar to that of the ultraviolet event except that the difference now increases with time and becomes large during the decay period. Donnelly states that most SFDs fit such cases. When the ultraviolet event is very slow and the rise time is much larger than the ionospheric time constant $(d/dt)(\Delta N_e)$ becomes a small term because now the loss term and the production term are nearly equal and as a result $\Delta f(t)$ is small, smooth and flat. It is thus seen that the SFDs are insensitive to slow and smooth ultraviolet events and are sensitive only to events in which the rise time is fast. As we have seen earlier, τ in the E and F region is typically between 15 and 40 s. Consequently, SFD observations are relatively insensitive to ultraviolet enhancements with rise times of 5 min or more.

Donnelly (1973) states that the peak enhancement in 10–1030 Å flux obtained from SFDs has a typical absolute accuracy of only within a factor of 4. This poor accuracy comes chiefly from insufficient information on the spectral distribution in the radiation and from insufficient information on the properties of the ionosphere along the path of the SFD probing radio wave at the time of the flare. The relative error increases with the time as the event progresses and may become large during the decay stage of the flare. Consequently, in studies of this kind the rapid rise and peak portions of the 10–1030 Å enhancements are the only ones used.

Donnelly (1973) has given, as an example of the test of accuracy of ultraviolet flux enhancements as deduced from the SFD data, the following diagram (Figure 125b). In this 10–1030 Å flux deduced from the SFD is compared with the broad band measurement by Kelly and Rense (1972) over the wavelengths 760–1030 Å from OSO-5 satellite for the flare of April 21, 1969. There is remarkable agreement between the two: the agreement even extends to the small fine structures even though some of these fine structures are about the size of the preflare noise in the 760–1030 Å data. Donnelly (private communication), however, finds that, in another case of comparison (March 12, 1969 flare), the agreement is not so good, although much better agreement occurred when electron loss rates lower than those considered as 'best estimates' were used.

Comparisons have also been made by Donnelly (Donnelly, 1969b) between the time variations of individual ultraviolet lines and the broadband flux deduced from the SFDs. We have indicated in Chapter 3 that the ultraviolet emission lines can be classified into three groups: (i) the lines of hydrogen and Si III (normally chromospheric lines) which precede the Hα maximum by 3–4 min; (ii) the lines of He II, C III, O V and O VI (normally chromosphere-corona transition region lines) which precede the

Hα maximum by about 2 minutes, (iii) lines of Ne VIII and Fe XVI (normally coronal lines) which maximise after the Hα maximum and the lines of Mg X, Si XII and Fe XV which follow the same behaviour. Donnelly found that the lines in group (i) and (ii) have the same time dependence as the flux enhancement in 10–1030 Å derived from SFDs except that they decay faster than the SFD radiation. The coronal lines have, however, a much slower time dependence than the SFD radiation. X-rays in 0.5–3 Å

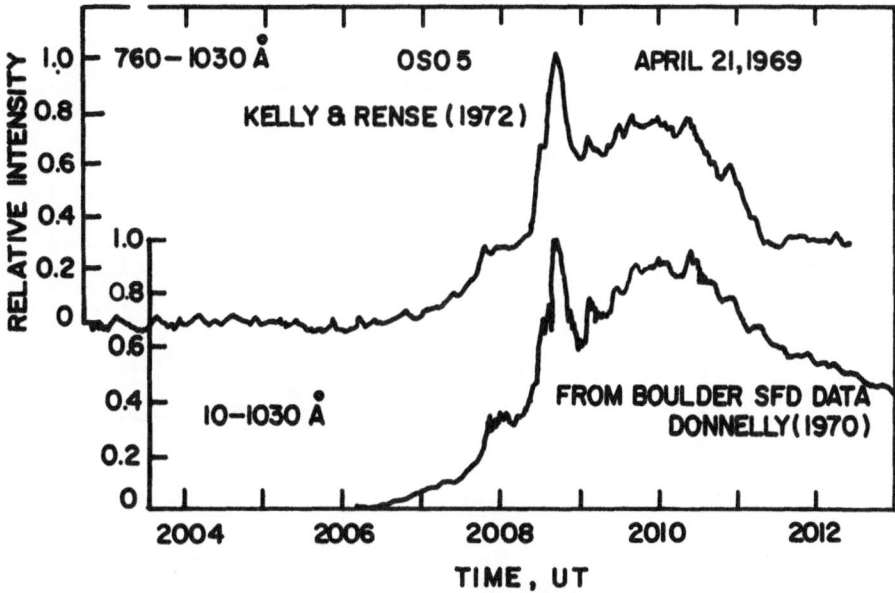

Fig. 125b. Effectiveness of the SFD in giving broadband EUV flux. This is exemplified by a comparison by Donnelly (1973) between the SFD deduced EUV flux in 10–1030 Å and direct measurement by Kelly and Rense (1972) of 760–1030 Å flux from OSO-5 for the solar flare event of April 21, 1969 (after Donnelly, 1973).

and 0–8 Å band were slower during the decay stage and X-rays in 8–20 Å band slower throughout the event than the fluxes in 10–1030 Å range. Figure 125c gives the comparison obtained by Donnelly between the SFD observed on 4 MHz at vertical incidence in Boulder for the event of March 15, 1966 when Explorer 30 observations in the bands 1–8 Å, 8–20 Å and 44–60 Å were available. The 44–60 Å detector is believed, however, to be dominated by 1–20 Å radiation and consequently its variation is likely to indicate the variation in 1–20 Å band. Δf computed from the time curve in this band is shown for comparison. In Figure 125d the sudden frequency deviation obtained on 13 MHz WWI Havana, Illinois to Boulder path for the flare of March 29, 1967 is compared with that derived from OSO-3 observations of He II 303.8 Å radiation (for $\tau = 30$ s). The agreement in the time variations is remarkably good.

In this early work, the number of lines from group (i) was very small. Subsequently a good deal of data on the EUV lines are available from OSO-4, OSO-5 and OSO-6.

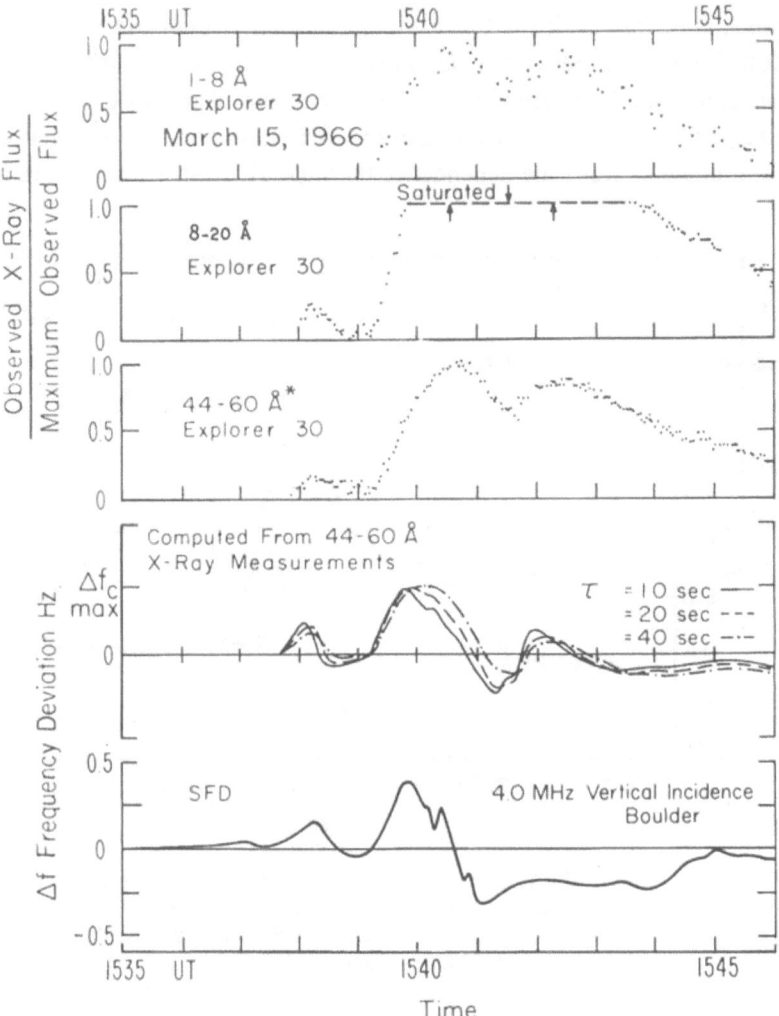

Fig. 125c.   Comparison of the time curve of 1–8 Å and 8–20 Å X-ray enhancements, as well as the normalised frequency deviation computed from 44–60 Å fluxes, with the observed frequency deviation for the flare of March 15, 1966 (after Donnelly, 1969b).

OSO-6 recorded a total of 205 EUV flares; its observations were mainly in C II 1335 Å line, C VI 1032 Å line and Mg X 625 Å line. These represent the chromosphere, chromosphere-corona transition region and lower corona, and are, therefore, quite representative.

In Figure 125e-h, we reproduce, after Donnelly *et al.* (1973), a comparison of 10–1030 Å radiation deduced from SFDs with OSO-6 observations of these ultraviolet lines. Figure 125e gives the time variation of C I continuum emission at 1098 Å and compares 10–1030 Å flux deduced from SFDs. Both have fast rise times. Figures 125f and g give similar comparisons with O IV 554 Å and O VI 1032 Å. Again the agree-

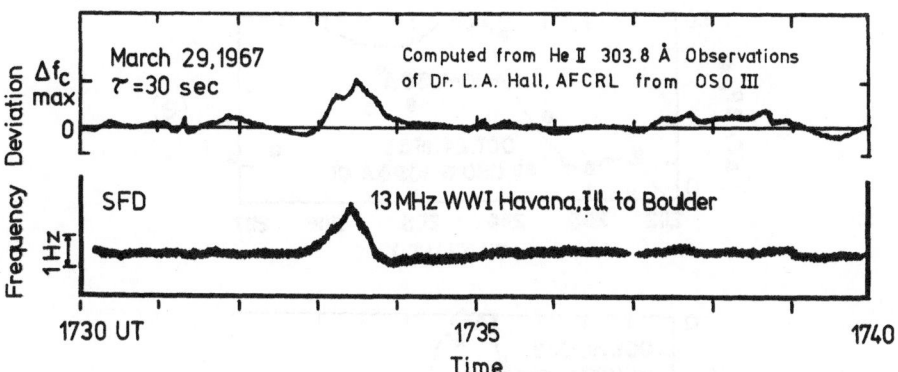

Fig. 125d. Comparison of the time dependence of the normalised frequency deviation for He II 303 Å flare measurements and observed frequency deviations for the flare event of March 29, 1967 (after Donnelly, 1969b).

ment is good during impulsive growth phase. The similarity in these observations indicate that the impulsive enhancements in the ultraviolet radiations from the chromosphere and from the chromosphere-corona transition region have similar rates of growth and are essentially identical with the growth curve of the 10–1030 Å within the limited accuracies and resolutions of the data. The result of comparison of the coronal line Mg x 625 Å with $F(10-1030 \text{ Å})$ is, however, surprising (Figure 125h). The coronal ultraviolet emissions, in general, show rather slow enhancements during flares. Part of the difference may arise from the fact that Mg x fluxes vary from flare to flare; some are dominated by impulsive bursts and others are a combination of slow and impulsive bursts.

We have so far discussed the time variation of satellite-observed EUV lines with those of $F(10-1030 \text{ Å})$ deduced from SFDs. We will now discuss to what extent the magnitudes of enhancements obtained directly from satellites and from SFDs agree. Donnelly (1969b) uses three parameters Rf, Φf and $R\Phi$ for such comparisons. Here $Rf(\lambda_i)$ indicates how much of the observed SFD is produced by the flare radiation at a particular emission line λ_i or wavelength range $\lambda_1 - \lambda_2$. The parameter Φf is used to make estimates of flare radiation enhancements at a particular wavelength or wavelength range from SFD observations. The last parameter $R\Phi$ indicates the intensity of the flare radiation enhancement at a particular wavelength or wavelength range relative to the total enhancement in the 1–1030 Å (or 10–1030 Å) range. Rf, Φf and $R\Phi$ were defined as follows:

$$Rf = 100 \, \frac{\Delta f_c \, (f_v = 5 \text{ MHz}, t_m)}{\Delta f \, (f_v = 5 \text{ MHz}, t_m)} \% \tag{100}$$

$$\Phi f = R_t R_\chi \, \frac{\Delta F \, (\lambda_i, t_g)}{\Delta f \, (f_v = 5 \text{ MHz}, t_m)} \text{ erg cm}^{-2} \tag{101}$$

$$R\Phi = 100 \, \frac{\Delta F(\lambda_i, t_g)}{\Delta F \, (10\text{–}1030 \text{ Å}, t_g)} \% = 100 \, \frac{\Phi f(\lambda_i)}{\Phi f \, (10\text{–}1030\text{Å})} \%, \tag{102}$$

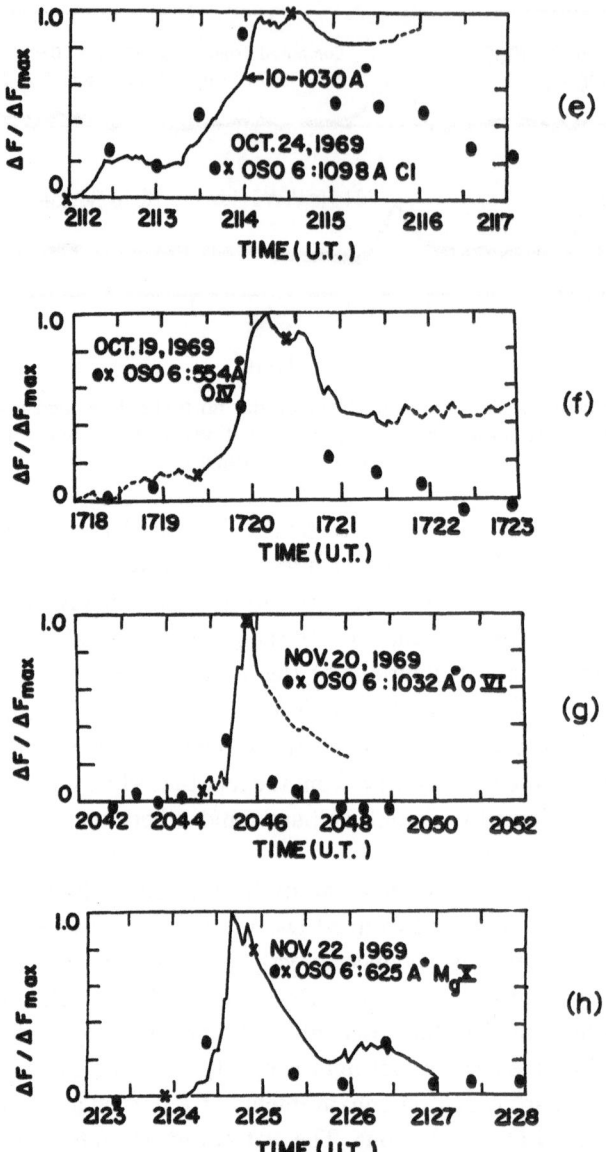

Fig. 125e–h. Comparison of flare emissions in C I continuum at 1098 Å, O IV 554 Å, O VI 1032 Å and Mg X 625 Å with 10–1030 Å flux deduced from SFDs. The flares are for October 24, 1969; October 19, 1969; November 20, 1969 and November 22, 1969 respectively (after Donnelly *et al.*, 1973).

where Δf_c was the frequency deviation *computed* from the radiation enhancement (ΔF) observed by satellite in the wavelength range $\lambda_1 - \lambda_2$ or at an emission line of wavelength λ_i, t_m and t_g were the times of SFD peak and peak net flux enhancements in the 1–1030 Å (or 10–1030 Å) range. R_t was a correction factor to account for the rise time of the ionizing radiation, and R_χ was a factor for removing the solar zenith angle (χ) dependence.

Rf and *Φf* were based on SFD observations at $f_v = 5$ MHz. The reference $f_v = 5$ MHz was chosen because Boulder SFD observations usually have some f_v values near 5 MHz and because 5 MHz is about the lowest value of f_v for which Δf is not highly sensitive to the pre-flare electron concentration height profile along the SFD propagation path.

Table XX gives the values of *Rf*, *Φf* and *RΦ* results obtained by Donnelly (1970) from a number of studies in different EUV lines and bands, starting from hard X-rays ($E > 10$ keV) to wavelengths in the radio spectrum. The remarks column indicates the nature of agreement or disagreement between the directly observed time variation in the wavelength region concerned and that deduced from SFD observations in the broadband 10–1030 Å.

One way of interpreting Table XX is that the largest contributors to SFDs on paths reflected from the *F* region are the EUV recombination continuum and line emissions from the more abundant solar constituents H, He, O, C and N. Another interpretation includes a bremsstrahlung continuum throughout the range 10–1030 Å. The contribution by any one line was not large. The total *Rf* for all wavelengths accounted for only half of the observed frequency deviation. The remaining half, Donnelly believes, may well have come from some type of continuum emission or by radiation enhancements at wavelengths for which no satellite observations are available. We shall see later that this inadequacy of the EUV emission lines in producing the observed ionization enhancements in the *F* region is also encountered in flare measurements with incoherent scatter facility.

Another check on the effectiveness of the SFD as an EUV detector would be to make a quantitative comparison between $\Delta F(\lambda_i)$ in specific EUV wavelengths as observed with satellites for different flares and the corresponding values of $\Delta F(10–1030$ Å$)$ obtained from SFDs. Donnelly *et al.* (1973) have given such a comparison. $\Delta F(\lambda_i)$ values were obtained from the spectroheliograph observations of EUV flares and SFDs mainly from Boulder observations. For this comparison they make the following adjustment in *F* values:

The peak flux enhancement from the *entire* flare region, $\Delta F_m(\lambda_i)$ erg cm^{-2} s^{-1}, was first derived by summing over the enhancements (erg cm^{-2} s^{-1} ster^{-1}) of each flaring spatial element ($35'' \times 35''$) (see for example, Section 3.2) during the raster having the largest enhancement of the brightest element of the flare and taking account of the effect of the field of view. ΔF_m included both slow as well as rapid rises in flux. Only the part that occurred by choosing two values of the brightest spatial element: one at the beginning of the rapid rise sequence $[F(\lambda_i, t_1)]$ and another near the end of the rapid rise or near the peak of the EUV burst $[F(\lambda_i, t_2)]$. Both are marked by × in Figure 125e–h taken from Donnelly *et al.* (1973). $\Delta F^*(\lambda_i)$ for the OSO-6 observations was then defined by:

$$\Delta F^*(\lambda_i) = \frac{F(\lambda_i, t_2) - F(\lambda_i, t_1)}{F_m(\lambda_i) - F_0(\lambda_i)} \Delta F_m(\lambda_i) \text{ erg cm}^{-2} \text{ s}^{-1} \text{ at 1 AU}.$$

Here $F_0(\lambda_i)$ gives the pre-flare flux and $F_m(\lambda_i)$ the peak for the brightest spatial

TABLE XX

Rf, Φf and $R\Phi$ as a function of wavelength (after Donnelly, 1970)

Radiation wavelength (Å)	Ion	Rf (%)	Φf (erg cm^{-2})	$R\Phi$ (%)	Number of events studied [a]	Remarks ($\Delta F(1-1030\ \text{Å}, t) \equiv \overline{\Delta F}$)
<1 (≥10 keV)		0	8.0×10^{-6}	~0.001	13t, 56s	
0.5–3		0.08	7.0×10^{-5}	0.09	4t	
1–8		2.4	6.3×10^{-3}	7.8	9t	ΔF(1–8 Å) decays slower than $\overline{\Delta F}$
8–20		<2.1	$\leqslant 4.6 \times 10^{-3}$	≤5.8	12t	ΔF(8–20 Å) rises, peaks, and decays slower than $\overline{\Delta F}$
303.8	He II	5.9	5.2×10^{-3}	6.6	8t, 1s	Good time dependence agreement with $\overline{\Delta F}$
335.3	Fe XVI	<0.6	$<4.0 \times 10^{-4}$	0.5	1t	Appears to be slower than $\overline{\Delta F}$
368.1	Mg IX	0.4	2.9×10^{-4}	0.4	1t, 1s	
465.2	Ne VII	0.3	2.0×10^{-4}	0.2	1s	
554	Group O IV	4.0	2.8×10^{-3}	3.4	1s	
584.3	He I	0.6	3.7×10^{-4}	0.5	1s	
629.7	O V	1.2	7.8×10^{-4}	1.0	4t,1 s	Good time dependence agreement with $\overline{\Delta F}$
760	Group O V	0.6	3.6×10^{-4}	0.5	1s	
765.1	N IV	0.8	4.7×10^{-4}	0.6	1s	
790	Group O IV	2.3	1.6×10^{-3}	2.0	1s	
834	Group O II and III	2.2	2.1×10^{-3}	2.7	1s	

680–911	H Cont	8.1	9.3×10^{-3}	12.0	1s	
949.7	H I Lδ	0.2	1.7×10^{-4}	0.2	1s	
972.5	H I Lγ	0	3.3×10^{-4}	0.4	1t, 1s	Good time dependence agreement with $\overline{\Delta F}$
977.0	C III	6.2	6.4×10^{-3}	8.0	2t, 1s	Good time dependence agreement with $\overline{\Delta F}$
990	Group N III	1.4	2.0×10^{-3}	2.5	1s	
1025.7	H I Lβ	2.3	1.3×10^{-3}	1.6	1s	
1031.9	O VI	0.1	3.2×10^{-3}	4.1	1s	
1085	Group N II	0	1.1×10^{-3}	1.4	1s	
1175	Group C III	0	1.7×10^{-2}	21.0	1s	
1206.5	Si III	0	1.4×10^{-2}	17.0	1s	
1215.7	H I Lα	<0.1	$\sim 8.0 \times 10^{-2}$	~ 100.0	2, 2t, 1s	Good time dependence agreement with $\overline{\Delta F}$
1080–1225		0	1.4×10^{-2}	17.0	1t	Good time dependence agreement with $\overline{\Delta F}$
1238.8	N V	0	6.1×10^{-4}	0.8	1s	
1225–1350		0	7.3×10^{-3}	9.1	1t	Good time dependence agreement with $\overline{\Delta F}$
6563	Hα	0	3.5×10^{-3}	$\leqslant 4.3$	3t	Good time dependence agreement with $\overline{\Delta F}$
3500–6500	White light	0		~ 100.0	4t	to within the time resolution of white light photographs
3–10 cm	radio	0	2.4×10^{-8}	3×10^{-5}	43t	

ᵃ t – denotes high time resolution; s – denotes low time resolution.

element of $35'' \times 35''$. Since

$$F(\lambda_i, t_1) \sim F_0(\lambda_i)$$

and

$$F(\lambda_i, t_2) \sim F_m(\lambda_i)$$
$$\Delta F^*(\lambda_i) \lesssim \Delta F_m(\lambda_i).$$

The corresponding parameter for the SFD-derived flux over 10–1030 Å was:

$$\Delta F^*(10\text{–}1030 \text{ Å}) = \Delta F(10\text{–}1030 \text{ Å}, t_2) -$$
$$- \Delta F(10\text{–}1030 \text{ Å}, t_1) \text{ erg cm}^{-2} \text{ s}^{-1} \text{ at 1 AU}.$$

The relationship between $\Delta F^*(\lambda_i)$ and $\Delta F^*(10\text{–}1030 \text{ Å})$ for O VI 1032 Å, CII 1335 Å and Mg x 625 Å, as obtained by Donnelly *et al.* (1973) are given in Figure 125i. For Mg x 625 Å, the impulsive enhancements fitted a power law:

$$\Delta F^*(625 \text{ Å}) = a[\Delta F(10\text{–}1030 \text{ Å})]^{0.93}.$$

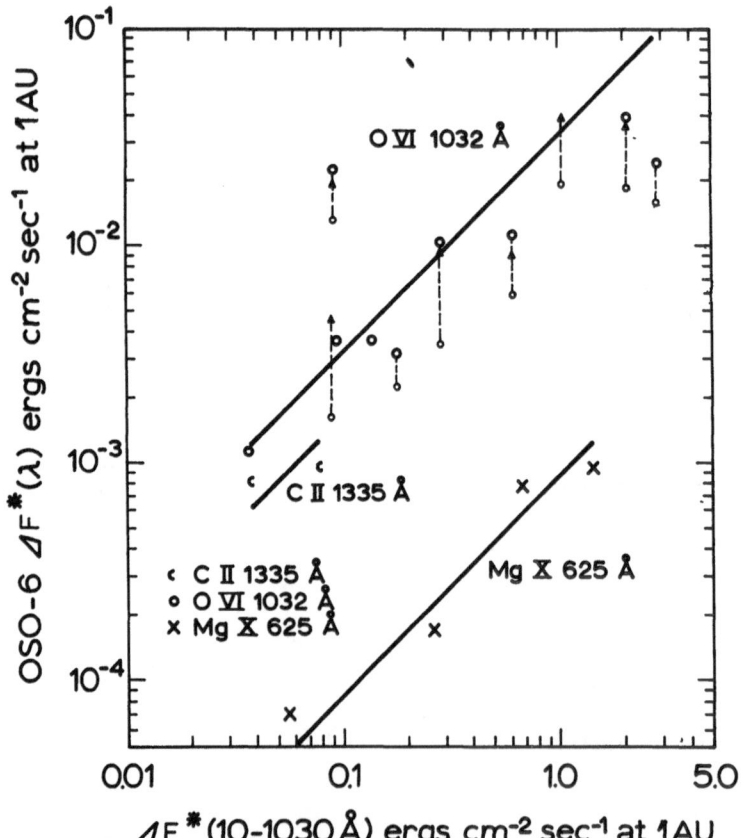

Fig. 125i.   Relationship between the fluxes for O vi 1032 Å, C ii 1335 Å and Mg x 625 Å obtained with OSO-6 satellite for different flare events during the fast rises with 10–1030 Å flux derived from the accompanying SFDs at corresponding times (after Donnelly *et al.*, 1973).

The impulsive emission in Mg X is thus clearly associated with the SFD-deduced EUV flux. For the impulsive O VI 1032 Å enhancement the relationship was less strong; the least square-power-law fit give:

$$\Delta F(1032 \text{ Å}) = a\left[\Delta F^*(10\text{–}1030 \text{ Å})\right]^{0.5}.$$

Donnelly (1970) summarised the EUV radiation detection characteristics of SFDs in the following way:

Wavelength Range:

1–1030 Å (or 10–1030 Å for F-region SFDs).

Dynamic Range:

Minimum detectable flux enhancement $\sim 10^{-2}$ erg cm^{-2} s^{-1}
Maximum measurable flux enhancement ~ 10 erg cm^{-2} s^{-1}.

Rise Time (t_r) Dependence of Sensitivity (for a linear rise):

$R_t \sim 1$ for $t_r \leqslant 10$ s
$R_t \sim \tau/t\left(1 - e^{-t/\tau}\right)$ for $t_r > 10$ s.

Time Resolution:

~ 1 s for Boulder observation (higher resolution possible)
Fine structure with rise times $t_r \geqslant 1$ s measurable.

Relative Intensity Resolution:

$\sim 10^{-2}$ erg cm^{-2} s^{-1} or 1% of maximum $\Delta F(10\text{–}1030 \text{ Å})$ for resolution of intensity of fine structure with $t_r \leqslant 10$ s.

Wavelength Resolution:

At best only in the following groups: 1–10 Å, 10–100 Å + 910–1030 Å, 100–400 Å + 800–910 Å and 400–800 Å.

Spatial Resolution:

None.

Absolute Intensity Accuracy:

Typically a factor of 4.

Advantages and Disadvantages:

Advantages are that the method is inexpensive, providing high time resolution and accuracy, good relative intensity resolution and good global coverage.
Disadvantages are poor wavelength and no spatial resolution, poor absolute intensity accuracy, high threshold detection level and ineffectiveness for non-impulsive events.

Considerable improvement likely in future if tied to incoherent measurements.

10.2.3. XUV FLUX DETERMINATIONS FROM INCOHERENT SCATTER OBSERVATIONS

The incoherent scatter observations provide a very powerful tool for spectrum building in the X-rays and the ultraviolet. It does not have the advantage of the SFD of being a routine patrol technique; but while with the SFD, the wavelength resolution of the derived EUV flux is rather poor, it is not so in the case of incoherent scatter observations. In this case, the limitation in the spectrum resolution is determined by the limitation of the height interval of the sampling heights. In the most favourable cases, flaretime enhancements have been obtained down to 80 km (Taylor, 1974) and upto 1000 km (Mendillo and Evans, 1974). At heights of 80 km, the dominating ionizing source is 1–8 Å, whereas as one moves up through 100 km to 150 km and through 150 km to 200, 250 and 300 km we are essentially sampling the time curves of different radiation bands.

A serious limitation is the lack of sufficient time resolution. The flare spectrum should not change significantly over the minimum resolvable time interval; furthermore, evaluation of dN_e/dt required for estimating q/q_n has a meaning only if Δt is small. Since the time resolution of most incoherent scatter measurements is 5–10 min, this can be a serious disadvantage, especially since the EUV flare radiation affecting the E and F regions often have rise and decay times shorter than this interval.

The key to successful spectrum building with such data is to divide the entire ionizing spectrum into a number of suitable wavelength groups with similar ionization and absorption cross section, and such that the heights of maximum electron production rate are well separated. The first work of this kind was done by Garriott et al. (1969) with the incoherent scatter $N_e(h, t)$ profiles of the large flares of May 21 and May 23, 1967 (reported in Chapter 6; also see Figure 38). The spectral groupings chosen by them are:

> Group 1 280–796 Å
> Group 2 205–280 Å
> Group 3 138–205 Å
>   796–911 Å
> Group 4   62–138 Å
>   911–1027 Å
> Group 5   1– 60 Å.

The intensity in each group was allowed to vary in such a way that the computed ratio $R_t\ (h) = q(h)/q_n(h)$ matched the observed ratio, R_{exp}, as closely as possible. The observed value of R was obtained through the equation:

$$q/q_n = \frac{1}{\alpha N_n^2} \frac{dN_e}{dt} + \frac{N_e^2}{N_n^2}$$

and on an assumed α-variation. It was observed that $R(h, t)$ increased at lower altitudes, indicating a hardening in the flare spectrum. Time histories of the derived

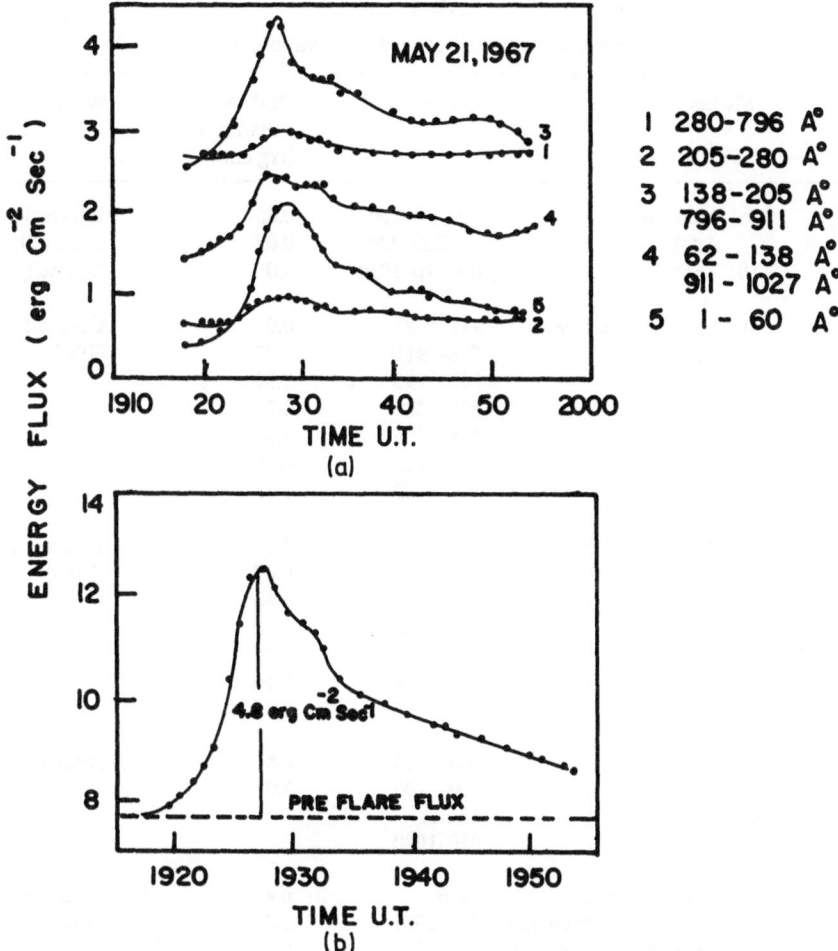

Fig. 126a–b.   Estimate of time histories of (a) EUV fluxes in selected bands and (b) total ionizing solar energy flux for the flare of 21 May 1967 from incoherent scatter measurements of electron density changes given in Figure 38 (after Garriott *et al.*, 1969).

flux in the five bands for the flare event of May 21, 1967 and corresponding to the ionization changes given in Figure 38 are shown in Figure 126a. The time history of the total ionizing flux is shown in Figure 126b. The flux rose rapidly to a peak value of 4.8 erg cm^{-2} s^{-1} at 1927 UT about 8 min after the beginning of the flare effect, and then decayed much more slowly. Half-an-hour after the beginning of the event, the ionizing radiation was still above normal. The enhancement of the fluxes in the different wavelength bands are given in Table XXI. The increasing enhancements with decreasing wavelength are given in Figure 126c.

The question that arises is how reliable are such determinations. Some idea about this can be formed from Figure 127. In Figure 127, the top curve is 1–60 Å flux variation deduced from ΔN_e changes determined by the incoherent scatter technique for May 23, 1967 event. This is compared with 2–12 Å flux changes observed directly

TABLE XXI

EUV flux enhancements as deduced from various F region SIDs

Flare event	Method	Bands ($\lambda\lambda$, Å)	Peak flux enhancements (erg cm^{-2} s^{-1})	Authors
July 7, 1966	(a) Ionosonde Analysis	(a) 44–100 200–900	0.85 0.02	Bhattacharyya et al. (1967)
	(b) SFD Analysis	(b) 10–1030	5.0	Donnelly (1968c)
May 21, 1967	Incoherent scatter data	911–1027 796– 911 280– 796 205– 280 138– 205 62– 138 1– 60	0.06 0.07 0.32 0.29 1.61 0.95 1.68	Garriott et al. (1969)
May 23, 1967	Incoherent scatter data	911–1027 796– 911 280– 796 205– 280 138– 205 62– 138 1– 60	0.06 0.07 0.40 0.16 2.25 1.16 3.35	Garriott et al. (1969)
August 28, 1966	SFD Analysis	100– 910 10– 100 and/or 910–1030	6.5 2.0	Donnelly (1968d)
March 27 1967	(a) Based on OSO-3 scanning spectrometer observations adjusted with 10–1030 Å flux deduced from SFDs, and other estimates	1216 911–1027 796– 911 280– 796	0.43 0.048 0.031 0.097	Donnelly and Hall (1973)
	(b) SFD analysis only	10–1030	0.49	

by the University of Iowa group. The curves for 1–20 Å and 1–60 Å bands are constructed on the basis of 2–12 Å curve and spot measurements in 8–20 Å and 44–60 Å bands assuming that the shape of the time curves in 1–60 Å and in 2–12 Å are about the same. From the very close agreement between the curve deduced from N_e changes (solid curve) and that expected from 2–12 Å measurement, it appears that reliability of this method can be quite good for the slow component of the flare emission.

Another example is shown in Figure 128 in which $\Delta N_e/N_e$ variation during the flare of February 20, 1970 as obtained by Taylor and Watkins (1970) is shown alongwith X-ray variations in the bands 0.3–3 Å, 1–8 Å and 1–20 Å bands observed with Vela satellite (Blocker et al., 1971). Here again the variations are similar.

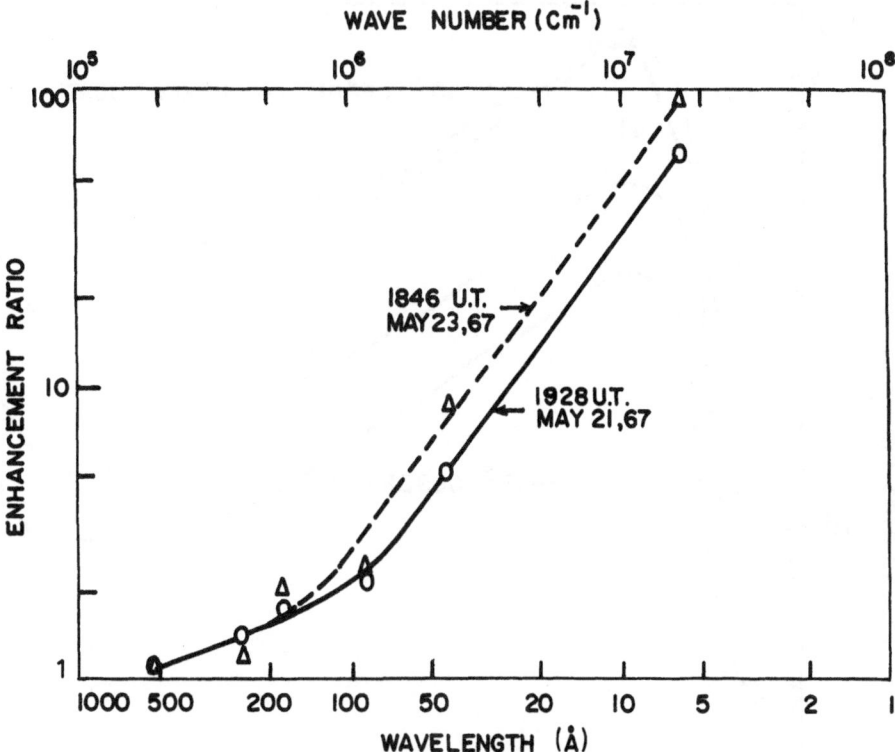

Fig. 126c. Ratio of flare to pre-flare solar flux as a function of wavelength for the flares of May 21 and 23, 1967, estimated from incoherent scatter measurements of electron density changes (after Garriott *et al.*, 1969).

The wavelength groupings selected by Garriott *et al.* are not exclusive. When observations are available mostly at heights in the *E* region and upper *D* region (such as those in Malvern reported by Taylor and Watkins, 1970 and Taylor, 1974) it is more appropriate to choose narrower bands in the X-ray region and those wavelengths in the ultraviolet which have their heights of maximum production below 150 km. It is not necessary, nor is it advisable, to use the same wavelength groupings for different heights. At 80 km, the sole contribution during the flare enhancement comes from X-rays of wavelengths below 20 Å; consequently $F(1\text{-}20 \text{ Å})$ is uniquely determined from this height. It is convenient to select as a first step such heights where the predominating ionizing contribution comes from one or two bands. Once these flux enhancements are reliably determined, the enhancements in the other bands can be determined more reliably.

As has been shown before, incoherent scatter measurements of flaretime ionization changes fall in the following categories:

(i) Measurements on ionization enhancements principally in the height range 90–150 km (e.g. Malvern observations).

(ii) Measurements on ionization enhancements principally between 120–200 km (e.g. Arecibo Observations).

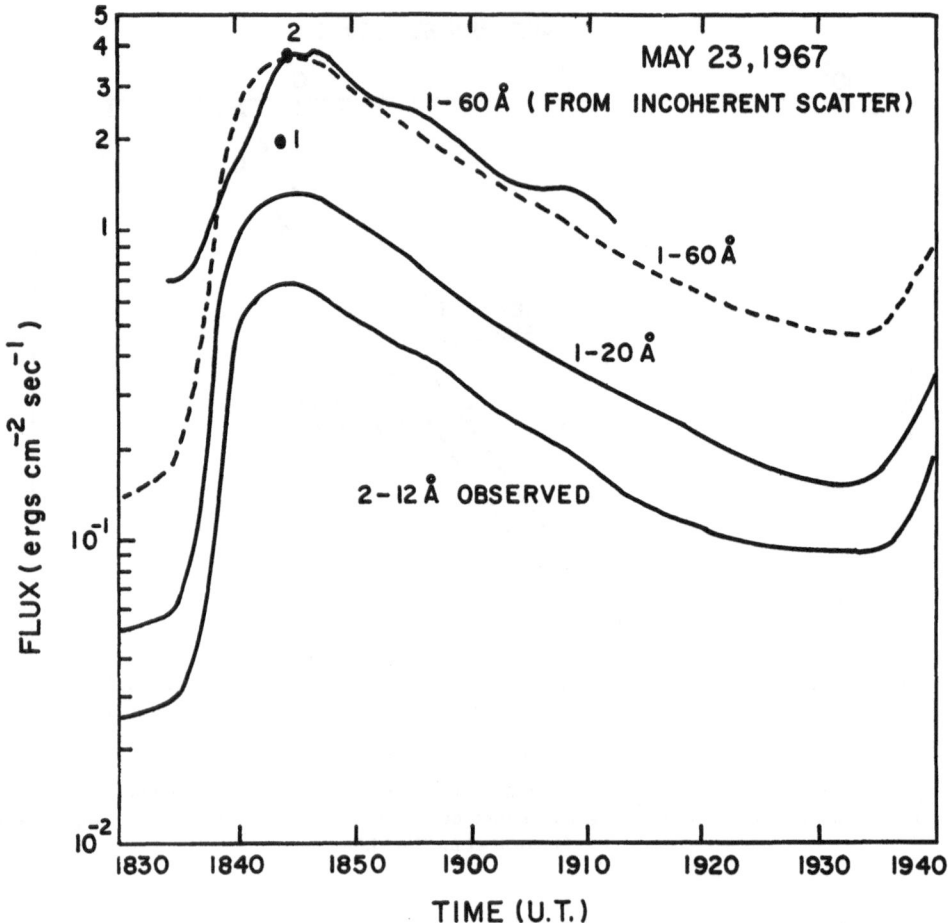

Fig. 127. Comparison of 1–60 Å flux derived from ΔN_e changes at 90 km from incoherent scatter observations made in Malvern, UK with time curves of 2–12 Å flux recorded by University of Iowa and estimated 1–60 Å changes based on 2–12 Å time curve and spot measurements in 8–20 Å and 44–60 Å bands (after Mitra, 1973).

(iii) Measurements on ionization enhancements extending into the topside ionosphere (e.g. Millstone Hill measurements).

Since the observed ionization changes in the topside ionosphere are a result of changes in the vertical drift and an increase in electron temperature (in which the two are believed to be related), observations above 300 km cannot be used effectively for any estimation of the EUV fluxes.

For the other heights, the points to keep in view are:

(a) That observations in (i) and (ii) require the use of different wavelengths groups.

(b) That the number of wavelengths groups (m) in which the XUV radiation is divided must be less than or equal to the number of heights (n) for which observations are essentially independent;

i.e. $m \leqslant n$.

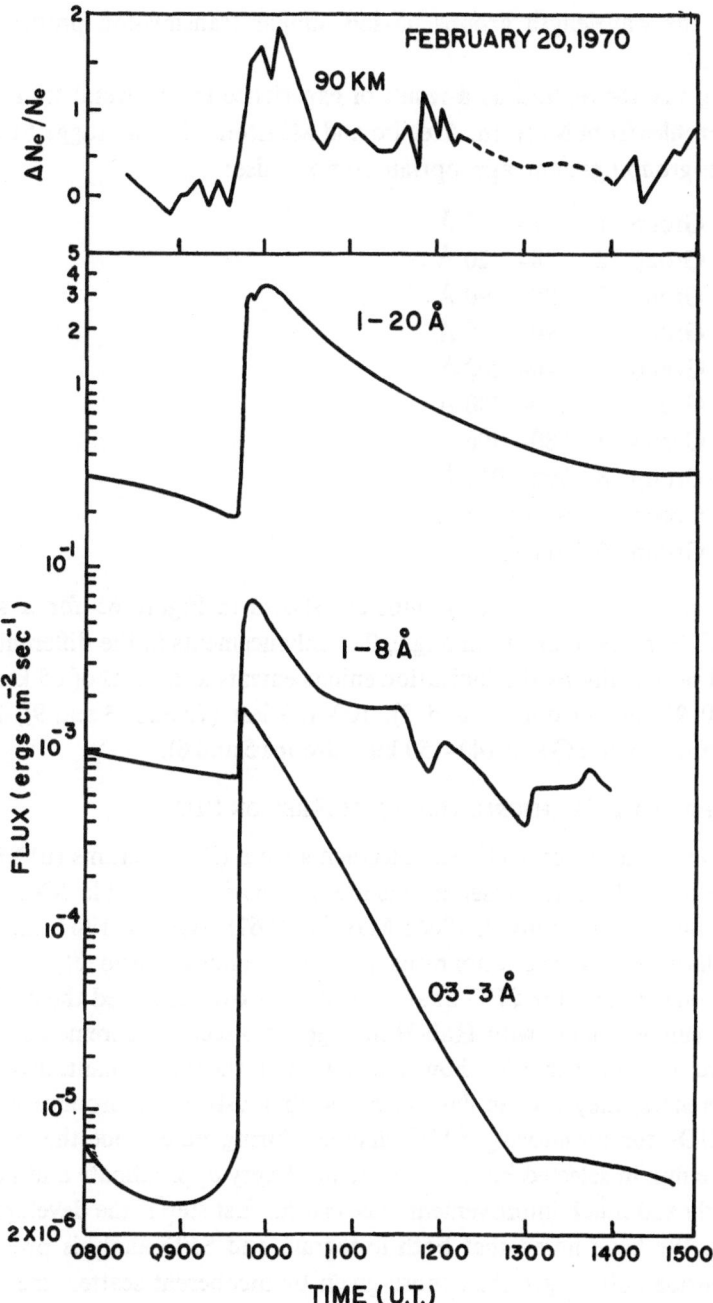

Fig. 128.   Time variations of $\Delta N_e/N_e$ for the solar flare of February 20, 1970, obtained at Malvern, UK by Taylor and Watkins (1970) shown along with X-ray variations in the bands 0.3–3 Å, 1–8 Å and 1–20 Å observed with Vela satellite by Blocker *et al.* (1971).

(c) That the wavelength groups having similar transmission profiles cannot be distinguished.

Considering the above, and as a result of experience from several trial calculations involving profiles from Malvern, Arecibo and Millstone Hill, we suggest the following wavelengths groupings as an appropriate compromise:

Group 1   1–   8 Å
Group 2   8– 20 Å
Group 3   20– 90 Å
Group 4   90– 140 Å
Group 5   140– 200 Å
Group 6   200– 280 Å
Group 7   280– 796 Å
Group 8   796– 911 Å
Group 9   911–1027 Å
Group 10 1216 Å .

The height sensitivities of these groups are shown in Figure 65 for a solar zenith angle of 32°. For this solar zenith angle, flux enhancements in the different groups are determined principally by the ionization enhancements at heights of 85 km (Groups 1 and 10), 90–95 km (Groups 2 and 3), 100–120 km (Groups 3 and 9), 120–130 km (Groups 4 and 8, 5 and 6) and 140–150 km (Groups 5 and 6).

10.2.4. COMBINED EUV INFORMATION FROM F-REGION SIDs

Combined information on EUV enhancements from (i) ionograms (ii) SFD analysis and (iii) incoherent-scatter observations are summarised in Table XXI. Flares considered in this table are July 7, 1966; May 21, 1967; May 23, 1967 and August 28, 1966 – all large events. The ionograms analysis, we notice, probably underestimated flux enhancements and the SFD analysis probably overestimated them. Comparison of the flux enhancements with Hall-Hinteregger's direct measurements for a class 3⁻ flare summarised in Table V show that although there are quantitative and qualitative differences, they are similar. There is obviously great promise in the use of F-region SIDs for monitoring EUV changes during flares, but the procedure for deriving the flux in selected bands is at the most very approximate and needs further critical study and much improvement. The crucial first step is the development of the N_e-profiles. We have noted that both ionogram and SFD analysis provide $\Delta N_e/N_e$ profiles considerably larger than those given by incoherent scatter, the first at least is grossly in error. Improvement in development of $\Delta N_e/N_e$ profiles is necessary and urgent. The second area of improvement concerns the determination of preflare α-h profile and of the solar ionizing flux in the selected wavelength bands so that the two are mutually consistent. The inadequacy of Hinteregger EUV flux in reproducing quiet day electron density profiles or thermal structure of the F region is well known. The general feeling is that the quiet day flux should be substantially enhanced, but to what extent is not clear. As long as this is uncertain, the α-h profile is also uncertain, and consequently any absolute determination of the fluxes will be correspondingly uncertain. One suggestion is to take a flux model (for quiet situation) of the following type:

Low solar activity	Mid solar activity	High solar activity
	Swartz (1972)	Taylor (1972)
$1\ H_{Qs}$	$2\ H_{Qs}$ at $\lambda < 350$ Å	$\approx 5\ H_{Qs}$
	$3\ H_{Qs}$ at $\lambda > 350$ Å	

and the use of laboratory rates for k_1, k_2 and $\bar{\alpha}_D$.

The most promising approach seems to be a combination of incoherent scatter and satellite beacon observations.

CHAPTER 11

POLAR CAP ABSORPTION EVENTS

In addition to the simultaneous effects, there are *delayed ionospheric effects* following a flare of two distinct kinds (Figure 1, Chapter 1):

(i) Polar Cap Absorption Events (PCAs) first observed with Riometers in the polar regions, and caused by MeV solar cosmic ray particles spiralling into the polar regions along the Earth's magnetic field lines some 20 min to 20 h after a flare.

(ii) Ionospheric (and magnetic) storms, the auroras, and the auroral absorption effects occurring between 20–50 h after a flare caused by lower energy particles (mostly protons and electrons). These are really clouds of solar gas blown out from the Sun, consisting of equal number of positive ions and electrons.

A sequence of the different flare-associated effects is indicated in Figure 129 with the example of August 7, 1972 event.

We notice, therefore, that the delayed effects are caused by ionization due to solar particles of different energies as they impinge upon the Earth's atmosphere. The actual mode of entry of these particles through the magnetospheric system is quite complex

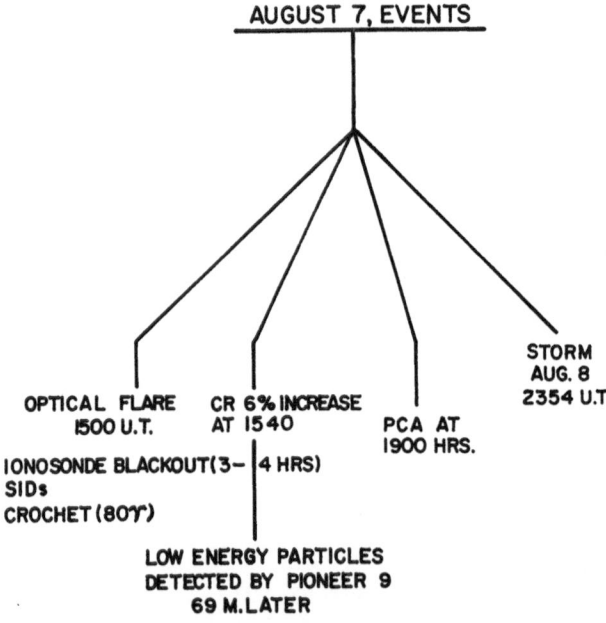

Fig. 129. Illustrative example giving sequence of flare-associated effects. Example shown is that of August 7, 1972 event.

and will not be discussed here. We will discuss here only the Polar Cap Absorption Events for the following reasons:

(i) PCAs, like SIDs, are major perturbations of the quiet ionosphere, with strong events entirely swamping it. This has a specific advantage. The uncertainties of the quiet ionosphere (such as those concerning the minor constituents, especially NO) are now unimportant.

(ii) The chemistry of the PCAs is in many ways similar to that of SIDs. Both provide important insight into the ionospheric processes of the lower ionosphere, including those obtaining during quiet times.

It is important to remember that the first realisation that the Sun provides an abundant supply of sub-relativistic particles came through the discovery of PCAs through riometer observations during the International Geophysical Year. This was before satellite monitoring became a major detector of the presence of solar protons. The particles are mainly protons in 1–100 MeV range, although in a few events alpha particles have been observed. Energetic solar electrons are also present. The basic physical factor, as in SIDs, is a great increase in ionization in the lower ionosphere, causing increased absorption of radiowaves in the HF and the VHF. A major value of these measurements is that arrival of such sub-relativistic particles is not detectable at the ground level, and can only be obtained either through PCA events or directly from space vehicles.

The influence of the Earth's magnetic field normally bars entry of these particles into latitudes below the auroral zone. PCA absorption is thus a characteristic feature of high latitudes only, above approximately 65 deg, and is practically uniform over the entire polar cap. Riometer observations showed that when the polar region is in permanent sunlight there is a rapid rise of the absorption to a peak in a matter of several hours and then the slow decay occurs over a period of several days due to the decay of the proton flux. As time progresses, the spectrum becomes softer, less penetrating and more stable. When observations are available through both day and night, absorption decreases markedly at night even though the particle influx may be of the same magnitude, causing a distinct modulation of the otherwise smooth form of the absorption profile.

11.1. Radio Measurement Techniques

The riometers have been the principal source of information in the HF and VHF since the IGY. Another method used extensively in the early days was the ionosonde where f_{min} provided a measure of the changing absorption. Subsequently, other techniques have also been employed, including those in the LF and VLF. In fact, all techniques used for SIDs are, or can be, employed. Those used with considerable success include (Reid, 1972a):

  (i) Riometers
 (ii) Normal sweep-frequency ionospheric sounders
(iii) Partial reflection experiments
(iv) LF and VLF measurements at both steep and oblique incidence

(v) Oblique HF propagation studies

(vi) VHF forward scatter signals.

11.1.1. Use of riometers

Sample records of PCAs as observed with riometer are shown in Figure 130. As in SIDs a frequency around 30 MHz is generally used. This represents a suitable compromise between the lower frequencies at which the absorption is more intense but

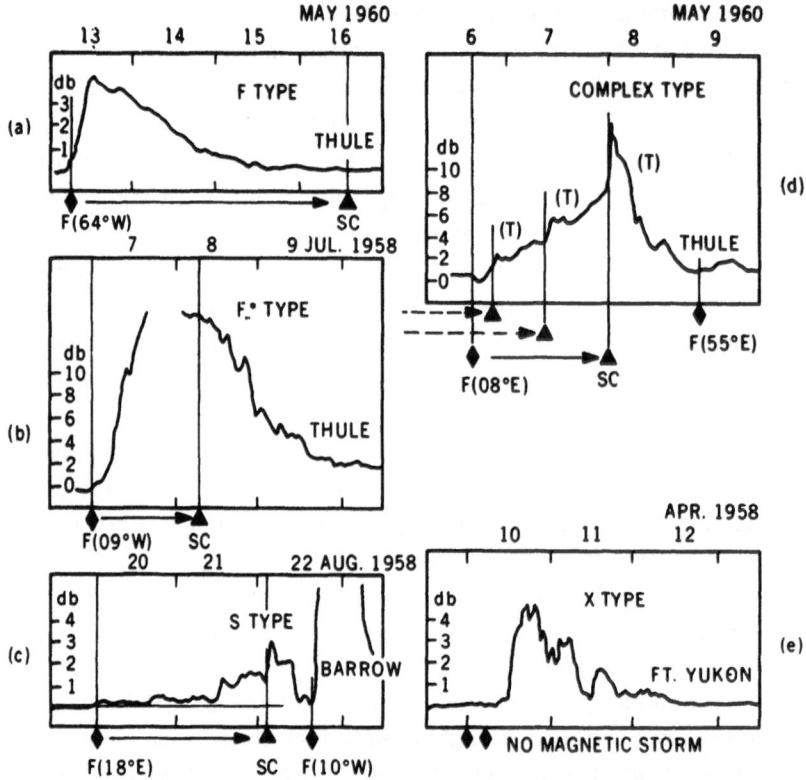

Fig. 130. Riometer absorption effects during PCA. Different types of PCA events are indicated (after Obayashi, 1964).

interference from radio transmissions and manmade noise is a serious handicap and higher frequencies at which the absorption is smaller (in accordance with f^{-2}) and more difficult to measure. Here again, quantitative analysis is hampered by the fact that the measurements represent only an integrated measure of the response of the entire lower ionosphere. For any meaningful study, it becomes necessary to make assumptions on the exact schemes of ionic reactions participating in the generation and loss of positive ions, negative ions and electrons. It is possible, however, to develop empirical relations between proton flux and absorption. Several such relations are available (Adams and Masley, 1966; Reid, 1970b; Parthasarathy and Venkentesan, 1964). Figure 131a shows an example of the relationship deduced by Reid (1970b).

This shows the time-history of the proton flux in four energy channels as observed by the satellite 1964-45A over the polar caps (solid lines) and the 30 MHz absorption measured by a riometer at the South Pole during the event of February, 1965. The dashed line is a least-square fit to the riometer data (excluding the brief enhancement on 7 February), and indicates that the time-history of the absorption is closer to that of the protons with energy greater than 20 MeV than to that of the other channels.

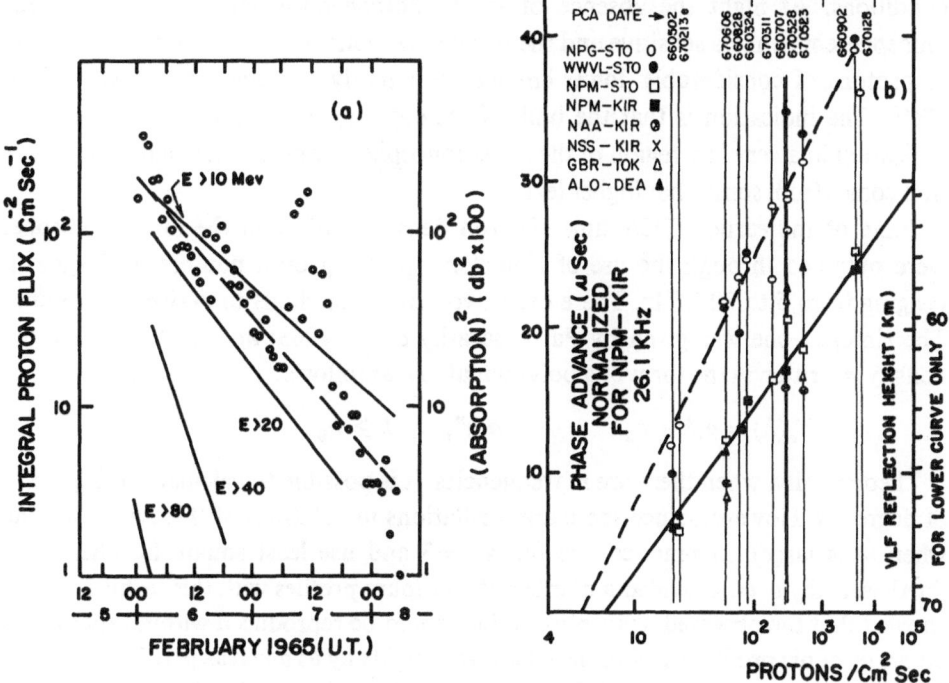

Fig. 131.   (a) Relationship between riometer absorption and flux of protons for proton flare event of February 1965 (after Reid, 1972b). (b) Relationship between VLF phase advance during PCAs and flux of protons (after Westerlund *et al.*, 1969).

This particular piece of data gives the following relationship between 30 MHz vertical absorption and proton flux:

$$J(> 20 \text{ MeV}) = 60\ A^2,\tag{103}$$

where A is the absorption in decibels and J is the 2π-omnidirectional proton flux in $\text{cm}^{-2}\ \text{s}^{-1}$. Reid uses this relation to examine the limits of usefulness of the riometer technique in an approximate way. He finds that the lower limit defined by an absorption of about 0.3 dB is a proton flux of above about 6 $\text{cm}^{-2}\ \text{s}^{-1}$ (or about 1 $\text{cm}^{-2}\ \text{s}^{-1}$ ster^{-1} in terms of directional flux). The upper limit is imposed by thermodynamic considerations, rather than by equipment sensitivity (see also Chapter 2, Section 5). The ionosphere emits thermal radiation in direct proportion to its power to absorb.

The temperature of the electrons in the lower ionosphere is of the order of 200 K. When absorption reduces the apparent cosmic noise temperature to 200 K, the riometer begins to measure the noise from the ionosphere itself. Ionospheric thermal radiation begins to be appreciable at 10 dB, and makes the absorption virtually unmeasurable much above 20 dB. The upper limit set by these considerations is a proton flux of about 2.4×10^4 cm^{-2} s^{-1} (directional flux of 4×10^3 cm^{-2} s^{-1} ster^{-1}). Thus, as a detector of solar protons, the riometer has an approximate dynamic range of 1 to 4000 protons cm^{-2} s^{-1} ster^{-1} for $E > 20$ MeV. These values refer to daytime conditions. At night the absence of solar photochemical effects make the lower ionosphere much less sensitive and the primary response is greatly reduced, and is due to protons of considerably lower energies apparently between 3 to 10 MeV (Reid, 1969). The indication is that the bulk of the daytime absorption comes from excess ionization in a very low region of the lower ionosphere, whereas the nighttime absorption comes from somewhat higher levels.

Some of the earliest PCA time electron density profiles in the lower ionosphere were obtained through the use of riometers operating on a number of frequencies ranging from 5 to 30 MHz. There exist two general methods to derive such profiles. The simplest one was given by Parthasarathy et al. (1963) and in this the electron density is written in the form of a polynomial in h as follows:

$$N_e(h) = a_1 h + a_2 h^2 + \cdots + a_n h^n, \qquad h \geqslant h_0.$$

Theoretically, when there are m frequencies, it is possible to calculate an expression of degree m. However, since there are oscillations in solutions of this type (especially when m is large), in practice one takes $n = 3$ and use least square fit when $m \geqslant 3$. Parthasarathy et al. provided a number of optimum profiles with this technique. They showed that the observed absorption values could be reproduced within experimental errors by cubic profiles varying the base level heights by as much as ± 10 km. However, it was shown that the choice of the base level had relatively little effect on the electron densities at heights greater than about 50 km. It was, therefore, considered that although the base level is not well defined by this method, the profiles are valid for $N_e \geqslant 100$ cm^{-3}. The second method (Hultqvist, 1966a) is more complicated algebraically, but is believed to give a better result.

Profiles obtained in these ways and through partial reflection were one of the earliest available for PCA times.

11.1.2. USE OF IONOSONDES

Use of the ionosonde during PCAs is identical with that during SIDs. The basic parameter is the abnormal increase in f_{min}. At these frequencies the magnitude of absorption is considerably larger than in the case of riometers. The technique is thus more sensitive to weak events. When this minimum frequency exceeds the F-region critical frequency, a complete blackout is recorded (as in the cases of large SIDs) and the method fails. The measurements have all the defects enumerated in Section 2.6 and is not quite suitable for quantitative information on PCA events. Nevertheless it has been effectively used to deduce information about the time of onset of events (Hakura, 1967a), their gross morphological features (Hakura, 1961), and their statistical properties, particularly before the more modern techniques became available. Some of the early PCA events of the IGY period were indeed recognized as a distinct phenomenon solely on the basis of high latitude ionosonde data.

11.1.3. PARTIAL REFLECTIONS

The measurement of partial reflections at lower frequencies is quantitatively more valuable (again as in SIDs), since here we have a method that provides information on the *distribution* of both electron concentration and collision frequency. Observations by Gregory (1963) using a frequency of 2.3 MHz at Scott Base, Antarctica, revealed the existence of several events in 1960 that were too weak to be recorded by riometers. Again, because of the relatively lower frequencies used, the technique is more sensitive to weak events than riometers. This, however, is a disadvantage for very strong events. Since the signals are heavily absorbed near the bottom of the absorbing region, the electron density profile can be measured only near the base. During the November 1969 PCA events (see Section 11.3) the Canadian Department of Communications operated partial reflection experiments at Ottawa, Churchill and Resolute Bay. Belrose (1972) has given a whole set of electron density profiles during these events, showing the nature of ionization growth and decay. While not as accurate as direct rocket measurements (Section 11.3) the obvious advantage is being able to watch the evolution of the ionization buildup and decay.

11.1.4. LF AND VLF OBSERVATION

Historically, the first VLF effect associated with a PCA was during the relativistic event of 23 February, 1956, which also marked the starting point for the entire field of polar cap absorption. Like the riometers it is also a sensitive detector of solar protons. Some believe it to be the most sensitive groundbased equipment for such detection. For February 1956 event Allan *et al.* (1957) recorded a phase advance of 110 μs over a long path (GBR – New Zealand) and also an amplitude decrease. Over a short path, GBR – Cambridge (England), Belrose *et al.* (1956) deduced a maximum decrease in phase height of 8 km, beginning about 2 h after the start of the flare.

Since VLF phase advances are almost linearly proportional to decreases in reflection height (Δh) and because Δh can be expressed in terms of relative solar radiation intensity, there should be an approximate linear relationship between the phase shift and the log log of the proton flux (recorded by satellites) that affects the region of propagation. Figure 131b, from Westerlund *et al.* (1969) shows that this relationship is fairly well obeyed (for $E > 25$ MeV), but the observations show two distinct straight lines for the VLF data from different paths. The authors show that the dashed line corresponds to paths that cross the Greenland ice-cap, where the ground conductivity is low ($\sigma_g = 0.01$ mmho m^{-1}), while the solid curve corresponds to paths that lie mainly over the oceans, with very high surface conductivity ($\sigma_g \sim \infty$). Writing the omni-directional flux as F and $F_0 = 1$ proton cm^{-2} s^{-1}, the experimental data can be represented respectively by:

$$[\varDelta\phi]_a = -12.9 + 24.5 \ln \ln F/F_0 \qquad F \geqslant 5.4.$$
$$[\varDelta\phi]_b = -10.9 + 16.5 \ln \ln F/F_0 \qquad F \geqslant 7. \tag{104}$$

Both curves indicate cut off at about 5 protons $cm^{-2} s^{-1}$ or about 1 $cm^{-2} s^{-1} ster^{-1}$ directional flux.

The much larger effects over the ice-cap paths are related to the change in the phase velocity of the waves in the Earth ionosphere waveguide, and implies that Antarctic VLF paths are more sensitive to PCA effects than most Arctic paths.

Thus it appears that VLF phase measurements have roughly the same sensitivity to solar proton fluxes as riometer measurements in the energy range to which riometer measurements apply, while their sensitivity to fluxes at both higher and lower energies is substantially greater than that of the riometer. The VLF effect is also ideal for recording the *onset* of an event.

The magnitude of the effects experienced on a long distance VLF path depends on the integrated properties of the entire portion of the Earth-ionosphere waveguide that affects the path, and different regions are usually experiencing different conditions of sunlight and of proton flux at any given time. Quantitative studies are thus very difficult, except in very broad terms. VLF measurements at vertical incidence are easier to interpret. VLF sounding has been carried out during PCA events by Helms and Swarm (1969) at Byrd Station, Antarctica, using a 34 km long dipole antenna buried in the ice. They were able to make direct measurements of reflection phase height and its changes, and found typical decrease in daytime reflection height from about 85 km under normal conditions to about 65 km during a PCA event.

In the LF major disturbance effects last for many days (upto 15 days), according to Belrose (1964). Observations on high-latitude paths at frequencies in the 70–80 kHz range during the November 1960 PCA events were reported by Belrose and Ross (1961), who found the daytime signals to be stronger than usual, and nighttime signals weaker. The transition between the two occurred quite suddenly at mid-path twilight. The normal diurnal variation disappeared entirely for strong events. For other events the diurnal variation was reduced, but did not disappear completely. A marked feature during the PCA was the absence of nighttime fading, indicating the presence of a homogeneous layer of ionization whose reflection properties were independent of the presence of sunlight.

11.1.5. OBLIQUE HF PROPAGATION

In the HF, riometer is the most sensitive PCA detector. *Oblique* HF propagation for reasons mentioned below shows a more subdued effect. Egan (1963) investigated the propagation of 12 MHz signals from Thule, Greenland, to College, Alaska, during the major PCA events of April and May 1960, and found that the absorption was considerably less (about 35 dB during the peak of the event) than the expected value (over 400 dB) from vertical observations with riometers. One interpretation is that the anomalously high signal strength is due to reflection from the D-region itself, where the electron density had considerably enhanced with only slight penetration of the radio waves into the absorbing region.

HF radio wave propagation during PCA events is actually helped by the substantial decrease in the background atmospheric noise level. In polar regions most of this noise is due to integrated atmospherics generated by distant (mainly tropical) thunderstorms and propagated to high latitudes by ionospheric reflection. The reduction in noise level keeps the signal-noise ratio on Arctic HF communications circuits relatively unaffected during PCAs.

Oblique HF propagation is, therefore, a poor monitor for PCAs.

11.1.6. VHF FORWARD SCATTER

PCAs have also been studied with forward scatter at upper HF and lower VHF frequencies. The pioneering work in this line was carried out by Bailey (1957, 1959), who made the first serious study of solar-proton-induced ionization on the basis of the major proton event of February 23, 1956. VHF forward scatter show similar effects to those observed by riometers, but quantitative interpretation is hampered by uncertainty in the height of the irregularities that are responsible for propagating the signal. Both *absorption* and *enhancement* of signals are observed. If there is a substantial portion of the absorbing region below the altitude of the irregularities, the signal is strongly absorbed, but if most of the absorbing region lies above the irregularities the signal is enhanced due to the increased electron concentration in the neighbourhood of the scatterers. The technique has, however, proved important as with ionosondes, in identifying events and in providing the basis for lists of past events (Bailey and Harrington, 1962).

11.2. PCA Morphology

11.2.1. CLASSIFICATION OF PCA EVENTS

The time curve of the ionospheric absorption recorded by riometers is a measure of the time curve of sub-relativistic solar cosmic ray flux. Different types of PCA time curves are observed (see Figure 130).

The PCAs may, in the first place, be divided into Single and Complex events. A Single event is an isolated one produced by a source flare with a single geomagnetic storm associated with it during the entire course of the solar cosmic ray event. A Complex event is one in which there is at least another source flare or geomagnetic storm, and the time variation is extremely complex. Single events are further classified into two types according to their time variations: F and S types. The F type represents a fast onset in the increase of flux, having a rapid risetime of 5 h or less between the flare and the beginning of PCA. The S type contains events with delaytimes exceeding 10 h. An intermediate type which has a fast onset but the flux does not reach its maximum until the beginning of the associated geomagnetic storm is labeled F^*.

The different types of events have been associated with different characteristics of solar disturbances. Hakura (quoted by Obayashi, 1964) found that F and F^* type events are followed by dm-IV and μ-IV solar outbursts while S-type events are related to strong m-IV outbursts and rather weak μ-IV component.

11.2.2. MICROWAVE-MILLIMETER WAVE SOLAR RADIO BURSTS AND PROTON FLUX

Another groundbased way of estimating proton flux has been given by Croom (1970a, b, 1971a, b). He drew attention to the use of impulsive single-frequency microwave bursts for providing warnings of the occurrence. He showed that there is a distinct relationship between the *mean duration* T_m in minutes of microwave-millimeter wave bursts in the range 8–20 GHz and the proton flux densities. The empirical relationship obtained by using the lower envelope of the available solar proton/microwave burst data is:

$$J(>10 \text{ MeV}) = 1.0 \times 10^{-2} T_m^5 \text{ protons cm}^{-2} \text{ s}^{-1} \text{ ster}^{-1}$$

in which T_m is given by:

$$T_m = \frac{1}{P_{\max}} \int_0^T P(t) \, dt, \tag{105a}$$

where $P(t)$ is the power of the flux density increase of the solar burst at time t and T is the overall duration of the burst.

Croom (1973) combined this with a modified version of Equation (103) to extend this to energies down to 10 MeV and found the following relationship to be valid:

$$A = 1.4 \times 10^{-2} T_m^{5/2} \text{ dB}, \tag{105b}$$

where A is the vertical *daytime* 30 MHz absorption.

Thus, the microwave-millimeter wave bursts can be used as a precursor prediction for the PCA events to follow, and, furthermore, the magnitude of this PCA can also be guessed. Croom finds that where this prediction failed – the so-called 'false alarms' – they were in fact associated with proton events, but conditions were not favourable for protons to reach the Earth.

11.2.3. TWILIGHT EFFECT

The day-night variation in riometer absorption during a PCA and the exact time of onset of the daytime condition (the so-called 'twilight effect') can be an important input in the understanding the mesospheric chemistry, especially of the role of photo-detachment from negative ions. Figure 132 shows early examples given by Hultqvist and Ortner (1959). The major virtue of PCA in such studies is that the ionization source exists during both day and night. One would expect that even if the proton flux remains the same, there will be less absorption at night due to decrease in free electrons from attachment to neutral particles to form negative ions. Hultqvist (1966b) found that the day/night absorption ratio was between 4–6. The sunrise or the sunset effect corresponded to the shadow of a solid Earth at about 45 km, or of an ozone layer

NEGATIVE IONS IN THE LOWEST IONOSPHERE

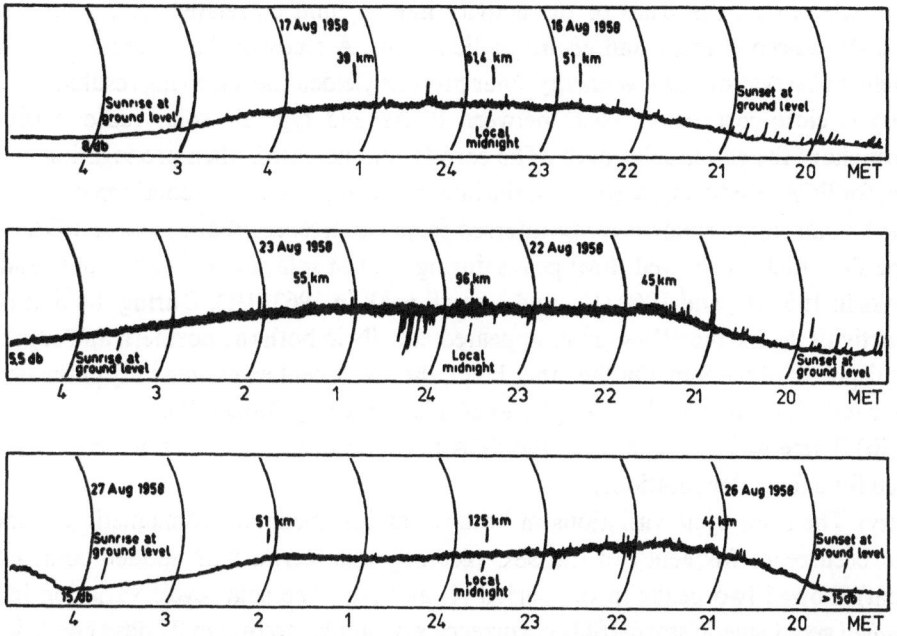

Fig. 132. Examples of records, after Hultqvist and Ortner (1959) of riometer absorption during PCAs at Kiruna showing sunrise and sunset effects. Values in km above the curves give the upper limit of the solid Earth's shadow vertically above Kiruna at midnight and estimated beginnings and ends of the periods of nighttime absorption.

producing a shadow at about 90 km. The former implies visible radiation; the latter, ultraviolet. Other results exist to show that the onset time for the twilight effect is generally in the vicinity $\chi = 90°$.

Another study of the variation of PCA in the twilight periods was published by Chivers and Hargreaves (1965). This involved measurement in both northern and southern polar caps. Their experimental results showed that absorption at the South Pole and Resolute Bay are identical in the middle of common daytime periods. They assumed that the flux was the same all the time in the two hemispheres and calculated the day-to-night ratio of the PCA in the periods when one station was in twilight and the other in darkness. Their results showed that after sunrise there was a time delay of several hours before the absorption reached the full daylight value. There was also some evidence for a time delay after sunset.

11.2.4. SOLAR CYCLE VARIATIONS IN THE PCAs

Obayashi (1964) pointed out that no relativistic solar cosmic rays were observed during the maximum sunspot activity. Švestka (1966), tracing PCA events back to 1938, showed that the sub-relativistic particles also tend to avoid the period of sun-spot activity peak during the last three sunspot cycles.

Hakura (1967b) has computed the daily indices of PCA-activity for years 1954–65, which covers the whole period of the Solar Cycle No. 19. Outstanding PCA events were selected on the basis of the activity indices, and correlated with solar flares, type IV radio outbursts, and geomagnetic storms. A study of these events along with satellite observations of low energy solar protons yielded the following results:

(i) A close correlation exists between PCAs and type IV solar radio outbursts throughout the whole solar cycle. The correlation increased when propagation condition for PCA-producing particles in the interplanetary space were considered.

(ii) Solar corpuscular activity inferred from occurrence frequencies of PCAs and type IV outbursts showed three peaks during the last solar cycle, i.e. two outstanding peaks in 1957 (I), and 1960 (II), and a small peak in 1963 (III). During the first peak of activity (I), the type IV-sources appeared equally in both the northern and southern hemispheres of the Sun. On the other hand, the active centers existed only in the northern hemisphere, during the later phases of solar activity (II) and (III).

(iii) There was a tendency for the PCA-flares occurring in the same active regions even for a few solar rotations.

(iv) The solar cycle variations in both the annual mean of geomagnetic ΣK_p index and occurrence frequency of the SSC (geomagnetic storms with sudden commencement) showed two peaks in the period (I) and (II). The solar cycle variation in the *gradual* geomagnetic storm (SG) occurrence was similar to that in 27 days coefficients. Both of them increased toward the end of the solar cycle, and had a prominent peak in the period (III).

(v) During the later phases of solar corpuscular activity (II) and (III), space vehicles detected a number of solar MeV protons which were detectable by the daily PCA index of higher sensitivity. A recurrent series of the MeV protons starting from the 5th–6th days lasted for about 15 solar rotations in 1963–64. This series coincided with a part of recurrent series of geomagnetic K_p index and galactic cosmic-ray variations, which could be traced for about 25 solar rotations from the end of 1962 to end of 1964. The result means that the MeV protons were confined within a region corotating with the Sun which caused enhancements of geomagnetic activity and at the same time modulated the galactic cosmic ray intensity at the orbit of the Earth with a 27 days recurrent period.

11.3. November 1969 PCA Events

There are certain similarities between a PCA and SID. In both, the effects are very pronounced and easily distinguished from the slowly varying components. The measurement techniques are very similar and both offer excellent opportunity to study atmospheric, chemical and physical processes occurring in the lower *D*-region provided the source intensity is measured. This is especially true during the decay period of the event when there is little change in the incoming energy spectrum. In the case of PCA there is another advantage in that since it lasts longer, on many occasions it covers conditions of varying sunlight. The difference in the absorption effect can be observed as a function of solar zenith angles during sunrise and sunset (the so-called twilight effect). If the effect covers both day and night, then we have the unusual advantage of exploring a situation of both large (night) and of low (day) negative ion concentrations, while the production rate remains essentially constant – a situation that does not exist during normal conditions since at night the normal ionizing source – the solar XUV radiation – vanishes. The 'Twilight Anomaly' has in the past been examined with the riometer absorption effect. A distinct advantage arises if one makes simultaneous measurement of the altitude variation of the negative and positive ion composition, electron and ion density, the proton and electron flux, neutral air density and temperature and other important parameters during a PCA by instrumented rocket flights. A coordinated PCA campaign in which all the inter-related parameters are simultaneously measured can be especially rewarding. Ideally one should measure during a PCA event the following:

(i) The primary source of atmospheric ionization: protons and alpha-particles, as well as X-rays, Lα and energetic electrons.

(ii) The effect of these ionizing particles on the atmosphere, indicating those of ionization, emission and heating.

After several false starts such a campaign was successfully achieved in the USA during the PCA events of November, 1969. The campaign involved firings of 47 scientifically instrumented rockets of which 11 were designed to provide certification and background information.

In this section we will discuss the results of this campaign in some detail, for these PCAs provided information in a way not otherwise available, and provide much of the basis of the PCA physical picture that follows.

The rocket program was conducted at the Churchill Research Range located in Manitoba, Canada within the auroral oval.

Figure 133 shows the absorption levels (relative to an average quiet day) for the 30 MHz polar riometer at Churchill during the event from 2 to 4 November, 1969 following the flare event of November 2 beginning at 1102 UT. The solar zenith angle variation as well as an indication of when each rocket was fired are shown at the top of the figure. The lower curves show the proton flux variation during this period obtained from Vela satellite (Singer, 1970) for $E > 25$ MeV, 3.2–5 MeV and 0.46–0.88 MeV. As shown, the main phase of the program began with the nighttime launches on 3 November, 1969 (0455 UT). The proton flux values indicate that conditions were fairly stable from these launches to the completion of the program. Table XXII gives a summary of the programs that were conducted during Operation PCA 1969.

Ionization measurements during this campaign was as complete as one could wish, considering the difficulties of organizing such a campaign. Both electron and positive ion densities were measured by probes carried in Black Brant rockets by Ulwick *et al.* (1970) on a certification day during daytime on November 2, 1969 and at night on November 3. Positive ions were also measured by Hale *et al.* (1972) with parachute-borne blunt probes, and by Conley (1972) with Gerdien condenser. There was also a deliberate multiplicity of electron density measurements with widely differing tech-niques: the Ballistic Research Laboratory used four-frequency propagation experi-

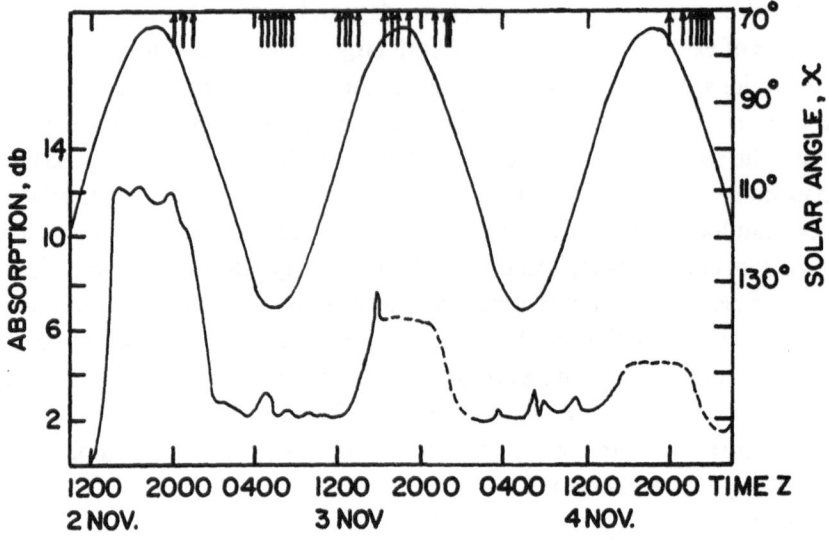

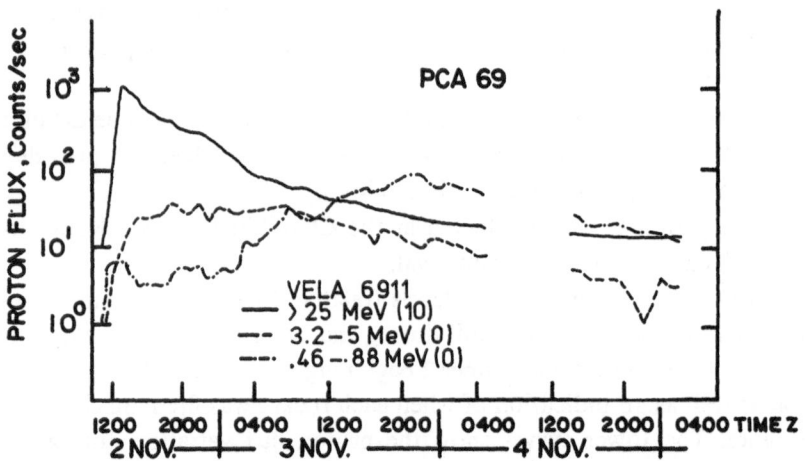

Fig. 133.   Riometer absorption during the PCA events of November 1969 with indication of the time of rocket firings from Fort Churchill. The lower curves give the proton fluxes recorded by Vela Satellite 6911 by Los Alamos Scientific Laboratories (after Ulwick and Blank, 1970).

ment (36, 72, 144 and 576 MHz) and propagation measurement for absorption experiment, and a Langmuir probe; the Boston College used Langmuir probes; AFCRL used an r.f. probe. What is still more valuable was the identification and measurement of concentrations of different species by Narcisi *et al.* (1972a, b). Of the neutral constituents, measurements were made of the total number density, of temperature and of ozone. An ultraviolet photometer at 2150 Å did not yield dependable results on nitric oxide concentration due to contamination from Rayleigh scattering which could not be eliminated, but indirect estimates were made during PCA day, night and sunrise by Narcisi *et al.* (1972c) from observations of the ionic ratio $[NO^+]/$

TABLE XXII

Summary of PCA 1969 observational programme
Results discussed in James C. Ulwick (ed.), *Proceedings of COSPAR Symposium on Solar Particle Event of November 1969*, AFCRL-72-0474 U.S.A.

Technique		Measurement
(A) Rockets	36 PCA 11 Certification and Background	Ionizing sources Minor Constituents Positive and Negative Ion Species Proton Flux Positive Ion and Electron Densities Neutral Density and Temperature Meteorological Parameters
(B) Aircraft	1 KC 135	Airborne Infrared Measurements
(C) Satellite	OVI-18 ESRO I and II Injun V Vela ATS	Energetic Particle Fluxes
(D) Groundbased Measurements	Partial Reflection Riometer VLF Ionosonde Magnetometer	*D*-Region Electron Density Cosmic Noise Absorption Phase and Amplitude Variations Absorption and Electron Density Variation in Magnetic Field

$[O_2^+]$. A Lα detector showed that the ambient daytime flux was 5 erg cm^{-2} s^{-1} (giving $q = 10$ cm^{-3} s^{-1} at 80 km and 1 at 100 km), and the nighttime flux was 1 kR (production rate $< 10^{-2}$ cm^{-3} s^{-1} and negligible at these heights). Proportional counters gave 1–10 Å spectral distribution. A radiometer operating at 1.27 μ yielded information of $O_2(^1\Delta_g)$ concentration, a competing, but normally weak source of ionization in the *D*-region, at midnight on 3rd November, and at sunset on 4 November. The 3914 Å (N_2^+) radiance was 15 kR at 75 km, 25 kR at 100 km and 10 kR at 125 km with two peaks, one at 70 km and the other at 100 km.

11.4. Principal Aeronomic Features of the Lower Ionosphere During a Major PCA

The most obvious feature is the very large enhancement in ionization in the lower ionosphere and the large production rates, compared to the quiet condition. For large PCAs, such as the November 1969 events the electron densities are even larger than those for strong flare conditions in the entire lower ionosphere. This is shown in Figure 134. In this the PCA (November 1969) profile given is that of Ulwick (1972). For comparison N_e profiles for a moderate flare and a strong flare are given as well as the quiet day ranges in electron density for midlatitude and moderate solar activity period. The major difference from the situation existing during a flare is that during

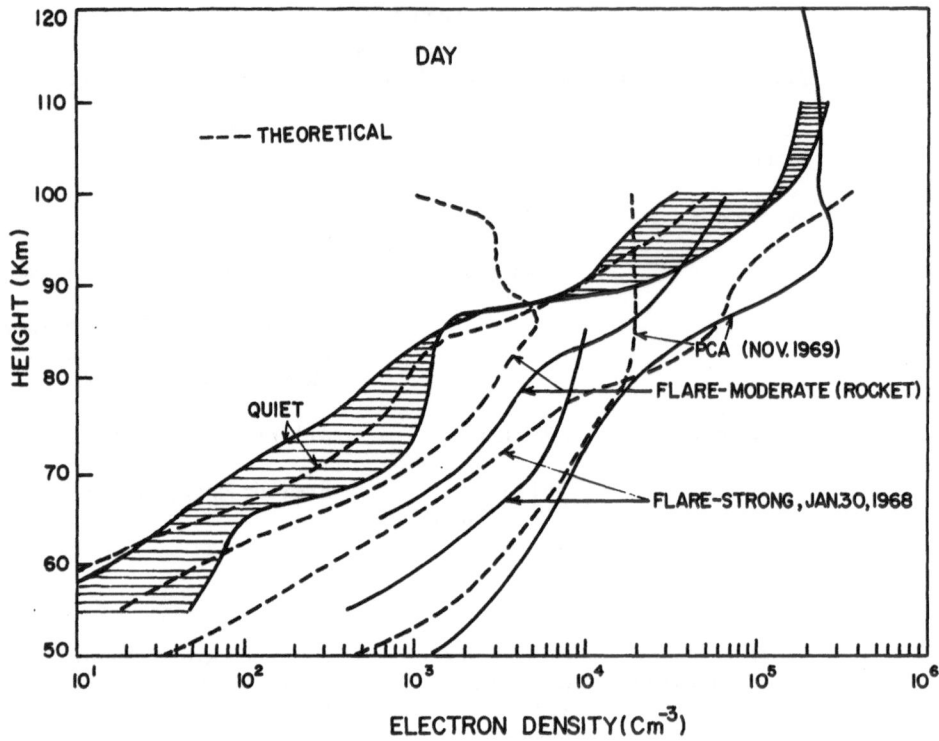

Fig. 134.   Comparison of electron density profiles during normal times, flares and PCAs.

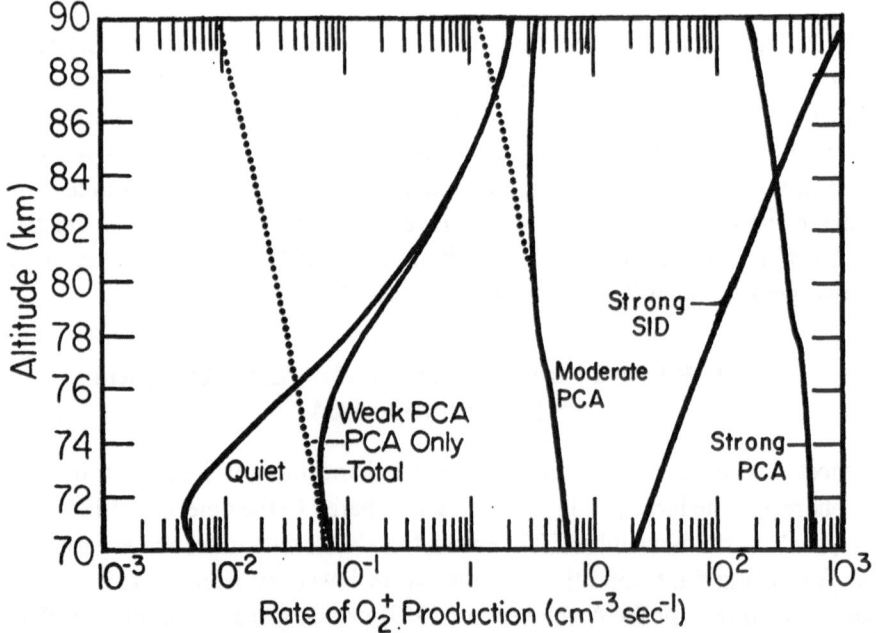

Fig. 135.   Representative profiles of the rates of O_2^+ production during weak, moderate and intense PCA events. Dotted curves show production due to protons alone (after Reid, 1972b). For comparison the altitude profile of production during a major flare is also shown (after Mitra and Deshpande, 1972).

PCA NIGHT ASCENT AG 7882 3 NOV 1969 0730 Z FORT CHURCHILL CAN.

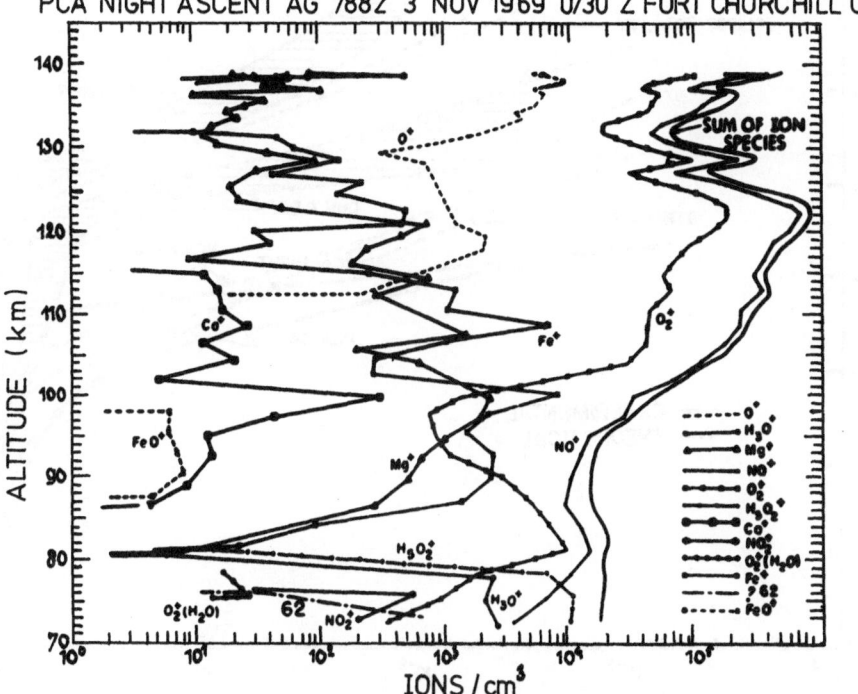

Fig. 136a. Positive ion composition during PCA night on November 3, 1969
(after Narcisi *et al.*, 1972a).

PCA DAY ASCENT AH 7886 3 NOV 69
1730 Z FORT CHURCHILL CANADA

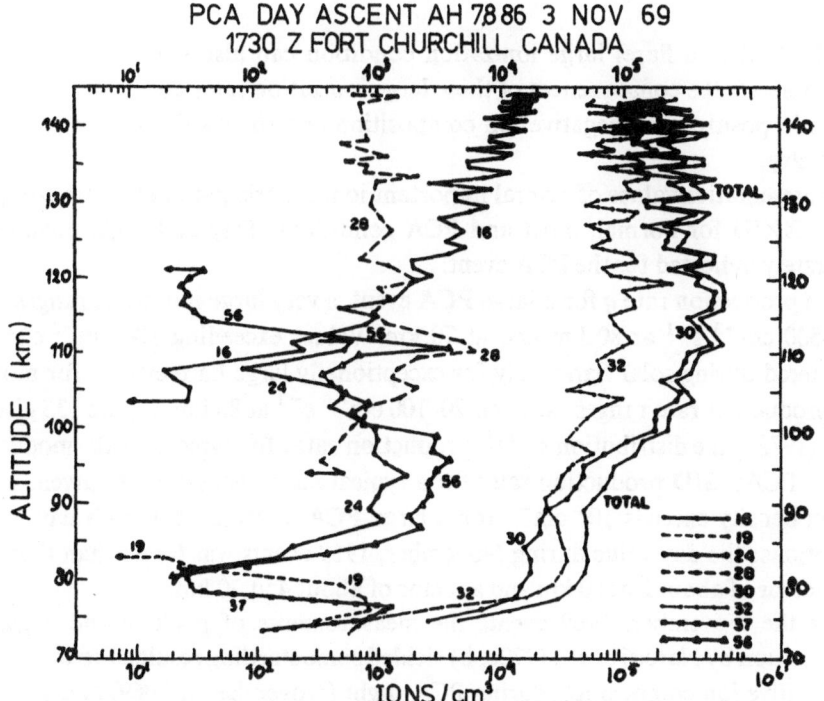

Fig. 136b. Positive ion composition during PCA day, November 3, 1969
(after Narcisi *et al.*, 1972a).

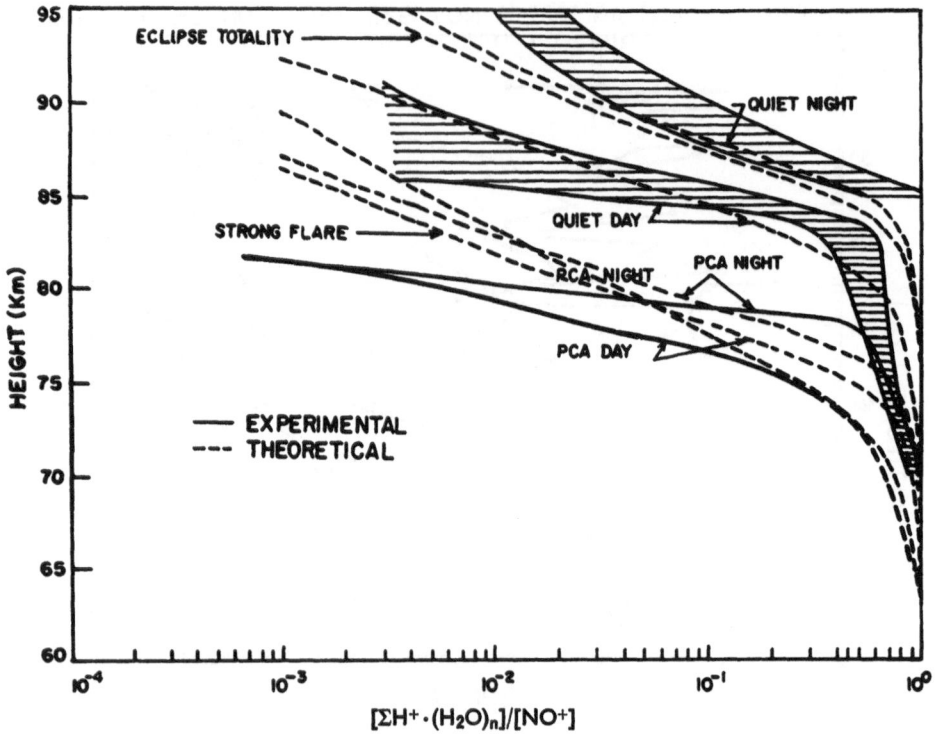

Fig. 136c. Changes during PCA in $[\Sigma H^+ \cdot (H_2O)_n]/[NO^+]$ ratio. Note the drastic decrease in percentage composition of water cluster ions during both PCA day and night conditions.

a PCA (unlike a flare) large ionization condition can also exist at night. More remarkable are the consequences of these large ionization rates and ionization enhancements in positive and negative ion composition and the resulting changes in the ion chemistry.

Representative values of several important ionospheric parameters are compared in Table XXIII for normal, quiet and PCA conditions. Day and night conditions are separately indicated for the PCA event.

The production rate q for a large PCA event is very large indeed. It ranges between 100–500 cm^{-3} s^{-1} at 80 km and at 70 km. Values exceeding 100 cm^{-3} s^{-1} are encountered during solar flares only for exceptionally large flare events: for most flares the production rate ranges between 20–100 cm^{-3} s^{-1} at 80 km. Figure 135 gives, after Reid (1972b) the distribution of O_2^+ production rates for typical weak, moderate and strong PCAs. SID production rates for a typical major flare are also given. The ionization density exceeds 10^4 cm^{-3} for a large PCA at 80 km for both day and night conditions; the day value during November, 1969 events was larger than that at night by a factor of about 2 at 80 km and a factor of about 4 at 70 km.

For the November, 1969 events the measurements of positive and negative ion composition by Narcisi *et al.* (1972a, b) yielded some startling results. Figure 136a shows the positive ion composition during PCA night (November 3, 1969) and Figure 136b

TABLE XXIII

Comparison of important ionospheric parameters during quiet, flare and PCA conditions

Height	Parameter	Normal day	Flare	PCA	
				Day	Night
80 km	$q(\text{cm}^{-3}\,\text{s}^{-1})$	5–10	100–200	300	300
	$N_e(\text{cm}^{-3})$	10^3	8×10^3	4×10^4	2×10^4
	$\dfrac{[\Sigma H^+ \cdot (H_2O)_n]}{N^+}$	0.9		0	0
	$\psi\,(\text{cm}^3\,\text{s}^{-1})$	$(1\text{–}2) \times 10^{-5}$	3×10^{-6}	5×10^{-7}	1.3×10^{-6}
	λ	$\ll 1$	$\ll 1$	$\ll 1$	0.7
75 km	q	1	50	500	500
	N_e	5×10^2	4×10^3	1×10^4	1×10^4
	$\dfrac{[\Sigma H^+ \cdot (H_2O)_n]}{N^+}$	~ 1	0.1	0	0.7
	ψ	$(1\text{–}4) \times 10^{-5}$	3×10^{-6}	5×10^{-7}	9×10^{-5}
	λ	$0.5 - 1$	–	0.4	10

that during PCA day (November 3, 1969). During the day flight, water cluster ions could not be seen down to 73 km, the limit of the measurement. This means that the cluster disappearance level has descended from its quiet day height of about 82 to at least 73 km and perhaps below. There was a similar lowering of the night cluster disappearance level from its normal value of about 87 km to about 77 km. $[NO^+]/[O_2^+]$ ratio increased from day to night as in normal conditions. The drastic decrease of water cluster ions relative to molecular ions can be seen in Figure 136c in which the altitude profiles of $[\Sigma H^+ \cdot (H_2O)_n]/[NO^+]$ are given for PCA day and night conditions along with quiet day and night conditions. The reduction in relative cluster condition is quite drastic for both day and night, but particularly drastic at night. There does not seem to be much difference in this ratio between day and night during a PCA.

The daytime situation, one should note, is similar to that expected for flare conditions, although for the latter the only evidence concerning disappearance of hydrated ions comes from the drastic decrease in effective recombination. The expected decrease in ψ for PCA conditions is also seen, quite as dramatically as one would have expected.

ψ-distributions deduced during PCAs from the use of the formula: $\psi = q/N_e^2$ have been obtained by several authors from measurements of electron density profiles during different PCAs and the production rates estimated from measurements of the incoming proton flux as measured with rockets and satellites. Some examples of $\psi - h$ profiles derived during PCAs are shown in Figure 88; in all cases the $\psi - h$ profiles are lower than those representing quiet day situation (e.g. Reid's or that given by Chakrabarty and Mitra). This is particularly apparent in Figure 137a, b where ψ obtained from q/N_e^2 by Ulwick (1972) for the November 1969 event is shown.

Another way of obtaining ψ is from ion composition measurements. For this, one may use the ratio of $[NO^+]/[O_2^+]$, the information that clusters have entirely vanished

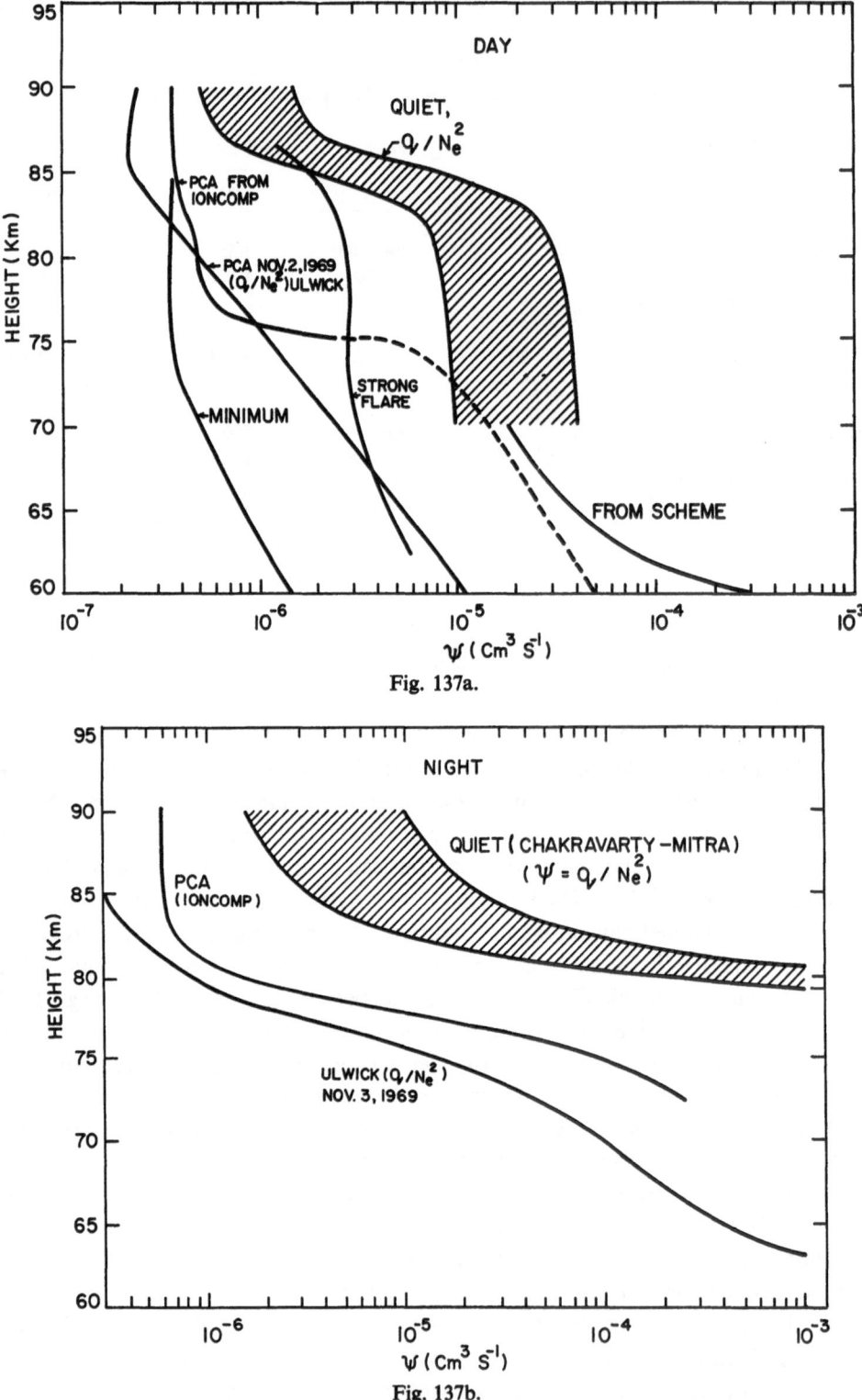

Fig. 137a.

Fig. 137b.

Fig. 137a–b. Altitude profile of the effective loss rate during a major PCA (deduced from ion composition information) compared with that for a major SID. The quiet time distribution of Cha-krabarty and Mitra is shown for comparison. Curves in 137a refer to daytime conditions and those in 137b to nighttime conditions.

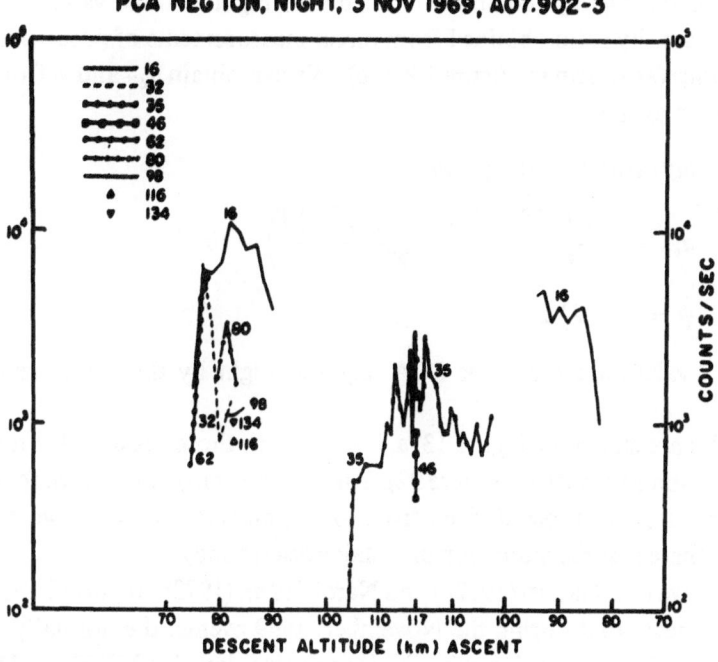

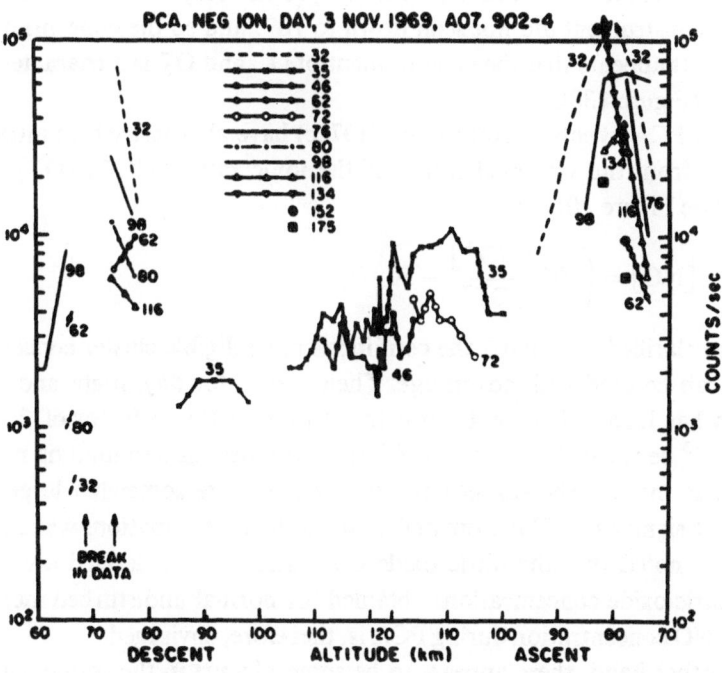

Fig. 138. Negative ion composition during the PCA of November 1969
(after Narcisi *et al.*, 1972b).

above 73 km during day and above 77 km during night, and the values of negative ion to electron density ratio obtained from direct measurements of positive ions and electron densities (as shown in Figure 136a, b). We can obtain α_{eff} and ψ from these using the following expressions:

$h > 73$ km (day) and > 77 km (night)

$$\alpha_{\mathrm{eff}} = \frac{3 \times 10^{-5}}{T} \left(\frac{5[NO^+] + 2[O_2^+]}{N^+} \right) + 10^{-7}\lambda \qquad (106)$$

$$\psi = (1 + \lambda)\,\alpha_{\mathrm{eff}}$$

in which λ values are given for PCA day and night by the curves given in Figure 99.

ψ profiles are shown in Figure 137a, b along with those deduced for major SIDs by Deshpande and Mitra (from Figure 92). The daytime ranges are shown for comparison. Ulwick's values and those derived from ion composition agree down to 75 km, but below this there is some indication of a major discrepancy.

For negative ions Narcisi (1972b) and Narcisi et al. (1972b) reported a very surprising result. It appears that during the November, 1969 events, the normally predominant heavy cluster ions were replaced by O_2^- ions during day (1147 CST, $\chi = 73.9°$) between 76 to 94 km and by O^- ions during night (0029 CST, $\chi = 136°$) at the lower altitudes. On the downleg O_2^- concentration became larger than O^- ions below 76 km, in contrast to quiet conditions when O_2^- ions are present only in small quantities. For the daytime the instrument did not scan down to 16^-, and so the concentration of O^- is not known. It appears that the enhancement of O^- and O_2^- is a characteristic of PCA conditions (Figure 138).

For these PCA events, Narcisi et al. (1972c) have also derived the nitric-oxide concentrations from the observed ratios of the ionic ratio $[NO^+]/[O_2^+]$ through the equation (see Figure 101 a):

$$[NO] = \left(900 \frac{[NO^+]}{[O_2^+]} - 50 \right) N_e. \qquad (107)$$

It has been clarified that under the conditions of negligible cluster concentrations this equation can be used with advantage. Their results for day, night and sunset PCAs are somewhat larger than, but close to Meira's, within a factor of 2 of the value 3×10^7 cm^{-3} between 77 and 88 km. All three profiles showed minima near 84 ± 1 km as the Meira model. The sunset and night values were somewhat larger, but this is perhaps not significant. The estimated error in the determinations was about $\pm 140\%$.

It will be noted that the nitric oxide concentrations so derived are very close to Meira's nitric oxide concentrations obtained for normal undisturbed mesosphere. No change in NO concentration during PCA is, therefore, envisaged.

On the other hand, there appears to be some change in the ozone concentrations. There appeared to be much less ozone at all altitudes (50–70 km) than at normal times: the difference was larger than the expected error. At 54 km the concentration was a factor of 2 lower, at 60 km a factor of 3 lower and at 67 km, a factor of 4 lower. This is in contrast to theoretical predictions where an *increase* in ozone is expected from dissociation of O_2 by energetic particles.

Some major aeronomical changes in PCA are outlined in Table XXIV.

TABLE XXIV

Major aeronomical features during PCAs

1. Water Cluster	– Down to 73 km during day (normally 82 km)
Disappearance	– Down to 77 km during night (normally 87 km)
Level Moves Downwards	

2. Presence of heavy negative ions: 70–90 km (NO_3^- or CO_3^- hydrates)
   O^- and O_2^- dominant at lower altitudes, where normally only heavy negative ions found

3. Drastic decrease in effective recombination coefficient
   λ unaffected during PCA day; $\lambda = 0$ level moves upto 85 km at night

4. Minor Constituents
   (a) Nitric oxide during day, night and sunrise PCA similar to Meira's
   (b) Decrease of ozone over 50–70 km: By factors of 2–4
   (c) $\lambda 5577$ Å intensity increased to $300R$ on November 2 PCA day
   $\lambda 4278$ Å (ionized nitrogen) increased to above $100R$ on November 2 PCA day
   $\lambda 6300$ Å increased to $200\ R$ on November 2 PCA day

11.5. Chemistry of the Lower Ionosphere During PCAs

The chemistry of the lower ionosphere during the PCA is almost entirely identical with that of the SID. As in SID there is creation of a large quantity of O_2^+ and N_2^+ ions, the only difference is that in this case these are produced by energetic protons, whereas in the SIDs these are produced by X-rays. The other difference is in the time scale. The peak production during SID is reached in 10–20 min; in PCAs the peak is reached in several hours and the level is maintained for hours, including conditions at night when photodetachment reactions will vanish, and atomic oxygen is reduced allowing a free run to negative ions to go essentially unhampered through the steps to the right in Figure 105.

If one looks at Figure 103, one can immediately understand why hydrated ions should decrease during a PCA. As in SIDs there are two reasons:

(i) One is that with increase in O_2^+ production, $O_2^+ \rightarrow O_4^+$ channel begins to emerge as an important channel and consequently one ends up on the path $O_2^+ \rightarrow O_4^+ \rightarrow O_2^+ \rightarrow \rightarrow NO^+$ in addition to direct path $O_2^+ \rightarrow NO^+$.

(ii) The second is that with increase in N_e, clusters will disappear more rapidly through dissociative recombination with electrons than the simple molecular ions.

Both will be effective, the relative contributions depending on the quantity $8 \times 10^{-21} \times [O_2]^2/[NO]$, and are consequently a function of altitude.

The chemistry of the positive and negative ions during a PCA has been examined by

Reid (1972b) and Narcisi (1972a, c) following essentially the schemes of Figures 102, 105 excepting that for NO^+ conversion to hydrates the faster process through N_2 was not used. Nevertheless, Narcisi *et al.* find the sharp cut off of the cluster ions above about 70 km.

In regard to negative ions Reid (1972b) points out that as in positive ions there should be a reversion to the earlier members of negative ion sequence with increase in ion-ion recombination rate as the ionization increases. His calculations are shown in Figure 139. An additional mechanism for throwing the negative ions to their initial stages (O^- and O_2^-) could be an increase in atomic oxygen concentration through dissociation of O_2 and O_3 by energetic particles. An increase in atomic oxygen would shunt off the heavy negative ion formation channel, reverting the ions to their basic forms O^- and O_2^-.

One particular aspect of the PCA needs a closer look. This is the Twilight Anomaly referred to before. The time of onset of the twilight variation gives an important insight on the negative ion chemistry. This occurs, as we have seen, near $\chi = 90°$. In the early days of PCA studies this was considered a major anomaly as it was then believed that O_2^- is the most dominant negative ion. It will be recalled that O_2^- ions are quite efficiently detached by the visible radiation (its electron affinity is 0.43 eV), and con-

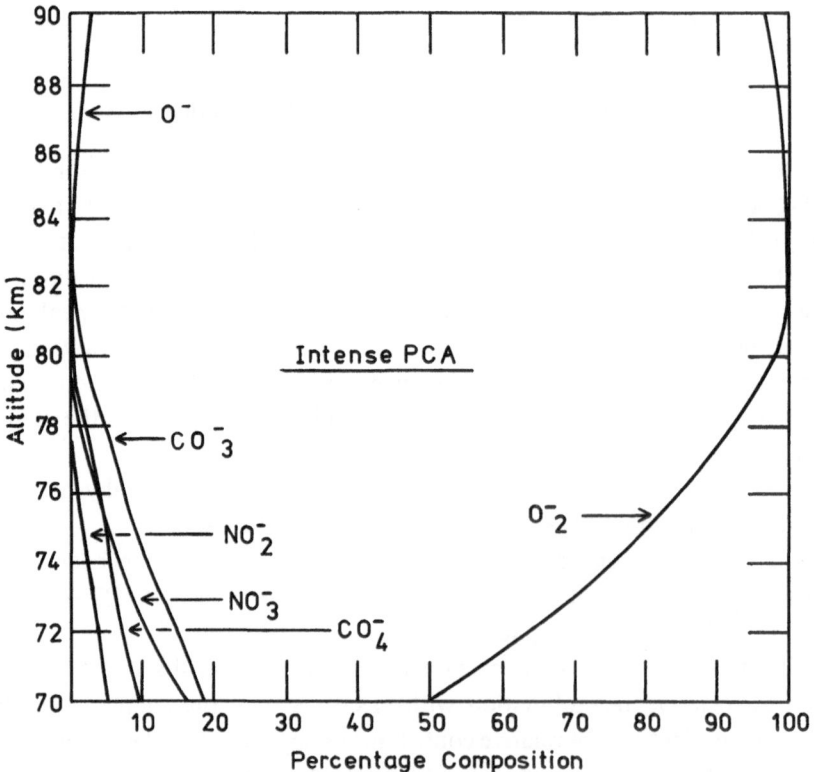

Fig. 139. Negative ion composition theoretically derived by Reid (1972b) for a PCA.

sequently the twilight effect would have been expected to begin only when the solid Earth shadow would reach the base of the D layer; this occurs at $\chi \approx 94°$. Later, when it was discovered that other negative ions exist, and that for several of these, the photo-detaching radiation is in the ultraviolet, an onset time at $\chi = 90°$ was expected.

It is known now, however, that neither of these give the correct or complete expla-nation. One of the most important detachment reactions is the associative detach-ment reaction described earlier (and shown in Figure 105):

$$O_2^- + O \rightarrow O_3 + e.$$

Those that survive the process and proceed to higher forms of negative ions are also subjected to similar, although sometimes less effective, detachment processes with atomic oxygen (such as $O_3^- + O$, $CO_3^- + O$). The day-night variation is then predom-inantly a result of day-night variation of atomic oxygen. The life time of atomic oxygen is short for heights below 75 km, and theoretical models give a sharp decrease of atomic oxygen from day and night at these heights. Now, atomic oxygen is a result of photodissociation of O_3 and O_2 both occurring in the ultraviolet, and conse-quently it is clear that if the detachment is controlled by atomic oxygen, the onset of the twilight effect should occur at $\chi = 90°$ as observed.

Even this cannot, however, be the full story. Firstly if there is any large scale formation of atomic oxygen from O_2 and O_3 by energetic particles, the expected day-night modulation in atomic oxygen is affected. Secondly theoretical calculations and some empirical deductions show that the photodetachment from the heavier negative ions is about 10^{-2} s^{-1} (γ_2 in Figure 103 and γ_2 to γ_9 in Figure 105) above 70 km and possibly as large as 0.1–1 s^{-1} around 60 km. This means that as the atomic oxygen concentration decreases with decreasing altitude, the atomic oxygen detach-ment processes are eventually replaced by photodetachment or other detachment processes from the heavy negative ions. The onset times, however, will, in all these cases occur near $\chi = 90°$. Observation of the twilight onset time alone cannot, there-fore, throw any further light on the relative importances of these processes. At higher altitudes, where the rocket observations have established negative ions to be in their basic forms of O^- and O_2^-, one would expect a change in the twilight onset pattern. Some evidence exists showing that this does in fact occur. There is, for example, the example quoted by Reid (1970b) giving the ratio of absorption observed at Dumont d'Urville, Antarctica, to that at South Pole which was in continuous darkness during the period May 24–25, 1967. Dumont d'Urville experienced a brief period of daytime absorption, which began and ended abruptly at $\chi \approx 99°$. The average ratio of absorpt-ion was closer to unity for $\chi > 99°$ both *before* sunrise and *after* sunset. For this zenith angle the visible-light shadow lay at about 80 km, and the ultraviolet shadow at about 110 km. Now that rockets have established the dominance of O^- and O_2^- at these heights, this result assumes new significance.

As we have indicated before, the chemistry of the PCA must not be considered in isolation. Any ion chemical scheme that we use should consider the changes that would occur when the normal electron production rate is *increased* (as in PCAs and solar flares) as well as when it is *decreased* (as during solar eclipses and at night). Or, in other words, one must provide a *unified* picture of the chemistry of the *normal* and the *disturbed D* region. Such a picture has recently been given by Mitra (1974), using the six-ion scheme of Mitra and Rowe given in Figure 103. This, as pointed out before, eliminates the uncertainties in the conversion of NO^+ to hydrated ions. He examined the changes that occur in the D region as we go to different types and degrees of disturbances. Mitra classified the D region conditions into two groups:

A. First group in which only: Quiet day, PCA day, Solar flares.
   $q(O_2^+)$ changes
B. Second group in which: Quiet night, Eclipse, PCA night.
   $q(O_2^+)$ changes and
   detachment processes are
   reduced or vanish

The parameters we will consider are: N_e, $[\Sigma H^+ \cdot (H_2O)_n]/N^+$, the concentrations of $H^+ \cdot (H_2O)_n$ and NO^+, and $[X^-]/[O_2^-]$.

In Figure 136c are given in dotted curves the theoretical altitude profiles of the quantity $[\Sigma H^+ \cdot (H_2O)_n]/[NO^+]$ derived by Mitra separately for quiet conditions, for the totality of an eclipse, for a strong flare and for PCA day and night conditions. These can be compared with the experimental values given by the solid ranges and the shaded curves. The calculations show that the cluster disappearance level is lowered during disturbed conditions and raised during conditions of $\gamma_2 = \gamma_1 = 0$. The theoretical values for these levels are:

Quiet night:	85 km
Eclipse totality:	84 km
Quiet day:	81 km
PCA night	76 km
Strong flare:	74 km
PCA day:	72 km.

Below 70 km, the dominance of water cluster ions is not visibly affected. The concentration of the total water cluster ions at 80 km decreases from its quiet day value of 60% to 10% during a strong flare and about 4% during the PCA day. The value increases to 90% during an eclipse but is still considerably depressed during PCA night (only 6%). At 70 km, the reduction is from 95% to 70% during PCA day and there is hardly any change during a flare. At night, there is complete dominance of cluster ions, its percentage composition approaching 100%.

Another way of looking at the problem is to concentrate at any one height and see what happens as we increase the O_2^+ production rate. In Figure 140 we have given the ratio $[\Sigma H^+ \cdot (H_2O)_n]/N^+$ as a function of $q(O_2^+)$ for the heights of 60, 70, 75, 80 and 85 km. The situations during normal conditions, a moderate solar flare, an outstanding solar flare and PCA day are marked on these curves by N, SF, OF and PCA respectively. For any selected level, the degree of destruction of the water cluster ions is in proportion to the O_2^+ production rate (i.e. to the ionizing capabilities of the flares

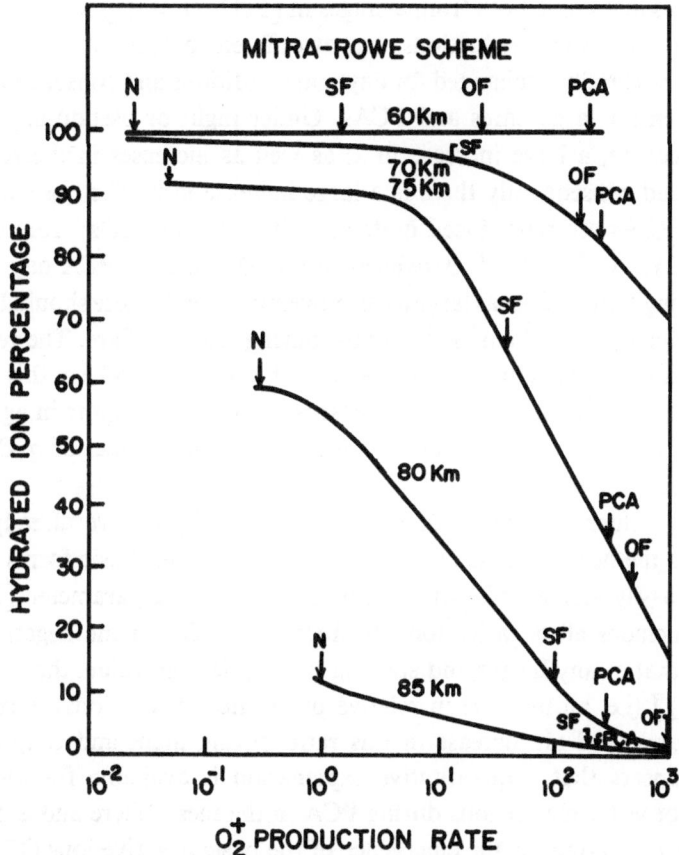

Fig. 140. Theoretical variation in hydrated cluster percentage with O_2^+ production rate. Note the large decrease in the percentage at heights below 70 km with increasing values of $q(O_2^+)$. N refers to normal condition, SF to the condition during a strong flare, OF that during an outstanding flare and PCA that during a polar cap absorption event (daytime).

or the PCAs). We note that at 60 km, there is no change in cluster composition even for very large disturbance conditions. At 70 km, measurable change in cluster composition occurs only if $q(O_2^+)$ at this height exceeds 10^1 cm^{-3} s^{-1}. At 75 km there is a rapid fall of the relative cluster composition for $q(O_2^+) > 3 \times 10^1$ cm^{-3} s^{-1}; this fall continues even when $q(O_2^+) = 10^3$ cm^{-3} s^{-1}. The fall is from a normal value of about 90% to as low as 15% for $q(O_2^+) = 10^3$ cm^{-3} s^{-1}. At 80 and 85 km, there are similar rapid and drastic decreases, but now there is a tendency of the reduction levelling off for values of $q(O_2^+) \geqslant 10^3$ cm^{-3} s^{-1}.

Consider now the parameter λ, the negative ion to electron ratio. The changes here are not as evident as in the cluster ion population. There is, however, a tendency of decrease in λ, at heights between 60 and 75 km during PCAs and flares by about a factor of 2 at 70 km and only a factor of 1.14 at 60 km. The values during PCA night show some reduction from quiet night values – although experimental results do not show any appreciable change.

The changes in ψ come both from changes in $[\Sigma H^+ \cdot (H_2O)_n]/N^+$ (changing $\bar{\alpha}_D$) and changes in λ. Both show reductions, but at different heights. At and below 60 km, both remain essentially unchanged for daylight conditions and consequently ψ remains unaltered even for large flares and PCAs. Under night or pseudo-night conditions. there is, of course, a large increase in λ, as well as increases in the relative cluster population and consequently there is a large increase in ψ. The observed reductions in ψ during PCAs are reproduced quite as well as during SIDs; from the quiet day value of $(1-2) \times 10^{-5}$ cm^3 s^{-1}, it reduces to 9×10^{-7} during PCA day and 7×10^{-6} during PCA night (for PCAs as large as of November 1969); these should be considered with a value of 1.6×10^{-6} cm^3 s^{-1} reached during a strong flare. The reduction in ψ levels off to a constant value for large values of $q(O_2^+)$: for 80 km this value should exceed 10^2 cm^{-3} s^{-1}. The mode of decrease is, however, not quite in agreement with observations which give a rapid fall at considerably lower values of $q(O_2^+)$ and then level off (see Figure 90).

We now consider the nature of the negative ions. We have already stated that during PCAs the negative ions tend to revert back to their basic forms O_2^- and O^-. This can be easily verified with the scheme by plotting the parameter $[X^-]/[O_2^-]$ in which X^- includes all negative ions produced from O_2^- in the negative ion chain. Mitra finds that at any height and starting from quiet day value, there is a *reduction* in $[X^-]/[O_2^-]$ (i.e. an increase in relative proportion of O_2^- ions) during PCA day and solar flares, and an increase in this ratio during night and eclipse conditions.

It thus appears that a quantitative explanation is available for the progressive destruction of water cluster ions during PCA in the mesosphere and a plausible, but qualitative, explanation for the emergence of the basic negative ions O^- and O_2^- and of the twilight anomaly. The smaller electron density at night is also understood in quantitative terms being a result of increase in recombination coefficient at night, firstly from the larger relative concentration of cluster ions at night, and secondly from the increase in λ. The need to extend the positive ion composition measurements below 70 km is felt not merely for PCA but also for quiet time – the expectation that at these heights the dominance of cluster ions is not largely affected must be checked. An important point concerns the fact that calculated from the ion composition data seem to break off from q/N_e^2 curve at heights below 75 km if one assumes that below this height $[\Sigma H^+ \cdot (H_2O)_n]/[NO^+]$ rapidly increases in magnitude. The trend in Figure 136c justifies this. The indication then would be that either cluster ions do not dominate even at these heights or that the production rate calculations from the observed proton fluxes are in error for these heights. It seems important that the different approaches currently attempted for q-calculations for PCAs are looked into thoroughly.

REFERENCES

Acton, L. W., Chubb, T. A., Kreplin, R. W., and Meekins, J.: 1963, *J. Geophys. Res.* **68**, 3335.
Adams, G. W. and Masley, A. J.: 1965, *J. Atmospheric Terrest. Phys.* **27**, 289.
Adams, G. W. and Masley, A. J.: 1966, *Planetary Space Sci.* **14**, 277.
Agy, V., Baker, D. M., and Jones, R. M.: 1965, NBS Tech. Note, No. 306.
Albee, P. R. and Bates, H. F.: 1965, *Planetary Space Sci.* **13**, 175.
Allan, A. H., Crombie, D. D., and Penton, W. A.: 1957, *J. Atmospheric Terrest. Phys.* **10**, 110.
Ananthakrishnan, S., Abdu, M. A., and Piazza, L. R.: 1973, *Planetary Space Sci.* **21**, 367.
Appleton, E. V. and Piggott, W. R.: 1954, *J. Atmospheric Terrest. Phys.* **5**, 141.
Arnold, F. and Krankowsky, D.: 1971, *J. Atmospheric Terrest. Phys.* **33**, 1693.
Arnoldy, R. L., Kane, S. R., and Winckler, J. R.: 1968, *Astrophys. J.* **151**, 711.
Bailey, D. K.: 1957, *J. Geophys. Res.* **62**, 431.
Bailey, D. K.: 1959, *Proc. Inst. Radio Engrs.* **47**, 255.
Bailey, D. K. and Harrington, J. M.: 1962, *J. Phys. Soc. Japan* **17**, Suppl. A-II, 334.
Bain, W. C.: 1953, *J. Atmospheric Terrest. Phys.* **3**, 141.
Bain, W. C. and Hammond, E.: 1974, private communication.
Bain, W. C. and Harrison, M. D.: 1972, *Proc. Inst. Elec. Engrs. London* **119**, 790.
Bain, W. C., Bracewell, R. N., Straker, T. W., and Westcott, C. H.: 1952, *Proc. Inst. Elec. Engrs. London*, Monograph No. 37.
Baker, D. C.: 1969, Sci. Rep. 334, Ionospheric Research Laboratory, Pennsylvania State University.
Baker, D. M., Chang, N., Davies, K., Donnelly, R. F., and Jones, J. E.: 1968, ESSA Tech. Rep. ERL 78-SDL 1.
Barletti, R. and Tagliaferri, G. L.: 1969, *J. Atmospheric Terrest. Phys.* **31**, 631.
Barletti, R. and Tagliaferri, G. L.: 1972, *J. Atmospheric Terrest. Phys.* **34**, 543.
Barth, C. A.: 1966, *Ann. Geophys.* **22**, 198.
Belrose, J. S.: 1957, Ph.D. Thesis, Cambridge University.
Belrose, J. S.: 1964, in W. T. Blackband (ed.), *Propagation of Radio Waves at Frequencies Below 300 kHz*, Pergamon Press, p. 149.
Belrose, J. S.: 1969, *Third Aeronomy Conf. Meteorological and Chemical Factors in D-Region*, Aeronomy Rept. No. 32, University of Illinois, p. 375.
Belrose, J. S.: 1970, *J. Atmospheric Terrest. Phys.* **32** 567.
Belrose, J. S.: 1972, AFCRL-72-0474, *Proc. COSPAR Symp. Solar Particle Event of November 1969*, Air Force Cambridge Research Laboratories, U.S. Air Force, p. 243.
Belrose, J. S. and Burke, M. J.: 1964, *J. Geophys. Res.* **69**, 2799.
Belrose, J. S. and Cetiner, E.: 1962, *Nature* **195**, 688.
Belrose, J. S. and Ross, D. B.: 1961, *Can. J. Phys.* **39**, 609.
Belrose, J. S., Devenport, M. H., and Weekes, K.: 1956, *J. Atmospheric Terrest. Phys.* **8**, 281.
Belrose, J. S., Bourne, I. A., and Hewitt, L. W. P.: 1967, in *Proc. Conf. on Groundbased Radiowave Propagation Studies of the Lower Ionosphere*, Defence Research Board, Ottawa, p. 125.
Belrose, J. S., Burke, M. J., and Coyne, T. N. R.: 1972, *J. Geophys. Res.* **77**, 4829.
Bennett, J. A.: 1967, *J. Atmospheric Terrest. Phys.* **29**, 887.
Best, A., Wernik, A., Gal'perin, Yu. I., Grafe, A., Knuth, R., Lashtovicka, Ya., Lippert, W., Mularchik, T. M., Nestorov, G., Samardjiev, D., Serafimov, K., Triska, P., Shuyskaya, F. K., and Entzian, G.: 1970, paper presented at *COSPAR Assembly*, Leningrad.
Bhattacharyya, J. C. and Balakrishnan, T. K.: 1967, *J. Atmospheric Terrest. Phys.* **29**, 1573.
Bibl, K.: 1951, *Ann. Geophys.* **7**, 208.
Biondi, M., Leu, M. T., and Johnson, R.: 1972, *COSPAR Symposium on D and E-region Chemistry*, Illinois, p. 266.

Blair, J. C.: 1959, NBS Rept. No. 6049.

Blocker, N. K., Chambers, W. H., Fehlau, P. E., Fuller, J. C., Kunz, W. E., and Milkey, R. W.: 1971, Los Alamos Scientific Laboratory, Rept. LA-4454, Vol. III.

Bracewell, R. N. and Straker, T. W.: 1949, *Monthly Notices Roy. Astron. Soc.* **109**, 28.

Bracewell, R. N., Budden, K. G., Ratcliffe, J. A., Straker, T. W., and Weekes, K.: 1951, *Proc. Inst. Elec. Engrs. London* **98**, 221 (Part III).

Brasseur, G. and Nicolet, M.: 1973, *Planetary Space Sci.* **21**, 925.

Briggs, B. H.: 1951, *J. Atmospheric Terrest. Phys.* **1**, 345.

Budden, K. G. and Ratcliffe, J. A.: 1937, *Nature* **140**, 1060.

Budden, K. G.: 1961, *The Waveguide Mode Theory of Wave Propagation*, Logos Press, London.

Bureau, R.: 1937, *Nature* **139**, 110.

Bureau, R.: 1947, *L'Onde Electrique*, No. 239.

Burgess, B. and Jones, T. B.: 1967, *Radio Sci.* **2**, 619.

Castelli, J. P., Barron, W. R., and Aarons, J.: 1973, AFCRL-TR-73-0086, Air Force Cambridge Research Laboratories, Bedford, U.S.A.

Castelli, John P., Aarons, J., and Michael, G. A., AFCRL Report 67-0622, November 1967.

Chakrabarty, D. K.: 1973, *Study of the D-Region of the Ionosphere During Quiet Times and During Solar Eclipses*, Ph.D. Thesis, Delhi University.

Chakrabarty, D. K., Jain, V. C., and Saha, A. K.: 1968, Scientific Report No. 40, Radio Propagation Unit, N.P.L., New Delhi.

Chakrabarty, D. K. and Mitra, A. P.: 1974, *Indian J. Radio Space Phys.* **3**, 76.

Chakrabarty, P.: 1974, Ph.D. Thesis, Delhi University.

Chilton, C. J., Steele, F. K., and Norton, R. B.: 1963, *J. Geophys. Res.* **68**, 5421.

Chilton, C. J., Steele, F. K., and Crombie, D. K.: 1964, *An Atlas of Solar Flare Effects Observed on Long VLF Path During 1961*, NBS Tech. Note No. 210, U.S.A.

Chilton, C. J., Conner, J. P., and Steele, F. K.: 1965, *Proc. Inst. Elec. Electron. Engrs.* **53**, 2018.

Chivers, H. J. A. and Hargreaves, J. K.: 1965, in B. Maehlum (ed.), *High Latitude Particles and the Ionosphere*, Logos Press, London, p. 257.

Conley, T. D.: 1972, *Proc. COSPAR Symp. on Solar Particle Event of Nov. 1969*, AFCRL-72-0474, Air Force Systems Command, U.S.A., p. 359.

Crombie, D. D.: 1965, *Proc. Inst. Elec. Electron. Engrs.* **53**, 2027.

Crombie, D. D.: 1966, in J. Frihagen (ed.), *Electron Density Profiles in Ionospheres and Exophere*, North-Holland Publishing Co., p. 118.

Croom, D. L.: 1970a, *Astrophys. Letters* **7**, 133.

Croom, D. L.: 1970b, *J. Geophys. Res.* **75**, 6940.

Croom, D. L.: 1971a, *Solar Phys.* **19**, 152.

Croom, D. L.: 1971b, *Solar Phys.* **19**, 171.

Croom, D. L.: 1973, *Planetary Space Sci.* **21**, 707.

Culhane, J. L.: 1969, *Monthly Notices Roy. Astron. Soc.* **144**, 375.

Culhane, J. L., Sanford, P. W., Shaw, M. L., Phillips, K. J. H., Willmore, A. P., Bowen, J. P., Pounds, K. A., and Smith, D. G.: 1969, *Monthly Notices Roy. Astron. Soc.* **145**, 435.

Culhane, J. L., Willmore, A. P., Pounds, K. A., and Sanford, P. W.: 1964, *Space Research IV*, North Holland Publishing Co., Amsterdam, p. 741.

Dasgupta, M. K. and Sarkar, S. K.: 1972, *Indian J. Radio Space Phys.* **1**, 256.

Dasgupta, M. K., Mitra, R. K., and Sarkar, S. K.: 1973, *J. Atmospheric Terrest. Phys.* **35**, 805.

Deshpande, S. D.: 1971, Ph.D. Thesis, Delhi University.

Deshpande, S. D. and Mitra, A. P.: 1972a, *J. Atmospheric Terrest. Phys.* **34**, 229.

Deshpande, S. D. and Mitra, A. P.: 1972b, *J. Atmospheric Terrest. Phys.* **34**, 243.

Deshpande, S. D. and Mitra, A. P.: 1972c, *J. Atmospheric Terrest. Phys.* **34**, 255.

Deshpande, S. D., Subrahmanyam, C. V., and Mitra, A. P.: 1972a, *J. Atmospheric Terrest. Phys.* **34**, 211.

Deshpande, S. D., Ganguly, S., Jain, V. C., and Mitra, A. P.: 1972b, *J. Atmospheric Terrest. Phys.* **34**, 267.

Dieminger, W. and Geisweid, K. H.: 1951, *J. Atmospheric Terrest. Phys.* **1**, 42.

Dodson, Helen, W. and Hedeman, E. Ruth, WDCA Report UAG-14, *An Experimental Comprehensive Flare Index and its Derivation for 'Major' Flares*, 1955–1969, July, 1971.

Donnelly, R. F.: 1968a, *Solar Phys.* **5**, 123.

Donnelly, R. F.: 1968b, ESSA Tech. Rept. ERL81-SDL2, ESSA Boulder, Colorado.
Donnelly, R. F.: 1968c, ESSA Tech. Rept. ERL86-SDL3, ESSA Boulder, Colorado.
Donnelly, R. F.: 1968d, ESSA Tech. Rept. ERL92-SDL6, ESSA Boulder, Colorado.
Donnelly, R. F.: 1969a, *J. Geophys. Res.* **74**, 1873.
Donnelly, R. F.: 1969b, ESSA Tech. Rept. ERL95-SDL7, ESSA Boulder, Colorado.
Donnelly, R. F.: 1969c, WDC-A, Upper Atmosphere Geophysics, Report UAG-5, Boulder, Colorado, p. 34.
Donnelly, R. F.: 1970, ESSA Tech. Rept. ERL169 – SDL14, ESSA Boulder, Colorado.
Donnelly, R. F.: 1971, *Solar Phys.* **20**, 188.
Donnelly, R. F.: 1973, in R. Ramaty and R. G. Stone (eds.), *Conference Proceedings on High Energy Phenomena on the Sun*, NASA X-693-73-193, p. 242.
Donnelly, R. F. and Hall, L. A.: 1973, *Solar Phys.* **31**, 411.
Donnelly, R. F. and Pope, J. H.: 1973, NOAA Tech. Rept. ERL276 – SEL25.
Donnelly, R. F., Wood, A. T., Jr., and Noyes, R. W.: 1973, *Solar Phys.* **29**, 107.
Drake, J. F., Sr., Gibson, J., and Van Allen, J. A.: 1968, *Iowa Catalog of Solar X-ray Intensity*, Iowa University.
Egan, R. D.: 1963, in G. J. Gassman (ed.), *The Effect of Disturbances of Solar Origin on Communication*, p. 47.
Ellison, M. A.: 1950, *Publ. Roy. Obs. Edinburgh* **1**, 53.
Ellison, M. A.: 1953, *J. Atmospheric Terrest. Phys.* **4**, 226.
Entzian, G.: 1964, *Vortragen der Sommerschule*, Kuhlungsborn/Heiligendaum.
Evans, W. F. J., Hunten, D. M., Llewellyn, E. J., and Vallance Jones, A.: 1968, *J. Geophys. Res.* **73**, 2885.
Fejer, J. A.: 1955, *J. Atmospheric Terrest. Phys.* **7**, 322.
Ferraro, A. J. and Lee, H. S.: 1967, *Groundbased Radiowave Propagation Studies of the Lower Ionosphere*, Conference Proceedings, Ottawa, 1967, p. 281.
Ferraro, A. J. and Lee, H. S.: 1968, *J. Geophys. Res.* **73**, 4427.
Ferraro, A. J., Lee, H. S., Rowe, J. N., and Mitra, A. P.: 1974, *J. Atmospheric Terrest. Phys.* **36**, 741.
Field, E. C.: 1970, *J. Geophys. Res.* **75**, 1927.
Findlay, J. W.: 1951, *J. Atmospheric Terrest. Phys.* **1**, 367.
Ganguly, S.: 1972, *J. Atmospheric Terrest. Phys.* **34**, 2009.
Gardner, F. F. and Pawlsey, J. L.: 1953, *J. Atmospheric Terrest. Phys.* **3**, 321.
Garriott, O. K., da Rosa, A. V., Davis, M. J., and Villard, O. G., Jr.: 1967, *J. Geophys. Res.* **72**, 6099.
Garriott, O. K., da Rosa, A. V., Davis, M. J., Wagner, L. S., and Thome, G. D.: 1969, *Solar Phys.* **8**, 226.
Gregory, J. B.: 1963, *J. Geophys. Res.* **68**, 3097.
Hachenberg, O. and Kruger, A.: 1959, *J. Atmospheric Terrest. Phys.* **17**, 20.
Heimerl, J. L., Venderhoff, J. A., Puckett, L. J., and Niles, F. E.: 1972, BRL Report No. 7570, Ballistic Research Laboratory, U.S.A.
Hakura, Y.: 1961, *Rep. Ionospheric Space Res. Japan* **15**, 1.
Hakura, Y.: 1966, *Rep. Ionospheric Space Res. Japan* **20**, 30.
Hakura, Y.: 1967a, Goddard Space Flight Centre, Report No. X-641-67-116.
Hakura, Y.: 1967b, Goddard Space Flight Center, Report X-640-67-258, June 1967.
Hakura, Y., Ohsiho, M., Yamashita, F., Énomé, S., Tanaka, H., Nagai, M., and Nakajima, H., *IAGA Assembly*, Kyoto, 1973.
Hale, L. C., Mentzer, J. R., and Nickell, L. C.: 1972, *Proc. COSPAR Symp. Solar Particle Event of Nov. 1969*, AFCRL-72-0474, p. 333.
Hale, L. C.: 1973, *COSPAR Assembly*, Konstanz.
Hall, L. A.: 1971, *Solar Phys.* **21**, 167.
Hall, L. A. and Hinteregger, H. E.: 1969, *Solar Flares and Space Research*, North-Holland, Amsterdam, p. 81.
Hall, L. A. and Hinteregger, H. E.: 1970, *J. Geophys. Res.* **75**, 6959.
Haubert, A.: 1959, *J. Atmospheric Terrest. Phys.* **13**, 379.
Haug, A. and Landmark, B.: 1970, *J. Atmospheric Terrest. Phys.* **32**, 405.
Helms, W. J. and Swarm, H. M.: 1969, *J. Geophys. Res.* **74**, 6341.
Hesstvedt, E.: 1969, in Willis L. Webb (ed.), *Stratospheric Circulation*, Academic Press, London, p.307.
Hinteregger, H. E.: 1961, *J. Geophys. Res.* **66**, 2367.

Hinteregger, H. E.: 1969, in *Annals of the IQSY*, MIT Press, Cambridge, Mass. **5**, 305.
Hinteregger, H. E. and Hall, L. A.: 1969, *Solar Phys.* **6**, 175.
Horan, D. M.: 1970, Ph. D. Thesis, Catholic University.
Hultqvist, B.: 1966a, *Space Sci. Rev.* **5**, 771.
Hultqvist, B.: 1966b, *Ann. Geophys.* **22**, 235.
Hultqvist, B. and Ortner, J.: 1959, *Planetary Space Sci.* **1**, 193.
Houston Jr. R. E.: 1957, Scientific Report No. 95, Ionosphere Research Laboratory, Pennsylvania State University, July 1.
Jacchia, L. G.: 1970, Smithsonian Astrophysical Observatory Special Report No. 313.
Jacchia, L. G.: 1971, Smithsonian Astrophysical Observatory Special Report No. 332.
Jean A. Glen., and Large, D. B.: 1967, ITSA Technical Memorandum No. 90, Institute for Telecommunicate Sciences and Aeronomy, Boulder.
Jones, R. E.: 1955, *J. Atmospheric Terrest. Phys.* **6**, 1.
Jones, R. M.: 1966, ESSA Tech. Rept IER17-ITSA 17.
Jones, T. B.: 1971, *J. Atmospheric Terrest. Phys.* **33**, 963.
Johnson, C. Y.: 1969, *Annals of the IQSY* V, p. 197.
Johnson, F. S., Purcell, J. D., and Tousey, R.: 1954, in R. Boyd and M. J. Sealon (ed.), *Rocket Exploration of the Upper Atmosphere*, Pergamon Press, p. 189.
Kamada, T.: 1961, *Proc. Res. Inst. Atmosphere.*, Nagoya University (Japan), **8**, 7.
Kaufmann, P., Schaal, R. E., Lopes, W., and Arakaki, L.: 1967, *J. Atmospheric Terrest. Phys.* **29**, 1443.
Kaufmann, P. and Pages de Barros, M. M.: 1969, *Solar Phys.* **9**, 478.
Kaufmann, P. and Mendes, A.: 1970, *J. Atmospheric Terrest. Phys.* **32**, 427.
Kebarle, P., Searles, S. K., Zolla, A., Scarborough, J., and Arshadi, M.: 1967, *J. Am. Chem. Soc.* **89**, 6393.
Kelly, P. T. and Rense, W. A.: 1972, *Solar Phys.* **26**, 431.
Keneshea, T. I., Narcisi, R. S., and Swider W. Jr., 1970, *J. Geophys. Res.* **75**, 845.
Kreplin, R. W., Chubb, T. A., and Friedman, H.: 1962, *J. Geophys. Res.* **67**, 2231.
Kreplin, R. W., Horan, D. M., Chubb, T. A., and Friedman, H.: 1968, paper presented at the *XI COSPAR Meeting*, Tokyo, Japan.
Kreplin, R. W., Horan, D. M., Chubb, T. A., and Friedman, H.: 1969, in C. de Jager and Z. Švestka (eds.), *Solar Flares and Space Research*, North-Holland Publishing Co., Amsterdam, p. 121.
Kreplin, R. W., Moser, P. J., and Castelli, J. P.: 1970, *Space Research X*, North-Holland Publishing Co., Amsterdam, p. 920.
Landini, M., Russo, D., and Tagliaferri, G. L.: 1966, *Space Research VI*, Spartan Book Washington, p. 1041.
Lee, H. S. and Ferraro, A. J.: 1969, *J. Geophys. Res.* **74**, 1184.
Lepine, D. and Hall, J. E.: 1972, *J. Atmospheric Terrest. Phys.* **34**, 1507.
Letfus, V., Apostolov, E. M., and Nestorov, G.: 1973, *J. Atmospheric Terrest. Phys.* **35**, 571.
Lindsay, J. C.: 1964, *Planetary Space Sci.* **12**, 379.
Little, C. G. and Leinbach, H.: 1959, *Proc. Inst. Radio Engrs.* **47**, 315.
Mandel'štam, S. L.: 1965, *Space Sci. Rev.* **4**, 587.
Manson, J. E.: 1972, *Solar Phys.* **27**, 107.
Matsoukas, D. A., Papagiannis, M. D., Aarons, J., and Klobuchar, J. A.: 1972, *J. Atmospheric Terrest. Phys.* **34**, 1275.
May, B. R.: 1966, *J. Atmospheric Terrest. Phys.* **28**, 553.
McIntosh, D. H.: 1951, *J. Atmospheric Terrest. Phys.* **1**, 315.
McNish, A.G.: 1937, *Terrest. Magn.* **42**, 107.
Mechtly, E. A., Bowhill, S. A., and Smith, L. G.: 1972, *J. Atmospheric Terrest. Phys.* **34**, 1899.
Meekins, J. F., Kreplin, R. W., Chubb, T. A., and Friedman, H., 1968, *Science* **162**, 891.
Meira, L. G.: 1971, *J. Geophys. Res.* **76**, 202.
Mendillo, M. and Evans, John, V.: 1974, *Radio Sci.* **9**, 197.
Mendillo, M., Klobuchar, J. A., Fritz, R. B., da Rosa, A. V., Kersley, L., Yeh, K. C., Flaherty, B. J., Rangaswamy, S., Schmid, P. E., Evans, J. V., Schödel, J. P., Matsoukas, D. A., Koster, J. R., Webster, A. R., and Chin, P.: 1974, *J. Geophys. Res.* **79**, 665.
Minnis, C. M. and Buzzard, G. H.: 1958, *Nature* **181**, 690.
Mitra, A. P.: 1954, Pennsylvania State University, Sci. Rept. No. 60, March 30.
Mitra, A. P.: 1958, Ionosphere Research Laboratory, Pennsylvania State University, Sci. Rept. No. 112.

Mitra, A. P.: 1966, *Space Research VI*, Spartan Press, p. 558.

Mitra, A. P.: 1968, *J. Atmospheric Terrest. Phys.* **30**, 1065.

Mitra, A. P.: 1969, *Third Aeronomy Conf. Meteorological and Chemical Factors in D-Region*, Aeronomy Rept. No. 32, University of Illinois, 174.

Mitra, A. P.: 1970, in E. R. Dyer (ed.) *Solar Terrestrial Physics*, Par. IV, p. 1.

Mitra, A. P.: 1971, *The Chemistry of the Ionosphere*, Radio Propagation Unit, National Physical Laboratory, New Delhi.

Mitra, A. P.: 1973, Symposium on Incoherent Scatter, Tromso.

Mitra, A. P.: 1974, Solar-Terrestrial Physics International Symposium, Brazil.

Mitra, A. P. and Banerjee, P.: 1971, *Space Research XI*, Akademic-Verlag, GDR, 1019.

Mitra, A. P. and Deshpande, S. D.: 1972, in S. A. Bowhill, L. D. Jaffe, and M. J. Rycroft (eds.), *Space Research XII*, Akademic-Verlag, Berlin, p. 1291.

Mitra, A. P. and Jones, R. E.: 1954, *J. Atmospheric Terrest. Phys.* **5**, 104.

Mitra, A. P. and Rowe, J. N.: 1972, *J. Atmospheric Terrest. Phys.* **34**, 795.

Mitra, A. P. and Rowe, J. N.: 1973, *IAGA Assembly*, Kyoto.

Mitra, A. P. and Shain, C. A.: 1953, *J. Atmospheric Terrest. Phys.* **4**, 1695.

Mitra, A. P. and Subrahmanyam, C. V.: 1966, *Indian Observations for the IQSY*, N.P.L. New Delhi, p. 194.

Mitra, A. P., Subrahmanyam, C. V., and Jain, V. C.: 1966, *Radio Sci.* **1** (New Series), 1188.

Mitra, A. P., Subrahmanyam, C. V., and Karabin Mirjana: 1964, *J. Atmospheric Terrest. Phys.* **26**, 1138.

Mitra, S. N.: 1964, *J. Atmospheric. Terrest. Phys.* **26**, 375.

Mogel, H.: 1930, *Telefunken-Zeitung* **11**, 14.

Montbriand, L. E.: 1973, Communications Research Center Report No. 1241, Dept. of Communications, Ottawa.

Montbriand, L. E. and Belrose, J. S.: 1972, *Radio Sci.* **7**, 133.

Morriss, R. W.: 1960, *Proc. Phys. Soc.* **76**, 79.

Narcisi, R. S.: 1972a, Aeronomy Rept. No. 48, *COSPAR Symposium on D and E-region Ion Chemistry*, June 1, 1972, p. 182.

Narcisi, R. S.: 1972b, Aeronomy Rept. No. 48, *COSPAR Symposium on D and E-region Ion Chemistry*, June 1, 1972, p. 221.

Narcisi, R. S.: 1972c, *Proc. COSPAR Symp. Solar Particle Event of Nov. 1969*, AFCRL-72-0474, p. 557.

Narcisi, R. S., Philbrick, C. R., Bailey, A. D., and Della Lucca, L.: 1969, Aeronomy Rept. No. 32, University of Illinois, U.S.A., p. 355.

Narcisi, R. S., Bailey, A. D., Della Lucca, L., Sherman, C., and Thomas, D. M.: 1971, *J. Atmospheric Terrest. Phys.* **33**, 1147.

Narcisi, R. S., Philbrick, C. R., Thomas. D. M., Bailey, A. D., Wlodyka, L. E., Baker, D., Federico, G., Wlodyka, R., and Gardner, M. E.: 1972a, AFCRL-72-0474, *Proc. COSPAR Symp. Solar Particle Event of November 1969*, Air Force Cambridge Research Laboratories, U.S. Air Force, p. 421.

Narcisi, R. S. Sherman, C., Philbrick, C. R., Thomas, D. M., Bailey, A. D., Wlodyka, L. E., Wlodyka, R. A., Baker, D., and Federico, G.: 1972b, AFCRL-72-0474, *Proc. COSPAR Symp. Solar Particle Event of November 1969*, Air Force Cambridge Research Laboratories, U.S. Air Force, p. 411.

Narcisi, R. S., Philbrick, C. R., Ulwick, J. C., and Gardner, M. E.: 1972c, *J. Geophys. Res.* **77**, 1332.

Narcisi, R. S., Bailey, A. D., Wlodyka, L. E., and Philbrick, C. R.: 1972d, *J. Atmospheric Terrest. Phys.* **34**, 647.

Nestorov, G.: 1971, *J. Atmospheric Terrest. Phys.* **33**, 119.

Nestorov, G. and Krivsky, L.: 1967, *Bull. Astron. Inst. Czech.* **18**, 143.

Neupert, W. M.: 1964, in W. N. Hess (ed.), *AAS-NASA Symposium on the Physics of Solar Flares*, NASA SP-50, p. 49.

Neupert, W. M.: 1967, *Solar Phys.* **2**, 294.

Neupert, W. M., Gates, W., Swartz, M., and Young, R.: 1967, *Astrophys. J.* **149**, L79.

Nicolet, M. and Aikin, C. A.: 1960, *J. Geophys. Res.* **65**, 1469.

Obayashi, T.: 1964, *Space Sci. Rev.* **3**, 79.

Ogawa, T.: 1970, in K. Davies (ed.), *Phase and Frequency Stabilities in Electromagnetic Wave Propagation*, Technivision, Slough, England, p. 478.

Ohle, K.-H., Knuth, R., Entzian, G., and Taubenheim, J.: 1974, *J. Atmospheric Terrest. Phys.* **36**, 513.

Ohshio, M.: 1964, *J. Radio Res. Lab. Japan* **11**, 377.

Ohshio, M.: 1968, *J. Radio Res. Lab. Japan* **15**, 307.

Ohshio, M.: 1971, *Nature* **229**, 239.

Ohshio, M., Maeda, R., and Sakagami, H.: 1966, *J. Radio Res. Lab. Japan* **13**, 245.

Ondoh, T. and Kishida, H.: 1966, *Rep. Ionosphere Space Res. Japan* **20**, 542.

Parthasarathy, R. and Venkentesan, D.: 1964, *J. Geophys. Res.* **69**, 549.

Parthasarathy, R., Lerfald, G. M., and Little, C. G.: 1963, *J. Geophys. Res.* **68**, 3581.

Park, J.: 1972, Ph.D. Thesis, University of Colorado.

Pearce, J. B.: 1969, *J. Geophys. Res.* **74**, 853.

Perona, G. E.: 1974, *J. Atmospheric Terrest. Phys.* **36**, 897.

Piggott, W. R., Pitteway, M. L. V., and Thrane, E. V.: 1965, *Phil. Trans. Roy. Soc.* **A257**, 243

Pontano, B. A.: 1970, Ionosphere Research Laboratory, Pennsylvania State University, Scientific Report No. 347.

Ramanamurthy, Y. V.: 1970, *Indian J. Pure Appl. Phys.* **8**, 569.

Reid, J. H.: 1969, *J. Atmospheric Terrest. Phys.* **31**, 859.

Reid, G. C.: 1970a, *J. Geophys. Res.* **75**, 2551.

Reid, G. C.: 1970b, in V. Manno and D. E. Page (eds.), *Intercorrelated Satellite Observations Related to Solar Events*, D. Reidel, Publishing Co., Holland, p. 319.

Reid, G. C.: 1972a, *Proc. COSPAR Symp. on the Nov. 1969 Solar Particle Event*, AFCRL-72-0474, p. 201.

Reid, G. C.: 1972b, in Kristen Folkestad (ed.), *Magnetosphere-Ionosphere Interactions*, Universitets-fortaget, Oslo, 39.

Richards, D. W.: 1971, Report-AFCRL-71-0392, Air Force Cambridge Res. Lab, Environmental Research Papers No. 363, AFCRL, U.S.A.

Ross, W. J.: 1960, *J. Geophys. Res.* **65**, 2601.

Rowe, J. N.: 1972, Ph.D. Thesis, IRL Rept. No. 406, Pennsylvania State University.

Rowe, J. N., Ferraro, A. J., Lee, H. S., Kreplin, R. W., and Mitra, A. P.: 1970, *J. Atmospheric Terrest. Phys.* **32**, 1609.

Rowe, J. N., Mitra, A. P., Ferraro, A. J., and Lee, H. S.: 1974, *J. Atmospheric Terrest. Phys.* **36**, 755.

Rowe, J. N.: private communication.

Sachdev, D. K.: 1958, *J. Sci. Industr. Res.* **17A**, 262.

Sakurai, K.: 1968, *J. Geomag. Geoelectr.* **20**, 271.

Sao, K., Yamashita, M., and Jindoh, H.: 1966, *J. Atmospheric Terrest. Phys.* **28**, 97.

Sao, K., Yamashita, M., Tanahashi, S., Jindoh, H., and Ohta, K.: 1970, *J. Atmospheric Terrest. Phys.* **32**, 1567.

Sarada, K. A. and Mitra, A. P.: 1962, *Proc. IGY Symp.*, CSIR (New Delhi) **1**, 158.

Schmeltekopf, A. L., Fehsenfeld, F. C., and Gilman, G. I.: 1967, *Planetary Space Sci.* **15**, 401.

Schwentek, H., Kreplin, R. W., and Hartmann, G.: 1970, presented at Solar-Terrestrial Physics Conference, Leningrad.

Sen, H. K. and Wyller, A. A.: 1960, *J. Geophys. Res.* **65**, 3931.

Sengupta, P. R.: 1968, Sci. Rept., Iowa University, Iowa.

Sengupta, P. R.: 1971, *J. Atmospheric Terrest. Phys.* **33**, 1953.

Sengupta, P. R. and Van Allen, J. A.: 1968, Sci. Rept. Dept. of Phys. and Astronomy, University of Iowa, Iowa.

Shain, C. A. and Mitra, A. P.: 1954, *J. Atmospheric Terrest. Phys.* **5**, 316.

Shapley, A. H. and Knecht, R. W.: 1957, *Inst. Radiat. Eng. Trans.* **AP5**, 326.

Singer, S.: 1970, Quoted by Ulwick and Blank in *Proc. of Meeting on Operation PCA 69*, AFCRL-70-0625, p. 10.

Smith, E. K. and Matsushita, S. (eds.): 1962, *Ionospheric Sporadic E*, Pergamon Press, Oxford.

Somayajulu, Y. V. and Aikin, C. A.: 1969, *Third Aeronomy Conf. Meteorological and Chemical Factors in D-region*, Aeronomy Rept. No. 32, University of Illinois, 373.

Somayajulu, Y. V. and Aikin, C. A.: 1970, Paper presented at *Xth COSPAR Assembly*, Prague.

Strauss, F. M., Papagiannis, M. D., and Aarons, J.: 1969, *J. Atmospheric Terrest. Phys.* **31**, 1241.

Strobel, D. F.: 1972, *J. Geophys. Res.* **77**, 1337.

Subrahmanyam, C. V.: 1967, Ph.D. Thesis, University of Delhi, Delhi, India.

Švestka, Z.: 1966, *Bull. Astron. Inst. Czech.* **17**, 262.
Swartz, W. E.: 1972, Ionosphere Research Laboratory Report No. 381, Pennsylvania State University, U.S.A.
Taubenheim, J.: 1962, *J. Atmospheric Terrest. Phys.* **24**, 191.
Taubenheim, J. and Entzian, G.: 1970, presented at Solar-Terrestrial Physics Conference, Leningrad.
Taubenheim, J., Entzian, G., Knuth, R., and Ohle, K. H.: 1974, *J. Atmospheric Terrest. Phys.* **36**, 525.
Taylor, G. N.: 1974, private communication.
Taylor, G. N. and Watkins, C. D.: 1970, *Nature* **228**, 653.
Taylor, R. G.: 1972, private communication.
Tellegan, B. D. H.: 1933, *Nature* **131**, 840.
Teske, R. C.: 1969, *Solar Phys.* **6**, 193.
Thomas, L., Gondhalekar, P. M., and Bowman, M. R.: 1973, *J. Atmospheric Terrest. Phys.* **35**, 385.
Thome, G. D. and Wagner, L. S.: 1971, *J. Geophys. Res.* **76**, 6883.
Thrane, E. V.: 1973, *COSPAR Symposium on Methods of Measurements and Results of Lower Ionosphere Structure*, Konstanz.
Thrane, E. V. and Piggott, W. R.: 1966, *J. Atmospheric Terrest. Phys.* **28**, 721.
Tisone, G. C.: 1973, *J. Geophys. Res.* **78**, 746.
Titheridge, J. E.: 1966, *J. Atmospheric Terrest. Phys.* **28**, 1135.
Titheridge, J. E.: 1972, in R. Leitinger (ed.), *Proc. Symp. Future Appl. Sat. Beacon Meas.*, Graz, p. 65.
Tříška, P. and Laštovička, J.: 1972, *J. Atmospheric Terrest. Phys.* **34**, 1065.
Ulwick, J. C.: 1972, AFCRL -72-0474 *Proc. COSPAR Symp. Solar Particle Event of November, 1969*, Air Force Cambridge Research Laboratories, U.S. Air Force, p. 395.
Ulwick, J. C. and Blank, C.: 1970, *Proc. of Meeting on Operation PCA 69*, AFCRL-70-0625, Air Force Cambridge Research Laboratories, U.S. Air Force.
Van Allen, J. A.: 1967, *J. Geophys. Res.* **72**, 5903.
Van Allen, J. A.: 1968, *Astrophys. J.* **152**, 185.
Van Allen, J. A.: 1969, WDC-A, Upper Atmosphere Geophysics, Report UAG-5, Boulder, Colorado, p. 46.
Volland, H.: 1966, *J. Atmospheric Terrest. Phys.* **28**, 409.
Wait, J. R.: 1962, *Electromagnetic Waves in Stratified Media*, Pergamon Press, New York.
Wait, J. R. and Spies, K. P.: 1964, NBS Tech. Note No. 300, Boulder, U.S.A.
Wakai, N., Fujii, S., and Ouchi, C.: 1973, *J. Radiat. Res. Lab. Japan* **20**, 67.
Watts, J. M. and Davies, K.: 1960, *J. Geophys. Res.* **65**, 2295.
Wende, C. D.: 1969, *J. Geophys. Res.* **74**, 4649.
Westerlund, S., Reder, F. H., and Abom, G.: 1969, *Planetary Space Sci.* **17**, 1329.
Wood, A. T., Jr. and Noyes, R. W.: 1972, *Solar Phys.* **24**, 180.
Zipf, E. C., Borst, W. L., and Donahue, T. M.: 1970, *J. Geophys. Res.* **75**, 6371.

SPECIAL DATA VOLUMES, BOOKS AND CONFERENCE PROCEEDINGS

1.   *Solar-Geophysical Data*, CRPL-F, Part B (Monthly).
2.   *Compilations of Solar-Geophysical Data*, WDCA for Upper Atmospheric Geophysics, Environmental Services Administration, Boulder, Colo. 80302, U.S.A.
3.   *IGY Solar Activity Data Series*, No. 24, WDCA for Solar Activity, High Altitude Observatory, Boulder, Colo., U.S.A.
4.   *Geophysikalische Beobachtungsergebnisse*, Heinrich-Hertz-Institut, Zentralinstitut für Solar-Terrestrische Physik, Berlin-Adlershof, G.D.R.
5   (i) *Ionosphere Data in Japan*, Radio Research Laboratories, Kokubunji, Japan (Monthly).
      (ii) *Report of Ionosphere and Space Research in Japan* (contains Catalogue of Disturbances).
6.   *Kiruna Geophysical Data*, Kiruna Geophysical Observatory, Sweden (Monthly).
7.   *Geophysical Observatory Reports*, Czechoslovak Academy of Sciences, Geophysical Institute, Czechoslovakia (Monthly).
8.   *Solar and Geophysical Data*, Part 2, Radio Propagation Unit, National Physical Laboratory, New Delhi, India (Monthly).
9.   *WDCA for Solar Terrestrial Physics Report UAG Series*. The following numbers in particular:

UAG-2   'A Reevaluation of Solar Flares, 1964–1966', by Helen W. Dodson and E. Ruth Hedeman of McMath-Hulbert Observatory, The University of Michigan, August 1968.

UAG-4   'Abbreviated Calendar Record 1966–1967', by J. Virginia Lincoln, Hope I. Leighton, and Dorothy K. Kropp of Aeronomy and Space Data Center, Space Disturbances Laboratory, ESSA Research Laboratories, January 1969.

UAG-5   'Data on Solar Event of May 23, 1967 and its Geophysical Effects', compiled by J. Virginia Lincoln, World Data Center A, Upper Atmosphere Geophysics, ESSA, February 1969.

UAG-6   'International Geophysical Calendars 1957–1969', by A. H. Shapley and J. Virginia Lincoln, ESSA Research Laboratories, March 1969.

UAG-8   'Data on Solar-Geophysical Activity October 24–November 6, 1968', Parts 1 and 2, compiled by J. Virginia Lincoln, World Data Center A, Upper Atmosphere Geophysics, ESSA, March 1970.

UAG-9   'Data on Cosmic Ray Event of November 18, 1968 and Associated Phenomena', compiled by J. Virginia Lincoln, World Data Center A, Upper Atmosphere Geophysics, ESSA, April 1970.

UAG-10   'Atlas of Ionograms', edited by A. H. Shapley, ESSA Research Laboratories, May 1970.

UAG-12   'Solar-Geophysical Activity Associated with the Major Geomagnetic Storm of March 8, 1970', Parts 1, 2 and 3, compiled by J. Virginia Lincoln and Dale B. Bucknam, World Data Center A, Upper Atmosphere Geophysics, NOAA, April 1971.

UAG-13   'Data on the Solar Proton Event of November 2, 1969 through the Geomagnetic Storm of November 8–10, 1969', compiled by Dale B. Bucknam and J. Virginia Lincoln, World Data Center A, Upper Atmosphere Geophysics, NOAA, May 1971.

UAG-14   'An Experimental, Comprehensive Flare Index and Its Derivation for 'Major' Flares, 1955–1969', compiled by Helen W. Dodson and E. Ruth Hedeman, McMath-Hulbert Observatory, The University of Michigan, July 1971.

UAG-19   'Reevaluation of Solar Flares 1967', by Helen W. Dodson and E. Ruth Hedeman, McMath-Hulbert Observatory, The University of Michigan, and Marta Rovira de Miceli, San Miguel Observatory, Argentina, June 1972.

UAG-24   'Data on Solar-Geophysical Activity Associated with the Major Ground Level Cosmic Ray Events of 24 January and 1 September 1971', Parts 1 and 2, compiled by Helen E.

Coffey and J. Virginia Lincoln, World Data Center A for Solar-Terrestrial Physics, December 1972.

UAG-28 'Collected Data Reports on August 1972 Solar-Terrestrial Events', Parts 1, 2 and 3, edited by Helen E. Coffey, World Data Center A for Solar-Terrestrial Physics, July 1973.

UAG-30 'Catalog of Data on Solar-Terrestrial Physics', prepared by Environmental Data Service, NOAA, Boulder, Colorado, October 1973.

10. *August 1972 Solar Activity and Related Geophysical Effects*, McKinnon, J. A., NOAA Technical Memorandum ERL SEL-22, NOAA TM ERL SEL-22, December, 1972.

11. *Ionospheric and Geophysical Data from India Related to the Solar Events of August 1972*, Scientific Report No. 85, Radio Science Division, National Physical Laboratory, New Delhi 12, India, August 1973.

12. *Geophysics and Space Data Bulletin*, Space Physics Laboratory, AFCRL, Mass., U.S.A.

13. *Comprehensive Tables of Sudden Ionospheric Disturbances Observed at the Radio Research Laboratories during the International Active Sun Years, January 1969–December 1971*, Tokyo, January 1972.

14. *An Atlas of Solar Flare Effects Observed on Long VLF Paths during 1961*, C. J. Chilton, F. K. Steele, and D. D. Crombie, NBS Technical Note No. 210, U.S. Department of Commerce, NBS, USA, March 13, 1964.

15. *An Atlas of Solar Flare Effects in the Ionosphere Observed with a High Frequency Doppler Technique, September 1960–December 1962*, NBS Technical Note No. 326, Donald M. Baker, U.S. Department of Commerce, NBS, Dec. 1, 1965.

16. *IAU Quarterly Bulletins of Solar Activity*.

17. *A Review of Some Ionospheric Studies Based on a High Frequency Doppler Technique*, Baker D. M., Chang N., Davies K., Donnelly R. F., and Jones J. E., ESSA Technical Report ERL 78-SDL 1 (Contains catalogue of SFDs from January 1963 through December 1967).

18. *Solar Terrestrial Activity Chart, Interdisciplinary Analysis Centre for Solar Terrestrial Activity*, National Committee on Solar Terrestrial Physics, Science Council of Japan.

19. *AFCRL Studies of the November 1960 Solar-Terrestrial Events*, Ed. J. Aarons and S. M. Silverman.

20. *Symposium on the July 1959 Events and Associated Phenomena*, IUGG, Helsinki, IUGG, 1960.

21. *Annals of the IQSY*
Vol. 1. Geophysical Measurements.
Vol. 2. Solar and Geophysical Events 1960–1965.
Vol. 3 The Proton Flare Project.
Vol. 4 Solar-Terrestrial Physics.
Vol. 5 Solar-Terrestrial Physics: Terrestrial Aspects.
Vol. 6 Survey of the IQSY Observations and Bibliography.
Vol. 7 Sources and Availability of IQSY Data.

22. *The Physics of Solar Flares*, AAS-NASA Symposium, NASA SP-50, Goddard Space Flight Center, 1963.

23. *Solar Flares and Space Research*, Ed. C. de Jager and L. Švestka, COSPAR, North-Holland Publishing Co., Amsterdam, 1969.

24. *Interrelated Satellite Observations Related to Solar Events*, Proceedings of the Third ESLAB/ESRIW Symposium, Noordwijk, Netherlands, September 16–19, 1969, D. Reidel Publishing Co., Netherlands, 1970.

25. *The Effect of Disturbances of Solar Origin on Communications*, Ed. G. J. Gassman, Pergamon Press, 1963.

26. *Solar Terrestrial Physics*, J. W. King and W. J. Newman, Academic Press, New York, 1967.

27. *Proceedings of COSPAR Symposium on Solar Particle Event of November 1969*, Ed. James C. Ulwick, Air Force Cambridge Research Laboratory, Bedford, Mass., U.S.A., Rep. AFCRL-72-0474, 11 August 1972.

28. *Phase and Frequency Instabilities in Electromagnetic Wave Propagation*, Ed. K. Davies, Technivision Services, Slough, England, 1970.

29. *Solar-Terrestrial Physics/1970*, Ed. E. R. Dyer, D. Reidel Publishing Co., Dordrecht-Holland, 1970.

INDEX OF NAMES

INDEX OF SUBJECTS

ASTROPHYSICS AND SPACE SCIENCE LIBRARY

Edited by

J. E. Blamont, R. L. F. Boyd, L. Goldberg, C. de Jager, Z. Kopal, G. H. Ludwig, R. Lüst,
B. M. McCormac, H. E. Newell, L. I. Sedov, Z. Švestka, and W. de Graaff

1. C. de Jager (ed.), *The Solar Spectrum. Proceedings of the Symposium held at the University of Utrecht, 26–31 August, 1963.* 1965, XIV + 417 pp.
2. J. Ortner and H. Maseland (eds.), *Introduction to Solar Terrestrial Relations, Proceedings of the Summer School in Space Physics held in Alpbach, Austria, July 15–August 10, 1963 and Organized by the European Preparatory Commision for Space Research.* 1965, IX + 506 pp.
3. C. C. Chang and S. S. Huang (eds.), *Proceedings of the Plasma Space Science Symposium, held at the Catholic University of America, Washington, D.C., June 11–14, 1963.* 1965, IX + 377 pp.
4. Zdeněk Kopal, *An Introduction to the Study of the Moon.* 1966, XII + 464 pp.
5. B. M. McCormac (ed.), *Radiation Trapped in the Earth's Magnetic Field. Proceedings of the Advanced Study Institute, held at the Chr. Michelsen Institute, Bergen, Norway, August 16– September 3, 1965.* 1966, XII + 901 pp.
6. A. B. Underhill, *The Early Type Stars.* 1966, XII + 282 pp.
7. Jean Kovalevsky, *Introduction to Celestial Mechanics,* 1967, VIII + 427 pp.
8. Zdeněk Kopal and Constantine L. Goudas (eds.), *Measure of the Moon. Proceedings of the 2nd International Conference on Selenodesy and Lunar Topography, held in the University of Manchester, England, May 30–June 4, 1966.* 1967, XVIII + 479 pp.
9. J. G. Emming (ed.), *Electromagnetic Radiation in Space. Proceedings of the 3rd ESRO Summer School in Space Physics, held in Alpbach, Austria, from 19 July to 13 August, 1965.* 1968, VIII + 307 pp.
10. R. L. Carovillano, John F. McClay, and Henry R. Radoski (eds.), *Physics of the Magnetosphere, Based upon the Proceedings of the Conference held at Boston College, June 19–28, 1967.* 1968, X + 686 pp.
11. Syun-Ichi Akasofu, *Polar and Magnetospheric Substorms.* 1968, XVIII + 280 pp.
12. Peter M. Millman (ed.), *Meteorite Research. Proceedings of a Symposium on Meteorite Research, held in Vienna, Austria, 7–13 August, 1968.* 1969, XV + 941 pp.
13. Margherita Hack (ed.), *Mass Loss from Stars. Proceedings of the 2nd Trieste Colloquium on Astrophysics, 12–17 September, 1968.* 1969, XII + 345 pp.
14. N. D'Angelo (ed.), *Low-Frequency Waves and Irregularities in the Ionosphere. Proceedings of the 2nd ESRIN-ESLAB Symposium, held in Frascati, Italy, 23–27 September, 1968.* 1969, VII + 218 pp.
15. G. A. Partel (ed.), *Space Engineering. Proceedings of the 2nd International Conference on Space Engineering, held at the Fondazione Giorgio Cini, Isola di San Giorgio, Venice, Italy, May 7–10, 1969.* 1970, XI + 728 pp.
16. S. Fred Singer (ed.), *Manned Laboratories in Space. Second International Orbital Laboratory Symposium.* 1969, XIII + 133 pp.
17. B. M. McCormac (ed.), *Particles and Fields in the Magnetosphere. Symposium Organized by the Summer Advanced Study Institute, held at the University of California, Santa Barbara, Calif., August 4–15, 1969.* 1970, XI + 450 pp.
18. Jean-Claude Pecker, *Experimental Astronomy.* 1970, X + 105 pp.
19. V. Manno and D. E. Page (eds.), *Intercorrelated Satellite Observations related to Solar Events. Proceedings of the 3rd ESLAB/ESRIN Symposium held in Noordwijk, The Netherlands, September 16–19, 1969.* 1970, XVI + 627 pp.
20. L. Mansinha, D. E. Smylie, and A. E. Beck, *Earthquake Displacement Fields and the Rotation of the Earth. A NATO Advances Study Institute Conference Organized by the Department of Geophysics, University of Western Ontario, London, Canada, June 22–28, 1969.* 1970, XI + 308 pp.
21. Jean-Claude Pecker, *Space Observatories.* 1970, XI + 120 pp.
22. L. N. Mavridis (ed.), *Structure and Evolution of the Galaxy, Proceedings of the NATO Advanced Study Institute, held in Athens, September 8–19, 1969.* 1971, VII + 312 pp.

23. A. Muller (ed.), *The Magellanic Clouds. A European Southern Observatory Presentation: Principal Prospects, Current Observational and Theoretical Approaches, and Prospects for Future Research. Based on the Symposium on the Magellanic Clouds, held in Santiago de Chile, March 1969, on the Occasion of the Dedication of the European Southern Observatory*. 1971, XII + 189 pp.

24. B. M. McCormac (ed.), *The Radiating Atmosphere. Proceedings of a Symposium Organized by the Summer Advanced Study Institute, held at Queen's University, Kingston, Ontario, August 3–14, 1970*. 1971, XI + 455 pp.

25. G. Fiocco (ed.), *Mesopheric Models and Related Experiments. Proceedings of the 4th ESRIN–ESLAB Symposium, held at Frascati, Italy, July 6–10, 1970*. 1971, VIII + 298 pp.

26. I. Atanasijević, *Selected Exercises in Galactic Astronomy*. 1971, XII + 144 pp.

27. C. J. Macris (ed.), *Physics of the Solar Corona. Proceedings of the NATO Advanced Study Institute on Physics of the Solar Corona, held at Cavouri-Vouliagmeni, Athens, Greece, 6–17 September 1970*. 1971, XII + 345 pp.

28. F. Delobeau, *The Environment of the Earth*. 1971, IX + 113 pp.

29. E. R. Dyer (general ed.), *Solar-Terrestrial Physics/1970. Proceedings of the International Symposium on Solar-Terrestrial Physics, held in Leningrad, U.S.S.R., 12–19 May 1970*. 1972, VIII + 938 pp.

30. V. Manno and J. Ring (eds.), *Infrared Detection Techniques for Space Research, Proceedings of the 5th ESLAB-ESRIN Symposium, held in Noordwijk, The Netherlands, June 8–11, 1971*. 1972, XII + 344 pp.

31. M. Lecar (ed.), *Gravitational N-Body Problem, Proceedings of IAU Colloquium No. 10, held in Cambridge, England, August 12–15, 1970*. 1972, XI + 441 pp.

32. B. M. McCormac (ed.), *Earth's Magnetospheric Processes. Proceedings of a Symposium Organized by the Summer Advanced Study Institute and Ninth ESRO Summer School, held in Cortina, Italy, August 30–September 10, 1971*. 1972, VIII + 417 pp.

33. Antonin Rükl, *Maps of Lunar Hemispheres*. 1972, V + 24 pp.

34. V. Kourganoff, *Introduction to the Physics of Stellar Interiors*. 1973, XI + 115 pp.

35. B. M. McCormac (ed.), *Physics and Chemistry of Upper Atmospheres. Proceedings of a Symposium Organized by the Summer Advanced Study Institute, held at the University of Orléans, France, July 31–August 11, 1972*. 1973, VIII + 389 pp.

36. J. D. Fernie (ed.), *Variable Stars in Globular Clusters and in Related Systems. Proceedings of the IAU Colloquim No. 21, held at the University of Toronto, Toronto, Canada, August 29–31, 1972*. 1973, IX + 234 pp.

37. R. J. L. Grard (ed.), *Photon and Particle Interaction with Surfaces in Space. Proceedings of the 6th ESLAB Symposium, held at Noordwijk, the Netherlands, 26–29 September, 1972*. 1973, XV + 577 pp.

38. Werner Israel (ed.), *Relativity, Astrophysics and Cosmology. Proceedings of the Summer School, held 14–26 August, 1972, at the BANFF Centre, BANFF, Alberta, Canada*. 1973, IX + 323 pp.

39. B. D. Tapley and V. Szebehely (eds.), *Recent Advances in Dynamical Astronomy, Proceedings of the NATO Advanced Study Institute in Dynamical Astronomy, held in Cortina d'Ampezzo, Italy, August 9–12, 1972*. 1973, XIII + 468 pp.

40. A. G. W. Cameron (ed.), *Cosmochemistry. Proceedings of the Symposium on Cosmochemistry, held at the Smithsonian Astrophysical Observatory, Cambridge, Mass., August 14–16, 1972*. 1973, X + 173 pp.

41. M. Golay, *Introduction to Astronomical Photometry*. 1974, IX + 364 pp.

42. D. E. Page (ed.), *Correlated Interplanetary and Magnetospheric Observations. Proceedings of the 7th ESLAB Symposium, held at Saulgau, W. Germany, 22–25 May, 1973*. 1974, XIV + 662 pp.

43. Riccardo Giacconi and Herbert Gursky (eds.), *X-Ray Astronomy*. 1974, X + 450 pp.

44. B. M. McCormac (ed.), *Magnetospheric Physics. Proceedings of the Advanced Summer Institute, held in Sheffield, U.K., August 1973*. 1974, VII + 399 pp.

45. C. B. Cosmovici (ed.), *Supernovae and Supernova Remnants. Proceedings of the International Conference on Supernovae, held in Lecce, Italy, May 7–11, 1973*. 1974, XVII + 387 pp.